Gamete Immunology

Hiroaki Shibahara • Akiko Hasegawa

Editors

Gamete Immunology

 Springer

Editors
Hiroaki Shibahara
Department of Obstetrics and Gynecology
School of Medicine, Hyogo Medical
University
Nishinomiya, Hyogo, Japan

Akiko Hasegawa
Department of Obstetrics and Gynecology
School of Medicine, Hyogo Medical
University
Nishinomiya, Hyogo, Japan

ISBN 978-981-16-9624-4   ISBN 978-981-16-9625-1   (eBook)
https://doi.org/10.1007/978-981-16-9625-1

This Springer imprint is published by the registered company Springer Nature Singapore Pte Ltd.
The registered company address is: 152 Beach Road, #21-01/04 Gateway East, Singapore 189721,
Singapore

Preface

I graduated from Kochi Medical University in 1984 and joined the department of Obstetrics and Gynecology, Hyogo College of Medicine, presided over by Prof. Shinzo Isojima. At that time, I was motivated and was interested in a new infertility treatment that was used in in vitro fertilization-embryo transfer (IVF-ET).

In 1968, Prof. Isojima developed a sperm-immobilization test and indicated that sperm-immobilizing antibodies are one of the causes of refractory female infertility and has long thought that IVF-ET is desirable as an ideal treatment method. But finally in 1986, he and his group applied IVF-ET to an infertile woman with sperm-immobilizing antibodies, and they succeeded in the world's first pregnancy and childbirth by this indication. In parallel, Prof. Isojima established H6-3C4, a monoclonal antibody derived from patient's peripheral lymphocytes, for the purpose of developing a contraceptive vaccine in 1987. Later, Prof. Koji Koyama and Dr. Akiko Hasegawa identified the corresponding antigen against H6-3C4 as CD52.

In 1990, I studied abroad at Eastern Virginia Medical School in the USA. Under the guidance of Prof. Nancy J. Alexander and Dr. Lani J. Burkman, I examined fertilization disorders caused by sperm-immobilizing antibodies using the hemizona assay (HZA) and obtained a degree in 1993. After that, I studied the clinical aspects of the antibodies in infertile women, but in 1996 I studied abroad again at the University of Virginia in the USA. Under the guidance of Prof. John C. Herr and Dr. Soren Naaby-Hansen, I joined the contraceptive vaccine development project team and conducted research until the end of 1997.

Soon after returning to Japan, I moved to Jichi Medical University in 1999 and began researching anti-sperm antibodies in infertile men. I have established guidelines for the treatment of infertility caused by anti-sperm antibodies possessed by men, which had been chaotic until then. In addition to studying the clinical significance, I pursued the corresponding antigen against sperm-immobilizing antibodies in men with Prof. Kiyotaka Toshimori at Chiba University.

In 2013, I returned to Hyogo College of Medicine as the chairman of the department of Obstetrics and Gynecology and resumed my research with Dr. Hasegawa in this area inherited from Prof. Isojima and Prof. Koyama. The time is ripe, and I decided to publish this textbook in order to preserve the knowledge I have acquired so far. I named this book *Gamete Immunology*, and I shared writing with Dr. Hasegawa, covering the field of anti-zona pellucida antibodies that Professors Isojima and Koyama have passionately researched.

It took 2 years for completing this project, and I would like to express my heartfelt gratitude to the members of Springer for their support during this time. From this text, I hope that research on the development of contraceptive vaccines in the future will become more and more active with the passion of young researchers.

Nishinomiya, Hyogo, Japan   Hiroaki Shibahara

Preface

Firstly, I would like to acknowledge the debt this book owes to two great Emeritus Professors, Dr. Shinzo Isojima and Dr. Koji Koyama. In 1977, I started working on gamete immunology at the department of Obstetrics and Gynecology, Hyogo College of Medicine, where Prof. Isojima was the chairperson. After his retirement, Prof. Koyama led the department from 1992 to 2009.

In 1981, Prof. Isojima succeeded in the establishment of a monoclonal antibody to zona pellucida (ZP) present in the pig ovary. Subsequently, Prof. Koyama and his colleagues determined the amino acid sequence of the pig ZP protein recognized by the antibody. Research into gamete immunology in our laboratory has been able to continue over the past 40 years based on these great predecessors' achievements.

Looking back, the meeting of immunology and the ovary occurred just 50 years ago, originating from a report that thymectomy of newborn mice induced ovarian disorders by self-reactive immunity (Nishizuka and Sakakura 1971). This finding opened the door of auto-immune research and led to the current knowledge that adult immune systems are harmoniously controlled, mediated by self- and not-self immunity.

In the 1970s, animal experiments and clinical research revealed that mammalian ZP in the ovary possessed a strong immunogenicity as well as unique functions. ZP is an extracellular matrix that surrounds the oocytes and forms markedly after puberty. Focusing on ZP, animal studies have explored immune contraceptive vaccine development and clinical studies have investigated pathogenic factors of infertility. Intensive efforts have been made by a huge number of researchers using biochemical analysis, DNA recombination, and modern database analysis. Parts IV–VI of this book reviews the history and leading edge of biological and immunological studies for ZP.

Finally, I would like to thank Prof. Hiroaki Shibahara, the current chairperson of the department of Obstetrics and Gynecology at Hyogo College of Medicine, for giving me the opportunity to contribute to this book as a second author.

Nishinomiya, Hyogo, Japan   Akiko Hasegawa

Professor Shinzo Isojima and Professor Koji Koyama

Dr. Akiko Hasegawa and Professor Hiroaki Shibahara

Contents

Fig. 1.1 Comparison of the conception rates between the female guinea pigs treated with homologous testis plus Freund's adjuvant and the controls. Fertility among the controls was 84% (37 of 44), while it was only 24% (7 of 29) in the group injected against homologous testis, which indicates that the active immunization of female guinea pigs with homologous testis or sperm using complete Freund's adjuvant markedly reduced pregnancy rates in immunized animals. There was a significant difference between the two groups ($P < 0.001$)

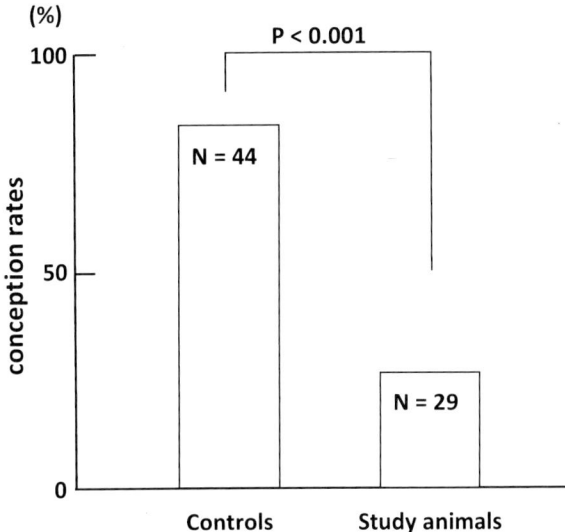

In 1959, Isojima et al. [7] demonstrated sterility in adult virgin female guinea pigs induced by injection with an emulsion of homologous adult testis and Freund's adjuvant before exposure to males. Fertility among the controls was 84% (37 of 44), while it was only 24% (7 of 29) in the group injected against homologous testis, which indicates that the active immunization of female guinea pigs with homologous testis or sperm using complete Freund's adjuvant markedly reduced pregnancy rates in immunized animals (Fig. 1.1). The testis-injected guinea pigs had also a high titer of antibodies against testis. The authors speculated that there were two possible mechanisms of sterility: one was the cellular immunity, as in Freund's experiment in aspermatogenesis [6], and the second was circulating antibodies which could appear in the vaginal fluid according to their observation or which could cause the uterus to contact with sperm, according to Katsh [8]. Katsh [9] also obtained a similar result by immunizing female guinea pigs with sperm using complete Freund's adjuvant. Later, Otani et al. [10] reevaluated and proved that immunization of female guinea pigs with homologous testis or sperm using complete Freund's adjuvant reduced pregnancy rates in immunized animals.

In mice, McLaren [11] showed the reduction of fertilization in females that received three intraperitoneal injections of live homologous sperm a week for 7 weeks. Females with high titers of anti-sperm agglutinins in their sera showed a marked reduction both in the size of the first litter and in total breeding performance over the 6-month period. They also showed that the reduced fertilization rate in immunized females was due to failure of the sperm to reach the site of fertilization.

Edwards [12] showed that, in his experiments, immunization of mice with sperm and adjuvant has not led to drastic reductions in infertility as found by Isojima et al. [7] in guinea pigs. It is difficult to make comparisons between species when the fertilization rate in the guinea pigs was unknown. The incidence of fertilization in his work was higher and more independent of the amount of circulating antibody than

Table 1.1 History of researches on the production of anti-sperm antibodies (ASA) and infertility in female animals

Year	Author	Ref. no.	Animal	Immunogen	Main results
1899	Landsteiner	[1]	Guinea pig	Sperm (bovine)	Sperm motility was lost
	Metchinikoff	[2]	Guinea pig	Various sperm	Sperm motility was lost
1923	MaCartney	[4]	Rat	Sperm	ASA production and infertility are related
1940	Henle	[5]	Guinea pig	Sperm	ASA production and infertility are not related
1959	Isojima	[7]	Guinea pig	Testis and FA	Reduced pregnancy rates with high ASA titers
	Katsh	[9]	Guinea pig	Sperm and FA	Reduced pregnancy rates with high ASA titers
1963	Otani	[10]	Guinea pig	Testis/sperm and FA	Reduced pregnancy rates with high ASA titers
1964	McLaren	[11]	Mouse	Sperm	Reduced fertilization rate with high ASA titers
	Edwards	[12]	Mouse	Sperm and FA	ASA production did not lead to infertility
1969	Bell	[13]	Mouse	Sperm	Systemic immunization resulted in infertility
					Intravaginal immunization did not result in infertility

FA complete Freund's adjuvant

was the incidence of infertility found in mice by McLaren [11]. Differences existed between the two investigations in the immunization procedure, and it is possible that a more effective immune response was produced in McLaren's work by the prolonged intraperitoneal immunization. He concluded that the high incidence of fertility remaining in mice after immunization indicates that an immunological method using sperm for the control of fertility has not yet been realized.

Bell [13] compared the efficacy of the methods for immunization of female mice between systemic and intravaginal sperm injection. He found that systemic iso-immunization of mice with sperm resulted in a significant reduction ($P < 0001$) of litter size. Re-immunization depressed fertility further to less than half the control level. Serum antibody levels were assayed both by a passive hemagglutination test adapted for detecting antibodies against mouse sperm and by a sperm-agglutination test (SAT). There was a significant regression of numbers of offspring on both sperm-agglutination and hemagglutination titers. Intravaginal iso-immunization with mouse sperm failed to alter the fecundity in short-term breeding periods or during long-term mating or to enhance the sterility induced by systemic injections.

The history of these researches on production of ASA and infertility in female animals is summarized in Table 1.1.

1.3   Immunization of Women Injected with Semen

In 1921, the editorials of the *JAMA* [14] commented that if sperm invade the female tissues and cause the formation of specific antibodies which are capable of preventing fertilization, may not such a process participate in the problem of sterility? They speculated that the traditional sterility of the prostitute depends sometimes on such a process rather than on inflammatory sealing of the tubes. In 1922, Mayer [15] and Vogt [16] stated that sperm overloading leads to sterility and that a natural cure occurs after prolonged abstinence from sexual intercourse.

In human, following some reports from birth control clinics of contraceptive efficacy of up to 20 months' duration by injecting human semen, Rosenfeld [17] was the first to suggest that a woman repeatedly injected with human semen became infertile. He attempted to immunize three women with multiple subcutaneous injections of semen and found serological evidence of immunity to sperm. Mild local reactions were noted but no systemic disturbance occurred. However, it seems to be a harmful and dangerous clinical trial as there might be common antigens against other organs except for reproduction in semen.

Following this report, Baskin [18] also reported infertility in women after immunization with human semen. He reported on 20 women who had received 3 intramuscular injections of 5–10 mL of fresh semen at 7-day intervals. The injections appeared harmless, and in all cases but one, humoral sperm cytotoxicity was induced and persisted for up to 12 months. Repeated testing of immune sera against sperm from different individuals indicated a lack of allospecificity of the reactions. No pregnancies occurred while positive serology persisted, and although the follow-up period was too short to allow a systematic assessment of subsequent fertility, one patient became pregnant 3 months after her serum reverted to negative following 9 months of immunity. Cervical secretions exhibited sperm cytotoxicity in about 50% of cases, but this showed no consistent relationship with serum activity. In 1937, Baskin was issued the US patent for a spermotoxic vaccine, which produced reversible sterilization in fertile women. Subsequent studies confirmed the induction of temporary infertility, without overt side effects in women injected with human sperm [19, 20].

Brunner reported a claim that 30 women had been immunized with the bovine sperm phospholipid and had remained sterile for 6 months in 1941 [21]. The history of these researches on production of ASA and infertility in human is summarized in Table 1.2.

1.4   Clinical Observations

Around World War II, studies for the relation between sperm immunity and infertility lost for a long time. In 1964, Franklin and Dukes [22, 23] reported their clinical observations on ASA in women. They found that 20.1% of 214 women undergoing

Table 1.2 History of researches on the production of anti-sperm antibodies (ASA) and infertility in human

Year	Author	Ref. no.	Immunogen	Main results
1926	Rosenfeld	[17]	Human semen	Repeated injection led to infertility
1932	Baskin	[18]	Human semen	No pregnancies occurred while positive serology persisted
1936	Escuder	[19]	Human sperm	Production of ASA and temporary infertility were induced
	Rodriguez	[20]	Human sperm	Production of ASA and temporary infertility were induced
1941	Brunner	[21]	Sperm phospholipid (bovine)	Temporary infertility was induced

infertility investigations had sperm-agglutinating activity in their serum. Within this group of patients, 31 (72.1%) of 43 women with unexplained infertility had a much higher incidence of sperm-agglutinating activity. The incidence was higher than those with organic causes of their infertility (8.4%) or fertile women (5.7%). They also reported that antibody titers in their sera declined markedly in all 13 women treated by condoms and/or abstained for 2–6 months. Ten women whose antibody titers dropped to undetectable levels were encouraged to resume unrestricted intercourse at the time of expected ovulation with the result that nine became pregnant. Later, they reported that 27 unexplained infertile women with sperm-agglutinating activity were treated by condoms and/or abstained. In 25 of 27 women, the antibody titers dropped to undetectable levels, and finally 20 of 25 women successfully conceived. There was a supportive appreciation for their clinical findings that they stimulated significant interest in the hypothesis that female immunological reactions to sperm could be involved in the etiology of otherwise unexplained infertility and in the concept of an anti-sperm contraceptive vaccine [24].

However, there was a serious problem for their studies that the phenomenon of sperm-agglutination is not always a specific event by ASA. It has been well known that sperm-agglutination frequently occurs naturally, not related with the existence of ASA. Therefore, after their reports, the relationship between sperm-agglutinating antibodies and infertility has been studied inconclusively by many other investigators [25–27]. One of the reasons may lie in the difficulty of discriminating true antibody-specific sperm-agglutination from nonspecific agglutination [28].

Because of the nonspecificity of the SAT for detecting ASA, a sperm-immobilization test (SIT), that utilized the function of sperm-immobilization in the presence of complement, was developed by Isojima et al. [25]. By using this method, it was found that the complement-dependent sperm-immobilizing antibodies were detected exclusively in 3 (12%) of 25 women with unexplained infertility. All of them showed a history of more than 5 years' infertility since marriage.

Later, Isojima et al. [29] compared the results of the SIT and Franklin-Dukes' and Kibrick's SAT by using sera of sterile, pregnant, and unmarried women. Positive reactions in the SIT were given by sera of 17.2% of the patients with sterility of unexplained cause and not by those of normal pregnant and unmarried women. However, Franklin-Dukes' and Kibrick's SAT gave rather higher percentages of positive sperm-agglutination reactions in normal pregnant women than in cases of sterility of unexplained cause but a very low percentage of positive reactions in unmarried women. They also showed that the sperm-immobilization value did not drop during condom therapy over a long period, while Franklin and Dukes reported a sharp drop in sperm agglutinin in the serum of cases after abstinence or condom therapy for several months [24].

Other studies also confirmed the reliability of SIT as an assay for detecting ASA closely related to female infertility. Ansbacher [30] reported that sperm-immobilizing antibodies appear to have a better correlation with infertility than sperm-agglutinating antibodies.

Jones et al. [31] used 2 tests, including SAT and SIT, to detect ASA in the female partners of 196 infertile couples and in a control group of 50 pregnant women. SAT demonstrated little difference in incidence of positive reactions in patients with unexplained and organic infertility and yielded an 18% incidence in the pregnant control patients. The results of the SIT showed a clear disparity in patients with unexplained infertility compared with those in the organic infertility and pregnancy groups. A comparison of these results with those obtained by other workers, using the same and other methods, suggests that the SIT provides the most reliable single method for the detection of circulating anti-sperm activity with possible clinical significance.

Petrunia [32] studied for the presence of anti-sperm factors in the sera from 102 women with infertility due to a variety of causes and from 40 pregnant women. Three techniques were used: sperm microagglutination, sperm-immobilization, and an indirect immunofluorescent technique for detection of sperm-bound immunoglobulins. There was no correlation between the results obtained using these three different techniques. Of the three, only the results of SIT correlated with primary unexplained infertility. The sperm microagglutination test appeared to measure nonspecific factors. Methanol fixation of sperm used in the indirect immunofluorescent technique apparently resulted in nonspecific binding of immunoglobulins.

Cantuaria [33] collected the sera and cervicovaginal secretions of 39 infertile women without abnormality except apparent cervical hostility and evaluated for the presence of sperm-immobilizing antibodies. The antibodies were found in 25.6% of sera and 20.5% of cervicovaginal secretions. No such antibodies were detected in controls.

In the recent review by Koyama [28], the incidences of sperm-immobilizing antibodies in the sera of infertile women or the control women were shown in Table 1.3. The incidences of the antibodies in unexplained and organic infertile women were found to be 13.2% and 1.4%, respectively, while the incidences of sperm-immobilizing antibodies in the controls, such as children, unmarried women, and pregnant women, were 0%, 0%, and 0.5%, respectively. In conclusion, the SIT proved to be one of the most reliable methods to detect ASA relevant to infertility in women.

Table 1.3 Incidences of sperm-immobilizing antibodies in the sera of infertile women and the controls

Subjects	Incidence
Infertility	
Unexplained	119/901 (13.2%)
Organic	31/2190 (1.4%)
Controls	
Children	0/1013 (0%)
Unmarried women	0/92 (0%)
Pregnant women	1/202 (0.5%)

References

1. Landsteiner K. Zur Kenntnis der spezifisch auf Blutkoreperchen wirkenden Sera. Zentr Bakteriol Parasitenk. 1899;25:546–9.
2. Metchnikoff E. Etudes sur la resorption de cellule. Ann Inst Pasteur. 1899;13:737–79.
3. Joel CA. Historical survey of research on spermatozoa from antiquity to the present. In: Joel CA, editor. Fertility disturbances in men and women. Basel: Karger; 1971. p. 3–13.
4. McCartney JL. Further observations on the antigenic effects of semen: mechanism of sterilization of female rats from injections of spermatozoa. Am J Physiol. 1923;66:404.
5. Henle W, Henle G. Spermatozoal antibodies and fertility: II. Attempt to induce temporary sterility in female guinea pigs by active immunization against spermatozoa. J Immunol. 1940;38:105.
6. Freund J, Thompson GE, Lipton MM. Aspermatogenesis, anaphylaxis, and cutaneous sensitization induced in the guinea pig by homologous testicular extract. J Exp Med. 1955;101:591–604.
7. Isojima S, Graham RM, Graham JB. Sterility in female guinea pigs induced by injection with testis. Science. 1959;129:44.
8. Katsh S. In vitro demonstration of uterine anaphylaxis in guinea pigs sensitized with homologous testis or sperm. Nature. 1957;180:1047–8.
9. Katsh S. Infertility in female guinea pigs induced by injection of homologous sperm. Am J Obstet Gynecol. 1959;78:276–8.
10. Otani Y, Behrman SJ, Porter CW, Nakayama M. Reduction of fertility in immunized guinea pigs. Int J Fertil. 1963;8:835–9.
11. McLaren A. Immunological controls of fertility in female mice. Nature. 1964;201:582–5.
12. Edwards RG. Immunological control of fertility in female mice. Nature. 1964;203:50–3.
13. Bell EB. Immunological control of fertility in the mouse: a comparison of systemic and intravaginal immunization. J Reprod Fertil. 1969;18:183–92.
14. Editorial. Have spermatozoa functions or effects other than fertilization? JAMA. 1921;77:42–3.
15. Mayer A. Increase of sterile marriage since the war. Klin Wchnschr. 1922;1:1142–4.
16. Vogt E. Sterility and sperm immunity. Klin Wchnschr. 1922;1:1144–6.
17. Rosenfeld SS. Semen injection with serologic studies: a preliminary report. Am J Obstet Gynecol. 1926;12:385–7.
18. Baskin MJ. Temporary sterilization by the injection of human spermatozoa. A preliminary report. Am J Obstet Gynecol. 1932;24:892–7.
19. Escuder CJ. La esterilizacion biologica temporaria la major poresperma humano. Arch Urug Med. 1936;8:484–6.
20. Rodriguez-Lopez MM. Esterilizacion biologica temporaria de la mujer. Arch Urug Med. 1936;121:373–6.
21. Katsh S. Immunology, fertility and infertility: a historical survey. Am J Obstet Gynecol. 1959;77:946–56.

22. Franklin RR, Dukes CD. Antispermatozoal antibody and unexplained infertility. Am J Obstet Gynecol. 1964;89:6–9.
23. Franklin RR, Dukes CD. Further studies on sperm-agglutinating antibody and unexplained infertility. JAMA. 1964;190:682–3.
24. Dukes CD, Franklin RR. Sperm agglutinins and human infertility: female. Fertil Steril. 1968;19: 263–7.
25. Isojima S, Li T, Ashitaka Y. Immunologic analysis of sperm-immobilizing factor found in sera of women with unexplained sterility. Am J Obstet Gynecol. 1968;101:677–83.
26. Tyler A, Tyler ET, Denny PC. Concepts and experiments in immunoreproduction. Fertil Steril. 1967;18:153–66.
27. Israelstam DM. The incidence of sperm-agglutinating antibodies in the serum of infertile women. Fertil Steril. 1969;20:275–8.
28. Koyama K. Gamete immunology: infertility and contraception. Reprod Immunol Biol. 2009;24:1–17.
29. Isojima S, Tsuchiya K, Koyama K, Tanaka C, Naka O, Adachi H. Further studies on sperm-immobilizing antibody found in sera of unexplained cases of sterility in women. Am J Obstet Gynecol. 1972;112:199–207.
30. Ansbacher R, Keung-Yeung K, Behrman SJ. Clinical significance of sperm antibodies in infertile couples. Fertil Steril. 1973;24:305–8.
31. Jones WR, Ing RMY, Kaye MD. A comparison of screening tests for anti-sperm activity in the serum of infertile women. J Reprod Fertil. 1973;32:357–64.
32. Petrunia DM, Taylor PJ, Watson JI. A comparison of methods of screening for sperm antibodies in the serum of women with otherwise unexplained infertility. Fertil Steril. 1976;27:655–61.
33. Cantuaria AA. Sperm immobilizing antibodies in the serum and cervicovaginal secretions of infertile and normal women. Br J Obstet Gynaecol. 1977;84:865–8.

Chapter 2
Detection of Anti-sperm Antibody (ASA) in Women

Hiroaki Shibahara

Abstract Several assay methods were developed to detect anti-sperm antibodies (ASAs). There are two ways to test for ASA in infertile couples. One is the assays for infertile men that directly detect ASA on the sperm membrane. For this purpose, the mixed antiglobulin reaction (MAR) test and direct immunobead test (direct IBT) are recommended. Direct IBT is widely used as a screening test for ASA, and the inhibitory effects by ASA on sperm motion and fertilizing ability are evaluated to make a decision for the strategy of infertility treatments.

The other tests for ASA involve assays that indirectly detect ASA in serum, cervical mucus, follicular fluid, peritoneal fluid, or seminal plasma. For indirect ASA testing, serological tests, such as sperm-agglutination test (SAT) and sperm-immobilization test (SIT), have been usually performed for clinical purposes in infertile women.

It is obvious that the most important consideration is the selection of the method. It has been recommended that clinically specific test for female infertility should be chosen for detecting circulating ASA. ASA with bioactivities, including sperm-immobilizing antibodies, sperm-agglutinating antibodies and fertilization-blocking antibodies are suitable for choosing as initial testing for infertile women.

2.1 Methods for Anti-sperm Antibody Testing

Anti-sperm antibodies (ASAs) in serum or cervical mucus have been implicated as an etiological factor of infertile women. Although the incidence of such antibodies is increased in subfertile individuals, the exact nature of their effects on sperm function has yet to be fully elucidated.

There are two ways to test for ASA in infertile couples. One is the assays for infertile men that directly detect ASA on the sperm membrane. For this purpose, the

H. Shibahara (✉)
Department of Obstetrics and Gynecology, School of Medicine, Hyogo Medical University, Nishinomiya, Hyogo, Japan
e-mail: sibahara@hyo-med.ac.jp

H. Shibahara, A. Hasegawa (eds.), *Gamete Immunology*,
https://doi.org/10.1007/978-981-16-9625-1_2

13

mixed antiglobulin reaction (MAR) test and direct immunobead test (direct IBT) are recommended by the World Health Organization (WHO) [1]. Direct IBT is widely used as a screening test for ASA [2–4], and the inhibitory effects by ASA on sperm motion and fertilizing ability are evaluated to make a decision for the strategy of infertility treatments [5].

The other tests for ASA involve assays that indirectly detect ASA in serum, cervical mucus, follicular fluid, peritoneal fluid, or seminal plasma. For indirect ASA testing, serological tests, such as sperm-agglutination test (SAT) and sperm-immobilization test (SIT), have been usually performed for clinical purposes in infertile women [6].

In this chapter, the indirect tests for ASA for infertile women are described.

2.2 Available Tests for the Detection of Anti-sperm Antibodies (ASAs)

Several assay methods were developed to detect ASA. However, it is obvious that the most important consideration is the selection of the method. Currently available tests for detecting ASA are summarized in Table 2.1. They are classified into three groups according to the characteristics of each test.

Group I contains tests for the detection of bioactivity of ASA, including SAT, SIT, and fertilization-blocking test (FBT) that detects fertilization-blocking antibodies. Group II contains tests for the detection of ASA bound for motile sperm, including IBT or immunospheres (IS), MAR test, and panning test. Group III contains tests for the detection of ASA against sperm or sperm extract, including radiolabeled antiglobulin test, indirect immunofluorescent test, fluorescence-

Table 2.1 Available tests for detecting anti-sperm antibody (ASA)

Group 1. Tests for detecting the bioactivity of ASA
(a)   Sperm-agglutination test (SAT)
(b)   Sperm-immobilization test (SIT)
(c)   Fertilization-blocking test (FBT)
Group 2. Tests for detecting ASA bound for motile sperm
(a)   Immunobead test (IBT) or immunospheres (IS)
(b)   Mixed antiglobulin reaction (MAR) test
(c)   Panning test
Group 3. Tests for detecting ASA against sperm or sperm extract
(a)   Radiolabeled antiglobulin test
(b)   Indirect immunofluorescent test
(c)   Fluorescent-activated cell sorter (FACS)
(d)   Enzyme-linked immunosorbent assay (ELISA)
(e)   Passive hemagglutination

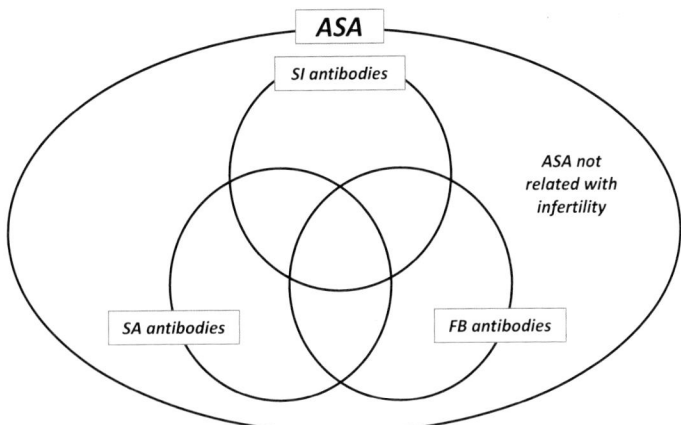

Fig. 2.1 The relationship of the various kinds of anti-sperm antibodies (ASAs) produced in immunological infertile women. There are various kinds of ASA that are produced in infertile women. They include at least two kinds of ASA. One is the ASA that only binds to sperm but not related with infertility. The other is the ASA, such as sperm-immobilizing (SI), sperm-agglutinating (SA), and fertilization-blocking (FB) antibodies, that is related with infertility to varying degrees. For example, some SI antibodies might have only SI activities (the zone that is not overlapped). Other SI antibodies might have all of SI, SA, and FB activities (the zone that is overlapped by the three circles)

activated cell sorter (FACS), enzyme-linked immunosorbent assay (ELISA), and passive hemagglutination for evaluation of ASA.

In these methods, it has been recommended that clinically specific test for female infertility should be chosen for detecting circulating ASA. The relationship between ASA with bioactivities, including sperm-immobilizing (SI) antibodies, sperm-agglutinating (SA) antibodies, and fertilization-blocking (FB) antibodies, and those without bioactivities is drawn in a scheme of Fig. 2.1. Therefore, the tests for detecting ASA that belonged to Group I are suitable for choosing as initial testing for infertile women.

2.2.1   Tests for the Detection of Bioactivity of Anti-sperm Antibodies

2.2.1.1   Sperm-Agglutination Test (SAT)

Detection of ASA has evolved considerably since the first serum agglutination assays were used in the 1950s. In 1952, Kibrick et al. [7] described a macroscopic procedure for identifying antibodies to mammalian sperm in serum. This test is known as gelatin agglutination test (GAT). Since the work of Rumke [8] in 1954, numerous investigators have used this assay to demonstrate the presence of ASA in the sera of infertile women. The GAT is performed by suspending semen from a donor without ASA with the complement-inactivated serum of the suspected subfertile patient in a gelatin mixture in a small glass tube. After incubation for 60 min at 37 °C, the formation of large agglutinated clumps of sperm in the upper portion of the tube indicates that ASAs were present in the serum sample. The GAT is a simple test to perform; however, it was suggested that agglutination may occasionally occur due to the factors that are unrelated with ASA.

In 1964, Franklin and Dukes [9, 10] introduced the tube slide agglutination test (TSAT).

In this assay, a mixture of donor semen and complement-inactivated patient serum is incubated at 37 °C for 60 min. An aliquot is pipetted onto a microscopic slide for observation of sperm-agglutination. They found that 20.1% of 214 women undergoing infertility investigations had detectable sperm-agglutinating activity in their serum. Women with unexplained infertility had a much higher incidence (72.1%) than women with organic causes for their infertility (8.4%) or fertile women (5.7%). It should be noted that this study found a very high incidence of ASA, and the results are not supported by the following reports. One of the reasons may lie in the difficulty of discriminating true antibody-specific sperm-agglutination from nonspecific agglutination.

After their reports, the relationship between sperm-agglutinating antibodies and infertility has been studied inconclusively by many other investigators. Tyler et al. [11] reported that the incidence of positive TSAT in 41 infertile women was 5%, while that in pregnant women was 8%. Isojima et al. [12, 13] reported that the sera from 38 (45.8%) of 83 pregnant women and those of 27 (37.5%) of 72 infertile women with unexplained cause gave a positive reaction with the use of Franklin-Dukes' TSAT [9, 10]. They also showed that the sera from 12 (37.5%) of 32 pregnant women and those of 10 (34.5%) of 29 infertile women with unexplained cause gave a positive reaction with the use of Kibrick's sperm-agglutination test. Isojima has shown that there are various types of sperm-agglutination by ASA as shown in Fig. 2.2 [14]. They include head-to-head (H-H, Fig. 2.2①), head-to-tail (H-T, Fig. 2.2②), and tail-to-tail (T-T, Fig. 2.2③) of sperm-agglutination patterns.

Rose et al. [15] suggested that sperm can also be agglutinated by some mycoplasmas, viruses, and non-antibody serum components. This means that control serum samples must always be included in each test series. Mandelbaum et al. [16] also reported that sperm-agglutination can occur with the presence of bacteria, fungi, or debris in seminal plasma and with non-immunoglobulin serum proteins or pregnancy-related steroids in serum. However, Franklin and Dukes' report was notable from a historical perspective in that it stimulated significant interest in the idea that female immunological responses to sperm could be involved in the development of otherwise unexplained infertility and in the concept of an anti-sperm contraceptive vaccine [17].

Later, Friberg [18] has introduced another microagglutination assay which is simpler to perform than the TSAT. This assay is called as the tray agglutination test (TAT). It uses microliter amounts of reagents and avoids the need for transfer of material from the tube to a microscope slide, which is required by the TSAT. In the TAT, donor sperm and patients' sera are incubated in a flat-bottom tissue-typing

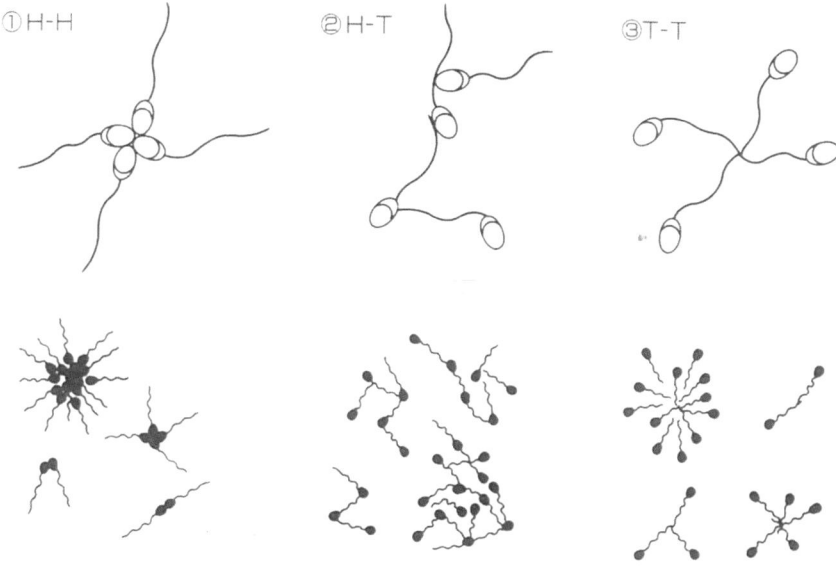

Fig. 2.2 Various types of sperm-agglutination by anti-sperm antibodies (ASAs) [14]. There are various types of sperm-agglutination by ASA. They include head-to-head (H-H, ①), head-to-tail (H-T, ②), and tail-to-tail (T-T, ③) of sperm-agglutination patterns

tray, and agglutination is observed using an inverted microscope. TAT enables a technician to test several hundred serum samples in a single day, using sperm from a single donor. The TAT is apparently more sensitive than the GAT technique, not only for detecting head-to-head agglutination but also tail-to-tail agglutination [19]. This difference in sensitivity is most prominent in the case of female sera, probably because the microscopic examination can detect small agglutinates commonly found with such sera [20].

2.2.1.2 Tests for the Detection of Sperm-Immobilizing Antibodies

2.2.1.2.1 Semiquantitative Sperm-Immobilization Test (SIT)

Because of the nonspecificity of the sperm-agglutination test (SAT), Isojima et al. developed a complement-dependent sperm-immobilization test (SIT) that detects sperm-immobilizing antibodies in the sera of infertile women [12, 13]. This assay utilized the function of sperm-immobilization in the presence of complement.

In their reports, positive reactions in the SIT were given by sera of 17.2% of the patients with infertility of unexplained cause and not by those of normal pregnant and unmarried women. Later, this observation was also confirmed by other studies [21–25]. Therefore, among the several assay methods detecting ASA, SIT has been

Table 2.2 Comparison of the incidence of sperm-immobilizing antibodies between two institutes in Japan

Institute	No. of patients		Incidence (%)
	Tested	Positive	
Jichi MU (*May 1999 to May 2008*)	1339	31	2.3[*]
Hyogo CM (*Jan. 1991 to Dec. 1995*)	1676	46	2.7[*]
Total	3015	77	2.6

[*]$P > 0.05$

shown to be the most reliable assay for aiding in the determination of female infertility because of its specificity.

The original assay method for the SIT [12] was slightly modified to improve the specificity and sensitivity [26]. In the modified test, a volume of 0.25 mL of heat-inactivated (56 °C, 30 min) test serum, 0.025 mL of human sperm suspension (40×10^6/mL), and 0.05 mL of guinea pig serum (ca. 10 C'H_{50} units as complement) are mixed in a small test tube and incubated with gentle shaking at 32 °C for 60 min. One drop of the mixture is put on a glass slide, and the number of motile sperm is counted under the microscope ($\times 200$). The sperm-immobilization value (SIV) is calculated by dividing the percentage of sperm motility in the control (C%) by that in the test serum with complement (T%). The ratio C/T is designated as the SIV. When the SIV is 2 or more, the test serum is judged as positive for sperm-immobilizing antibodies. The higher the SIV, the stronger the sperm-immobilizing antibody activity; however, it should be noted that this result is only to be semi-quantitative. The recent incidences of the SIT in the two institutes that this author belonged were 2.3–2.7% (Table 2.2).

As the ordinary tube assay method is unsuitable for the examination of sperm-immobilizing antibodies in a small volume of specimens like cervical mucus, a microtechnique of the SIT was developed [27, 28]. To improve its specificity, sensitivity, and feasibility, a slightly modified method was developed. As shown in Fig. 2.3, 10 μL of inactivated patient's serum, 1 μL of human sperm suspension (40×10^6/mL), and 2 μL of guinea pig serum as complement source (C'$H_{50} = 200$) or heat-inactivated guinea pig serum as control are mixed on a Terasaki microplate and incubated at 32 °C for 60 min with shaking on a plate mixer. A 10 μL of inactivated unmarried women's serum with complement and 10 μL of inactivated standard patient's serum containing sperm-immobilizing antibodies with complement are used as negative and positive controls. The sperm motility is measured directly in the plate under the microscope ($\times 200$).

2.2.1.2.2 Quantitative Sperm-Immobilization Test (SIT)

In the qualitative sperm-immobilization method, the very high titer of SIV introduces errors, and high titers of SIV are not quantitative. As the SIV cannot quantify the antibody exactly, Isojima et al. developed the quantitative method for this antibody [29].

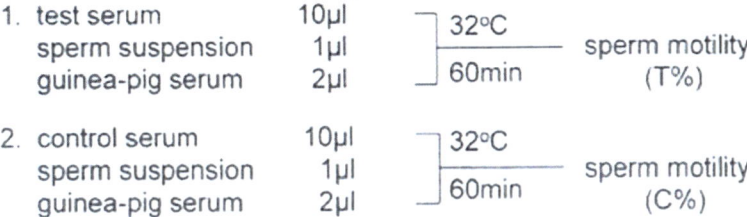

$$\text{SIV (Sperm Immobilization Value)}$$
$$\text{SIV} = C / T \quad \cdots \quad \geq 2 : \text{positive}$$
$$< 2 : \text{negative}$$

Fig. 2.3 Sperm-immobilization test (micro-method). 10 μL of inactivated patient's serum, 1 μL of human sperm suspension (40×10^6/mL), and 2 μL of guinea pig serum as complement source ($C'H_{50} = 200$) or heat-inactivated guinea pig serum as control are mixed on a Terasaki microplate and incubated at 32 °C for 60 min with shaking on a plate mixer. 10 μL of inactivated unmarried women's serum with complement and 10 μL of inactivated standard patient's serum containing sperm-immobilizing antibodies with complement are used as negative and positive controls. The sperm motility is measured directly in the plate under the microscope ($\times 200$)

For quantitative assay for SIT, each test serum is serially diluted twofold with control serum. The sperm motility in the test serum (T%) and in the control serum (C %) were calculated according to the semiquantitative assay method. The antibody activity for sperm-immobilization is calculated by the formula (C-T/C $\times$ 100), and the value is plotted against the dilutions of the test serum on a semi-logarithmic chart to obtain a sigmoid dose-response curve. The dilution of the test serum at which the sigmoid curve crossed the value of 50 for the antibody activity is determined on the dose-response line and designated as 50% sperm-immobilization unit (SI_{50}), as shown in Fig. 2.4. Thus, the amount of sperm-immobilizing antibodies in various sera can be compared. This method is very useful for routine examination of sperm-immobilizing antibody in the sera of infertile women.

Isojima and Koyama found by using the scanning electron microscope that the sperm-immobilizing antibodies opened holes at the acrosome lesion of sperm head and also around the midpiece in collaboration with the complement (Fig. 2.5, unpublished data). This phenomenon is the reason for the sperm-immobilization due to the lysis of sperm membrane by the antibodies and the complement.

Koyama et al. applied this quantitative method to a follow-up study of the SI_{50} titers in the sera of infertile women [30]. When the SI_{50} titers were followed over 3 years in infertile women with sperm-immobilizing antibodies, the SI_{50} titers were found to be unstable and undulated over a period of several months (Fig. 2.6). Later, Kobayashi et al. reported that they found the SI_{50} titers associated with pregnancy rates [31]. Patients with high SI_{50} titers (Fig. 2.6, Group A, greater than 10 units) did not conceive by intrauterine insemination (IUI) except when they were treated with in vitro fertilization-embryo transfer (IVF-ET). Patients with relatively low SI_{50} titers (Fig. 2.6, Group C, less than 10 units) could conceive by IUI, though the

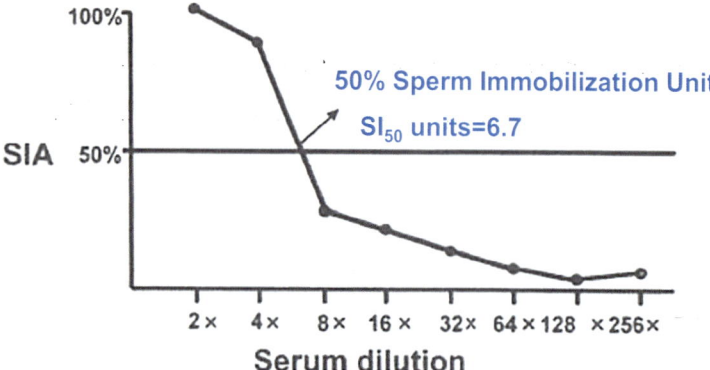

Fig. 2.4 Sperm-immobilization test (quantitative method). Sperm-immobilization in various dilutions of a standard human serum with the sperm-immobilizing antibody is presented. A certain standard patient's serum was serially diluted twofold with control serum. The sperm motility in the test serum (T%) and in the control serum (C%) were calculated according to the semiquantitative assay method. The sperm-immobilization activity (SIA) was calculated by the formula (C-T/C × 100), and the value was plotted against the dilutions of the test serum on a semi-logarithmic chart to obtain a sigmoid dose-response curve. The dilution of the test serum at which the sigmoid curve crossed the value of 50 for the antibody activity was determined on the dose-response line and designated as 50% sperm-immobilization unit (SI_{50}). The SI_{50} titer of the standard serum was calculated as 6.7 on this dose-response line

Fig. 2.5 Electron microscopic findings of immobilized human sperm reacted with sperm-immobilizing antibodies and complement. By using the scanning electron microscope, it was found that the sperm-immobilizing antibodies opened holes at the acrosome lesion of sperm head and also around the midpiece in collaboration with the complement (Fig. 2.5, unpublished data). This phenomenon is the reason for the sperm-immobilization due to the lysis of sperm membrane by the antibodies and the complement

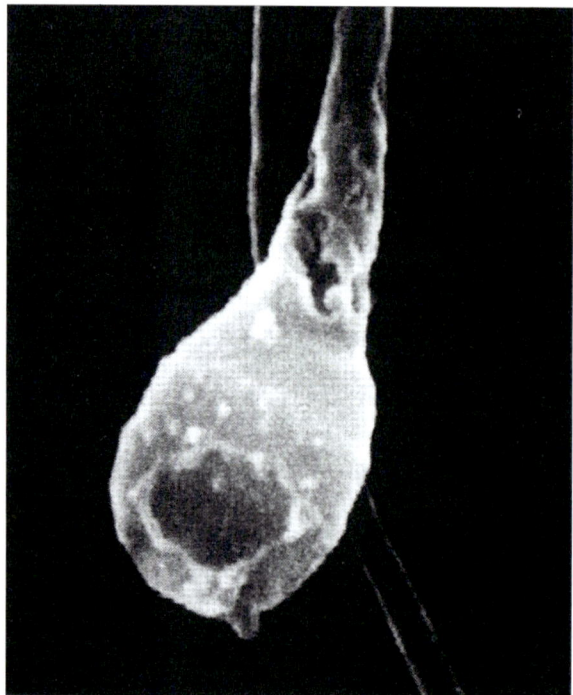

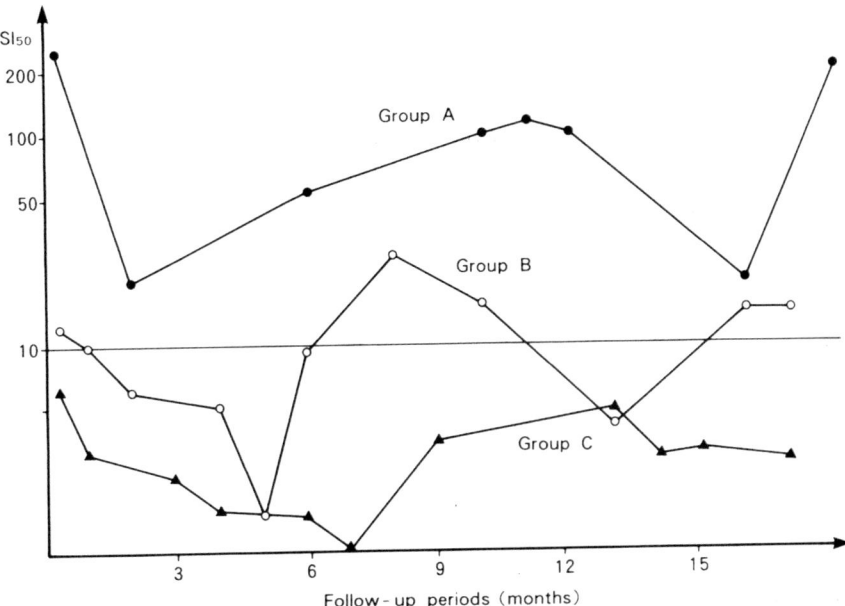

Fig. 2.6 Three undulation patterns (Groups A, B, and C) of SI_{50} titers in the sera of infertile women [31]. Patients were retrospectively divided into three groups (A, B, and C) according to their SI_{50} titers, and the typical patterns of SI_{50} titers in these three groups are shown. Group A included patients whose SI_{50} titers were constantly >10 units, Group B included patients whose SI_{50} titers varied around 10 units, and Group C included patients whose SI_{50} titers were constantly <10 units

success rates were lower than by IVF-ET. They concluded that it is important to assess the SI_{50} titers by the quantitative method to select treatments for infertile women with sperm-immobilizing antibodies.

2.2.1.2.3 Objective Measurement of the Sperm-Immobilization Test (SIT)

Because the judgment of sperm motility is subjective, the results may vary depending on the experience and ability of the operator. To improve it, Komori et al. [32] applied and found the usefulness of Sperm Quality Analyzer (SQA, United Medical Systems Inc., Santa Ana, CA, USA) [33–39] to estimate objective sperm motility for the detection of sperm-immobilizing antibodies. They showed that there was a significant correlation in comparison with SI_{50} titers between the traditional method and the digital method using the SQA. The SQA is a simple and inexpensive commercial unit; they concluded that the results obtained by the method were in good agreement with those obtained by the conventional method.

Recently, Wakimoto et al. [40] applied the computer-aided sperm analysis (CASA) (SMAS, DITECT CO., Tokyo, Japan) for the detection of sperm-immobilizing antibodies. CASA systems that can identify and track human sperm have been developed to revolutionize the research of the movement of human sperm [37, 39, 41–46]. The CASA has the advantages of providing objective semen analysis data on sperm kinetics. Our group has been using CASA to predict sperm fertilizing ability in IVF [43], comparing its functions with other devices such as SQA [34], and analyze the association with multinucleate formation in IVF [44]. We have also shown that three semen parameters including normal morphology before sperm separation and rapid and VCL after sperm separation were identified as predictors of pregnancy by IUI. These variables would be helpful when counseling patients before they make the decision to proceed with in vitro fertilization (IVF)/ intracytoplasmic sperm injection (ICSI)-ET [46].

Wakimoto et al. developed a novel method using CASA, and the results were compared with those obtained by the traditional method. The results were identical, and 25 of 78 samples tested were positive and 53 samples were negative for sperm-immobilizing antibodies based on both methods. For the SIT-positive samples, the values of SI_{50} obtained using the two methods correlated closely with high co-efficiency. They concluded that the novel method using CASA will make it possible to objectively evaluate SIT and SI_{50} data at several clinical facilities and will increase the convenience of using the SIT as a clinical indicator. Furthermore, we recently developed a modified method of SIT using CASA by substitution of cryopreserved sperm in case fresh sperm are unavailable [47].

2.2.1.3 Tests for the Detection of Fertilization-Blocking Antibodies

There are various stages of the normal fertilization process including sperm capacitation, sperm-zona pellucida binding, acrosome reaction, sperm penetration through zona pellucida, sperm-egg fusion, and post-fusion events. Several diagnostic tests were developed to predict the fertilization potential of sperm, including the hemizona assay (HZA) for sperm-zona pellucida tight binding [48], zona pellucida penetration assay for sperm penetration through zona pellucida [49], and zona-free hamster egg-sperm penetration assay for sperm-egg fusion [50]. It has been shown that the sperm-immobilizing antibodies can exert inhibitory effects on sperm-egg interaction at the level of the acrosome reaction [51] and zona pellucida recognition and penetration [52–60].

2.2.1.3.1 Test for the Detection of Anti-sperm Antibodies that Inhibit the Hemizona Assay (HZA)

In 1988, Burkman et al. developed the hemizona assay (HZA) to predict the fertilizing potential of sperm in human [48]. The HZA uses the matching halves of a human zona pellucida from a non-fertilizable and non-living oocyte, providing an internal control on zona-to-zona variability. Maximal binding of human sperm to the hemizona usually occurred after 4–5 h of coincubation. Sperm from fertile men exhibited significantly higher binding capacity to the hemizona compared with sperm from men who had fertilization failure during IVF treatment. The hemizona index (HZI) is calculated as follows: (bound sperm from subfertile male) ÷ (bound sperm from fertile male) × 100. They concluded that the HZA may be a useful diagnostic tool in male infertility evaluations (Fig. 2.7).

We applied the procedure of the HZA in human to detect the inhibitory effects of sperm-immobilizing antibodies on sperm-zona pellucida tight binding [5, 56, 59–61]. Briefly, human oocytes obtained from excised ovarian tissues were stored at −70 °C in a 2.0 M dimethyl sulfoxide solution in phosphate-buffered saline. The frozen oocytes were thawed and cut almost in half using the micromanipulators (Narishige, Tokyo, Japan) mounted on a phase-contrast microscope (Nikon, Garden City, NY, USA). After discarding the degenerated ooplasma, the two matched hemizona were placed overnight at 4 °C in a droplet of medium under mineral oil. Swim-up human sperm were incubated with patient's serum or test solution containing monoclonal anti-sperm antibodies against human sperm before exposure to hemizona at 37 °C for 1 h. One hemizona was placed in a 100 μL drop of swim-up sperm suspension with test sample, while the matched hemizona was placed in a drop of control serum or diluent of solution containing monoclonal anti-sperm antibodies confirmed to be non-inhibitory to sperm-zona binding. After 4 h of coincubation, each hemizona was removed and rinsed vigorously to detach loosely associated sperm. Then the number of sperm tightly bound to the outer hemizona surface was counted. Each zona and each sample were tested twice. The HZI was the number of sperm bound to the hemizona in the test sample divided by that in the control, all multiplied by 100 (Fig. 2.8). When the HZI was 50% or less, the sperm binding to zona pellucida was considered to be inhibited in the test sample as compared with the control sample, according to the criterion of Mahony et al. [57, 58].

To evaluate the in vitro effects of sperm-immobilizing antibodies on sperm-zona pellucida (ZP) tight binding, the HZA was used to study the inhibitory effects of infertile women's sera with and without sperm-immobilizing antibodies on sperm-zona pellucida tight binding. These results were compared with those of monoclonal sperm-immobilizing antibodies. Sera from 40 infertile women (24 with and 16 without sperm-immobilizing antibodies) and 2 postpartum women as control were used. Of 24 patients' sera with sperm-immobilizing antibodies, 23 (96%) showed significant inhibitory effect, whereas none of 16 patient's sera without sperm-immobilizing antibodies exhibited any inhibitory effect. However, there was no correlation between the antibody titers of sperm-immobilizing antibody and the hemizona index. The vast majority of sperm-immobilizing antibodies reduce sperm-zona pellucida binding even without the presence of complement [59].

2.2.1.3.2 Test for the Detection of Anti-sperm Antibodies that Inhibit the Zona Pellucida Penetration Assay (ZPA)

To assess the fertilizing capacity of human sperm, the zona pellucida penetration assay (ZPA) was developed by Yanagimachi et al. [49]. We used the ZPA to test the

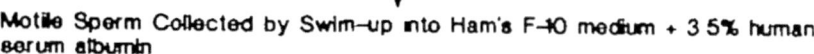

Semen from proven fertile men

Motile Sperm Collected by Swim—up into Ham's F-10 medium + 3 5% human serum albumin

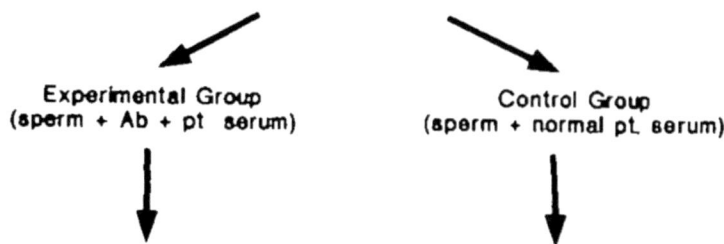

Experimental Group
(sperm + Ab + pt serum)

Control Group
(sperm + normal pt. serum)

Two sperm drops prepared under oil (2 million motile sperm/ml) Oocytes are bisected   One hemizona from a matching pair added to each 100ul drop

After 4 hour coincubation, each hemizona was rinsed to dislodge loosely attached sperm. Assessment of sperm bound to outer surface of each hemizona.

Calculation of Hemizona Index (HZA)

$$HZI = \frac{\text{\# sperm bound in the presence of antibody}}{\text{\# sperm bound in the absence of antibody}} \times 100$$

Fig. 2.7 Procedure for the hemizona assay (HZA) to detect the inhibitory effects of anti-sperm antibodies on sperm-zona pellucida tight binding [5]. The frozen human oocytes were thawed and cut almost in half using the micromanipulators mounted on a phase-contrast microscope. After discarding the degenerated ooplasma, the two matched hemizona (HZ) were placed overnight at 4 °C in a droplet of medium under mineral oil. Swim-up human sperm were incubated with patient's serum or test solution containing monoclonal anti-sperm antibodies (ASAs) against human sperm before exposure to HZ at 37 °C for 1 h. One HZ was placed in a 100 μL drop of swim-up sperm suspension with test sample, while the matched HZ was placed in a drop of control serum or diluent of solution containing monoclonal ASA confirmed to be non-inhibitory to sperm-zona binding. After 4 h of coincubation, each HZ was removed and rinsed vigorously to detach loosely associated sperm. Then the number of sperm tightly bound to the outer HZ surface was counted. Each zona and each sample were tested twice. The HZI was the number of sperm bound to the HZ in the test sample divided by that in the control, all multiplied by 100 (Fig. 2.8). When the HZI was 50% or less, the sperm binding to zona pellucida was considered to be inhibited in the test sample as compared with the control sample

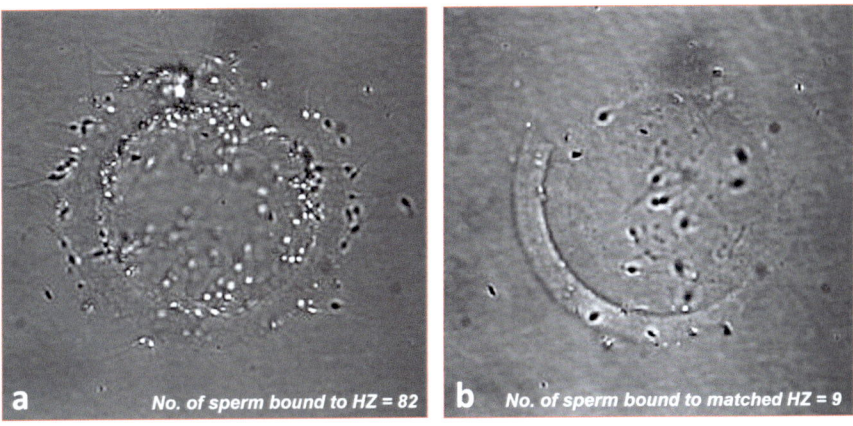

Control sample   **Test sample**

Fig. 2.8 Inhibitory effect of sperm-immobilizing antibodies in the sera of infertile women on hemizona assay (HZA) [5]. The hemizona index (HZI) was the number of sperm bound to the hemizona (HZ) in the test sample divided by that in the control, multiplied by 100. When the HZI was 50% or less, the sperm binding to the zona pellucida was considered to be inhibited in the test sample as compared with the control sample. The number of sperm bound to the HZ was (**a**) 82 when inseminated with control sperm and (**b**) 9 when inseminated with test sperm with ASA. The HZI in this patient was 11.0

blocking effects of anti-sperm antibodies on sperm penetration through the human zona pellucida. Human immature oocytes obtained from excised ovarian tissue were incubated in Menezo's B2 medium (France) containing 1% bovine serum albumin (BSA) for 48 h and then stored at 4 °C in a highly concentrated salt solution containing 0.5 M ammonium sulfate (Wako-junyaku, Osaka, Japan), 1 M magnesium chloride (Wako-junyaku), and 0.1% dextran (Wako-junyaku) until use. Swim-up sperm were incubated with test solution containing monoclonal antibodies against human sperm or control solution containing NS-1 (mouse myeloma cells) before exposure to salt-stored human zona pellucida at 37 °C for 24 h. Three eggs were used in each assay, and each monoclonal antibodies was tested three times. The zona pellucida penetration index (ZPI) was calculated by dividing the number of sperm penetrated into the perivitelline space in the test sample by that in the control sample and multiplying by 100. A value <50% was considered to show inhibition of zona pellucida penetration by sperm.

Four monoclonal antibodies (2C6, 1G12, 3B10, and H6-3C4) against human sperm were used to investigate the diversity of the ASA on fertilization by using the sperm functional assays including HZA, ZPA, and the zona-free hamster egg-sperm penetration assay (SPA), as described below. Both of the two monoclonal antibodies, 2C6 and 1G12, had strong sperm-immobilizing activities and showed significant inhibitory effects in all the three assays. One monoclonal antibody, 3B10, which did not have sperm-immobilizing activity, never inhibited sperm binding to the zona pellucida but showed blocking effects on sperm penetration through both the zona pellucida and the ooplasm. A human monoclonal antibody, H6-3C4, had strong sperm-immobilizing activities but did not show any inhibitory effects in any of the assays [60].

2.2.1.3.3   Test for the Detection of Anti-sperm Antibodies that Inhibit
            the Zona-Free Hamster Egg-Sperm Penetration Assay (SPA)

To assess the fertilizing capacity of human sperm, the zona-free hamster egg-sperm penetration assay (SPA) was developed by Yanagimachi et al. [50]. We used the SPA to test the blocking effects of anti-sperm antibodies on sperm penetration through zona-free hamster egg. Female golden hamsters at the age of 8 weeks were induced to ovulate by an intraperitoneal (i.p.) injection of 30 IU of pregnant mare serum (PMS) on the morning of post-estrous vaginal discharge. At 48 h later, each animal received an i.p. injection of 30 IU human chorionic gonadotropin (HCG, Mochida Pharmaceutical Co., Ltd., Tokyo, Japan). Each hamster was killed 15–17 h later, its oviducts were removed, and the cumulus mass was freed; 0.1% hyaluron-idase (Sigma) and 0.1% trypsin (Sigma) were used to free the ova from the cumulus and to remove the zona pellucida. From 10 to 20 ova were added to 0.1 mL of capacitated human sperm plus patient's serum, covered with mineral oil, and allowed to incubate for 3 h at 37 °C. The ova were then mounted on a slide, compressed with a coverslip, and examined at ×400 magnification with a phase-contrast microscope. Penetration was determined by the presence of a swollen sperm head and attached tail within the cytoplasm (Fig. 2.9) [62]. Sperm samples with no serum added served as a control and an indicator of the integrity of the SPA on any given day. Each sample was tested twice in different experiments. The sperm penetration index (SPI) was calculated by dividing the number of eggs penetrated in the test sample by that in the control sample and multiplying by 100. A value <50% was considered to show inhibition of sperm penetration.

We compared the hemizona index and sperm penetration index in infertile patients to study the diversity of the blocking effects of ASA on fertilization in human. Sera from 22 infertile women with sperm-immobilizing antibodies and 13 women with unexplained infertility without the antibodies were tested for their effects on sperm-zona pellucida binding by using HZA and on sperm penetration into the ooplasm by using SPA. Of 22 patient sera with sperm-immobilizing antibodies and 13 patient sera without the antibodies, 21 (95.5%) and none showed inhibitory effects on HZA, respectively. Of 22 patient sera with sperm-immobilizing antibodies and 13 patient sera without the antibodies, 19 (86.4%) and 8 (61.5%) inhibited SPA, respectively. There was no correlation between the HZI and the SPI in either group. No statistical correlations were found between SI_{50} titers and HZI, nor between SI_{50} titers and SPI [59].

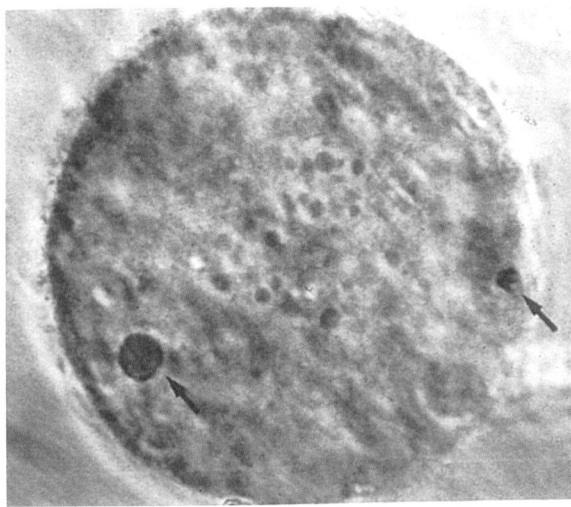

Fig. 2.9 Assessment of the swollen human sperm head in the zona-free hamster egg-sperm penetration assay (SPA) [62]. The zona-free hamster egg-sperm penetration assay (SPA) was used to test the blocking effects of anti-sperm antibodies on sperm penetration through zona-free hamster egg. The oviducts were removed from a hamster, and the cumulus mass was freed with hyaluronidase, and the zona pellucida was removed with trypsin. The zona-free oocytes were added to capacitated human sperm plus patient's serum, covered with mineral oil, and allowed to incubate for 3 h at 37 °C. Penetration was determined by the presence of swollen sperm heads (arrow) and attached tail within the cytoplasm

2.2.2   Tests for the Indirect Detection of Anti-sperm Antibodies Bound for Motile Sperm

The identification of patients with immunological infertility has traditionally depended on tests for ASA in circulating blood serum. However, some individuals possess antibodies in local secretions such as cervical mucus or seminal plasma without the presence of detectable systemic antibodies [63]. Therefore, assays such as the IBT have been developed to allow for the demonstration of membrane-bound antibodies, reflective of local immunity to sperm antigens [2–4]. The IBT, used directly, can detect and characterize sperm membrane-bound antibodies; indirectly (see Chap. 8), it can identify and provide semiquantitative data on ASA in reproductive tract secretions.

The IBT is considered the gold standard of sperm antibody assays. This test uses polyacrylamide beads labeled with antiglobulins (anti-IgG, anti-IgA, and anti-IgM), which bind to the corresponding antibody on the sperm surface. The indirect IBT can be performed on serum, cervical mucus, and follicular fluid and also on seminal plasma of semen with too few motile sperm for a direct IBT [64]. Samples for the indirect IBT may be stored frozen until analyzed. Before use, the sample is heated to 56 °C for 30 min to inactivate the complement. Aliquots of the sample are incubated with donor sperm previously proven negative for membrane-bound antibodies by the direct IBT. These sperm are washed to eliminate unreacted immunoglobulin and mixed with immunobeads, as in the direct IBT. ASA in the test sample will bind to the donor sperm, which then can interact with the anti-human immunoglobulin on the immunobeads.

We collected serum samples from 23 infertile patients with sperm-immobilizing antibodies and 1 pregnant patient to screen sera to determine whether they contained factors to inhibit sperm-zona pellucida tight binding by using the HZA to test for this binding [56]. The HZA showed that all 23 serum samples inhibited sperm-zona pellucida tight binding. The HZI ranged from 3 to 53 with a mean of 18.1 compared to a normal HZI of 100. All 23 serum samples bound to the surface of sperm plasma membrane after 1 h coincubation as evidenced by the fact that they all demonstrated 50% IgG beads bound. Further the results of the indirect IBT for IgG showed that positive sera significantly inhibited binding more than negative sera (HZIs = 12.4 vs 24.4, $p < 0.05$). Yet serum with positive I-IBT for IgM did not affect sperm-zona binding (HZIs = 17.1 vs 19.4, $p > 0.05$). No association existed between HZI and site of immunobead binding. These results might indicate that sera with both sperm-immobilizing antibodies and antibodies recognized by indirect IBT for IgG and IgA may play a significant role to inhibit the sperm-zona pellucida tight binding. In conclusion, physicians should expect patients with low HZI to have more problems conceiving than those with normal HZI. IVF using heat-inactivated human cord serum or donor serum may help them to conceive.

In our laboratory, we have been adopting immunospheres (IS) assay since the IBT was discontinued of its production [65]. The IS uses color-coded latex beads of uniform 3.0 micron size coated with the antiglobulins which can be viewed with bright-field light microscopy. Centola GM et al. compared the IBT and IS in an indirect test using human serum specimens ($n = 42$), which were tested for the presence of antibody isotypes IgG, IgA, and IgM. Donor sperm was washed in BWW with 5% BSA and diluted to a final concentration of 50×10^6 motile sperm/mL. The sperm were incubated with a 1:10 dilution of test serum for 30 min to 1 h at 37 °C and then washed by three cycles of centrifugation. The sperm and beads (IBT, IS) were mixed on a glass slide, covered with a coverslip, and observed within 5 min. At least 100 motile sperm were counted and scored for bead binding. A specimen was considered positive if 20% or more of the sperm were coated with one or more beads. The IS was able to detect 94% of IgG antibodies, 91% of IgA antibodies, and 100% of IgM antibodies. One serum specimen was IgG negative by IS (14% binding), but positive by IBT (20%). A second serum specimen was IgA negative by IS (16%) yet positive by IBT (29%). There were no false positives with the IS assay. Of the IgM positives (five of six) occurred alone and not with IgG or IgA, suggesting the necessity for testing all specimens also for IgM. They concluded that the ASA test results obtained by the IS assay are in agreement with the results obtained with the IBT test [65]. The IS are monodispersed and color-coded and can be visualized with bright-field microscopy.

2.2.3 Tests for the Detection of Anti-sperm Antibodies Against Sperm or Sperm Extract

The ELISA technique has been adapted to quantitatively assess the presence of ASA. ELISA combines the specificity of the antigen-antibody reaction with the continuous degradation of chromogenic substrate by an enzyme to amplify the sensitivity of the reaction. Numerous materials and methods have been used as variations for the ELISA procedure: solid-phase materials (silicone rubber, glass, polyvinyl chloride, polystyrene), carriers (test tubes, beads, disks, microtitration plates), enzymes (alkaline phosphatase, horseradish peroxidase, glucose oxidase, galactosidase), substrates (p-nitrophenyl phosphate, o-phenylenediamine), and a wide range of wash solutions and incubation conditions [66]. Many different methods are used, differing mainly in the type of antigen. Most researchers use whole sperm as antigen, while some use a sperm membrane extract. Other variables include sperm concentration, type of sperm fixation, blocking agents, serum, and seminal plasma dilutions [66]. Said et al. described that the complexity, instrumentation, and expense of the ELISA have prevented its widespread use in the workup of immunological infertility [67].

The use of flow cytometry has been reported to detect sperm-bound antibodies and to quantitate the sperm antibody load (antibody molecules/spermatozoa). Following staining of the washed sperm samples, dead sperm are excluded with fluorescein isothiocyanate-conjugated F(ab')2 fragments of anti-IgG and IgA antibodies by the use of calibration standards. Flow cytometry has the potential reliability and objectivity to quantitate sperm antibodies; therefore, the sperm antibody load can be used to compare different patients or to follow up the progression of the same patient [68]. Similar to ELISA, flow cytometry is not currently widely used for the detection of ASA due to its complexity, expense, and instrumentation requirement [69].

In the same context, the agglutinin radiolabeled antibody assay for the detection and quantitation of ASA is of limited use. This method is also limited by an inability to determine specific ASA location, expense, and reliance on highly skilled labor [69].

References

1. World Health Organization, editor. WHO laboratory manual for the examination and processing of human semen. 5th ed. Geneva: WHO; 2010.
2. Clarke GN, Baker HWG. Treatment of sperm antibodies in infertile men: sperm antibody coating and mucus penetration. In: Bratanov K, editor. Proc Int Symp immunology of reproduction. Bulgaria: Bulgarian Academy of Sciences Press; 1982. p. 337.
3. Clarke GN, Stojanoff A, Cauchi MN. Immunoglobulin class of sperm-bound antibodies in semen. In: Bratanov K, editor. Proc Int Symp immunology of reproduction. Bulgaria: Bulgarian Academy of Sciences Press; 1982. p. 482.

4. Bronson RA, Cooper GW, Rosenfeld DL. Correlation between regional specificity of antisperm antibodies to the spermatozoan surface and complement-mediated sperm immobilization. Am J Reprod Immunol. 1982;2:222–4.
5. Shibahara H, Shiraishi Y, Suzuki M. Diagnosis and treatment of immunologically infertile males with antisperm antibodies. Reprod Med Biol. 2005;4:133–41.
6. Shibahara H, Koriyama J, Shiraishi Y, Hirano Y, Suzuki M, Koyama K. Diagnosis and treatment of immunologically infertile women with sperm-immobilizing antibodies in their sera. J Reprod Immunol. 2009;83:139–44.
7. Kibrick S, Belding DL, Merrill B. Methods for the detection of antibodies against mammalian spermatozoa. I. A modified macroscopic agglutination test. Fertil Steril. 1952;3:419–29.
8. Rumke P. The presence of sperm antibodies in the serum of two patients with oligozoospermia. Vox Sang (Basel). 1954;4:135.
9. Franklin RR, Dukes CD. Antispermatozoal antibody and unexplained infertility. Am J Obstet Gynecol. 1964;89:6–9.
10. Franklin RR, Dukes CD. Further studies on sperm-agglutinating antibody and unexplained infertility. JAMA. 1964;190:682–3. ·
11. Tyler A, Tyler ET, Denny PC. Concepts and experiments in immunoreproduction. Fertil Steril. 1967;18:153–66.
12. Isojima S, Li TS, Ashitaka Y. Immunologic analysis of sperm immobilizing factor found in sera of women with unexplained sterility. Am J Obstet Gynecol. 1968;101:677–83.
13. Isojima S, Tsuchiya K, Koyama K, Tanaka C, Naka O, Adachi H. Further studies on sperm-immobilizing antibody found in sera of unexplained cases of sterility in women. Am J Obstet Gynecol. 1972;112:199–207.
14. Isojima S. Anti-sperm antibodies detected in women. In: Isojima S, editor. Clinical immunology of reproduction. Tokyo: Medical View; 1996. p. 18–33. (in Japanese).
15. Rose NR, Harper MK, Hjort T, Vyazov O, Rumke P. Techniques for detection of iso- and auto-antibodies to human spermatozoa. Clin Exp Immunol. 1976;23:175–99.
16. Mandelbaum SL, Diamond MP, DeCherney AH. The impact of antisperm antibodies on human infertility. J Urol. 1987;138:1–8.
17. Clarke GN. ASA in the female. In: Krause WKH, Naz RK, editors. Immune infertility. Impact of immune reactions on human fertility. 2nd ed. Springer; 2017. p. 161–172.
18. Friberg J. A simple and sensitive micro-method for demonstration of sperm-agglutinating activity in serum from infertile men and women. Acta Obstet Gynecol Scand Suppl. 1974;36: 21–9.
19. Ingerslev HJ, Ingerslev M. Clinical findings in infertile women with circulating antibodies against spermatozoa. Fertil Steril. 1980;33:514–20.
20. Alexander NJ, Ackerman SB, Windt M-L. Immunology. In: Acosta AA, Swanson RJ, Ackerman SB, Kruger TF, van Zyl JA, Menkveld R, editors. Human spermatozoa in assisted reproduction. Philadelphia: Lippincot Williams & Wilkins; 1990. p. 208–22.
21. Ansbacher R, Keung-Yeung K, Behrman SJ. Clinical significance of sperm antibodies in infertile couples. Fertil Steril. 1973;24:305–8.
22. Jones WR, Ing RMY, Kaye MD. A comparison of screening tests for anti-sperm activity in the serum of infertile women. J Reprod Fertil. 1973;32:357–64.
23. Petrunia DM, Taylor PJ, Watson JI. A comparison of methods of screening for sperm antibodies in the serum of women with otherwise unexplained infertility. Fertil Steril. 1976;27:655–61.
24. Cantuiria AA. Sperm immobilizing antibodies in the serum and cervicovaginal secretions in infertile and normal women. Br J Obstet Gynecol. 1977;84:865–8.
25. Blumenfeld Z, Gershon H, Makler A, Stoler J, Brandes JM. Detection of antisperm antibodies: a cytotoxicity immobilization test. Int J Fertil. 1986;31:207–12.
26. Isojima S, Koyama K. Techniques for sperm immobilization test. Arch Androl. 1989;23:185–99.
27. Isojima S, Koyama K. Microtechnique of sperm immobilization test. In: Immunology of reproduction (Proc 3rd Int Symp). Sofia: Bulgarian Academy of Sciences; 1975. p. 215–9.

28. Koyama K. Gamete immunology: infertility and contraception. Reprod Immunol Biol. 2009;24: 1–17.
29. Isojima S, Koyama K. Quantitative estimation of sperm immobilizing antibody in the sera of women with sterility of unknown etiology; the 50% sperm immobilization unit (SI_{50}). Excerpta Medica Intl Cong Series. 1976;370:10–5.
30. Koyama K, Kubota K, Ikuma K, Shigeta M, Isojima S. Application of the quantitative sperm immobilization test for follow-up study of sperm-immobilizing antibody in the sera of sterile women. Int J Fertil. 1988;33:201–6.
31. Kobayashi S, Bessho T, Shigeta M, Koyama K, Isojima S. Correlation between quantitative antibody titers of sperm immobilizing antibodies and pregnancy rates by treatments. Fertil Steril. 1990;54:1107–13.
32. Komori S, Hamada Y, Hasegawa A, Tsubamoto H, Horiuchi I, Tanaka H, Kasumi H, Shigeta M, Koyama K. A digital method of sperm immobilization test: comparison to the conventional method. Am J Reprod Immunol. 2003;50:481–4.
33. Bartoov B, Ben-Barak J, Mayevsky A, Sneider M, Yogev L, Lighyman A. Sperm motility index: a new parameter for human sperm evaluation. Fertil Steril. 1991;56:108–12.
34. Johnston RC, Clarke GN, Liu DY, Baker HW. Assessment of the sperm quality analyzer. Fertil Steril. 1995;63:1071–6.
35. Shibahara H, Naito S, Hasegawa A, Mitsuo M, Shigeta M, Koyama K. Evaluation of sperm fertilizing ability using the sperm quality analyzer. Int J Androl. 1997;20:112–7.
36. Shibahara H, Hamada Y, Hasegawa A, Wakimoto E, Toji H, Shigeta M, Koyama K. Relationship between the sperm motility index assessed by the sperm quality analyzer and the outcome of intracytoplasmic sperm injection. J Assist Reprod Genet. 1999;16:540–5.
37. Suzuki T, Shibahara H, Tsunoda H, Hirano Y, Taneichi A, Obara H, Takamizawa S, Sato I. Comparison of the sperm quality analyzer IIC variables with the computer-aided sperm analysis estimates. Int J Androl. 2002;25:49–54.
38. Shibahara H, Suzuki T, Obara H, Hirano Y, Onagawa T, Taneichi A, Takamizawa S, Sato I. Accuracy of the normal sperm morphology value by sperm quality analyzer IIC: comparison with the strict criteria. Int J Androl. 2002;25:45–8.
39. Hirano Y, Shibahara H, Shimada K, Yamanaka S, Suzuki T, Takamizawa S, Motoyama M, Suzuki M. Accuracy of sperm velocity assessment using the sperm quality analyzer V. Reprod Med Biol. 2004;2:151–7.
40. Wakimoto Y, Fukui A, Kojima T, Hasegawa A, Shigeta M, Shibahara H. Application of computer-aided sperm analysis (CASA) for detecting sperm-immobilizing antibody. Am J Reprod Immunol. 2018;79(3). https://doi.org/10.1111/aji.12814.
41. Sundqvist T, Fjällbrant B, Magnusson KE. Computer-aided counting with the coulter counter of low numbers of spermatozoa in human semen. Int J Androl. 1981;4:18–24.
42. Aitken RJ, Sutton M, Warner P, Richardson DW. Relationship between the movement characteristics of human spermatozoa and their ability to penetrate cervical mucus and zona-free hamster oocytes. J Reprod Fertil. 1985;73:441–9.
43. Hirano Y, Shibahara H, Obara H, Suzuki T, Takamizawa S, Yamaguchi C, Tsunoda H, Sato I. Relationships between sperm motility characteristics assessed by the computer-aided sperm analysis (CASA) and fertilization rates in vitro. J Assist Reprod Genet. 2001;18:213–21.
44. Shibahara H, Yamanaka H, Hirano Y, Kikuchi K, Yamanaka S, Suzuki T, Takamizawa S, Suzuki M. Analysis of factors associated with multinucleate formation in human IVF. Int J Androl. 2003;26:211–4.
45. Shibahara H, Onagawa T, Ayustawati JS, Hirano Y, Suzuki T, Takamizawa S, Suzuki M. Clinical significance of the acridine orange test performed as a routine examination: comparison with the CASA estimates and strict criteria. Int J Androl. 2003;26:236–41.
46. Shibahara H, Obara H, Ayustawati HY, Suzuki T, Ohno A, Takamizawa S, Suzuki M. Prediction of pregnancy by intrauterine insemination using CASA estimates and strict criteria in patients with male factor infertility. Int J Androl. 2004;27:63–8.

47. Wakimoto Y, Fukui A, Wakimoto G, Sugiyama Y, Hasegawa A, Shibahara H. Sperm immobilization test and quantitative sperm immobilization test using frozen-thawed sperm preparation applied with computer-aided sperm analysis. Reprod Med Biol. 2021;20(3):321–6.
48. Burkman LJ, Coddington CC, Franken DR, Kruger TF, Rosenwaks Z, Hodgen GD. The hemizona assay (HZA): development of a diagnostic test for the binding of human spermatozoa to the human hemizona pellucida to predict fertilization potential. Fertil Steril. 1988;49:688–97.
49. Yanagimachi R, Lopata A, Odom CB, Bronson RA, Mahi CA. Penetration of biologic characteristics of zona pellucida in highly concentrated salt solution the use of salt stored eggs for assessing the fertilizing capacity of spermatozoa. Fertil Steril. 1979;35:562–74.
50. Yanagimachi R, Yanagimachi H, Rogers BJ. The use of zona-free animal ova as a test-system for the assessment of the fertilizing capacity of human spermatozoa. Biol Reprod. 1976;15:471–6.
51. Bandoh R, Yamano S, Kamada M, Daitoh T, Aono T. Effect of sperm-immobilizing antibodies on the acrosome reaction of human spermatozoa. Fertil Steril. 1992;57:387–92.
52. Bronson RA, Cooper GW, Rosenfeld DL. Sperm-specific isoantibodies and autoantibodies inhibit the binding of human sperm to the human zona pellucida. Fertil Steril. 1982;38:724–9.
53. Alexander NJ. Antibodies to human spermatozoa impede sperm penetration of cervical mucus or hamster eggs. Fertil Steril. 1984;41:433–9.
54. Kamada M, Daitoh T, Hasebe H, Irahara M, Yamano S, Mori T. Blocking of human fertilization in vitro by sera with sperm-immobilizing antibodies. Am J Obstet Gynecol. 1985;153:328–31.
55. Tsukui S, Noda Y, Fukuda A, Matsumoto H, Tatsumi K, Mori T. Blocking effect of sperm immobilizing antibodies on sperm penetration of human zonae pellucidae. J In Vitro Fert Embryo Transf. 1988;5:123–8.
56. Shibahara H, Shigeta M, Koyama K, Burkman LJ, Alexander NJ, Isojima S. Inhibition of sperm-zona pellucida tight binding by sperm immobilizing antibodies as assessed by the hemizona assay (HZA). Acta Obstet Gynaecol Jpn. 1991;43:237–8.
57. Mahony MC, Fulgham DL, Blackmore PF, Alexander NJ. Evaluation of human sperm-zona pellucida tight binding by presence of monoclonal antibodies to sperm antigens. J Reprod Immunol. 1991;19:269–85.
58. Mahony MC, Blackmore PF, Bronson RA, Alexander NJ. Inhibition of human sperm-zona pellucida tight binding in the presence of antisperm antibody positive polyclonal patient sera. J Reprod Immunol. 1991;19:287–301.
59. Shibahara H, Burkman LJ, Isojima S, Alexander NJ. Effects of sperm-immobilizing antibodies on sperm-zona pellucida tight binding. Fertil Steril. 1993;60:533–9.
60. Shibahara H, Shigeta M, Inoue M, Hasegawa A, Koyama K, Alexander NJ, Isojima S. Diversity of the blocking effects of antisperm antibodies on fertilization in human and mouse. Hum Reprod. 1996;11:2595–9.
61. Shibahara H, Shiraishi Y, Hirano Y, Suzuki T, Takamizawa S, Suzuki M. Diversity of the inhibitory effects on fertilization by anti-sperm antibodies bound to the surface of ejaculated human sperm. Hum Reprod. 2003;18:1469–73.
62. Hasegawa A, Isojima S. Methods for the detection of anti-sperm antibodies. In: Isojima S, editor. Clinical immunology of reproduction. Tokyo: Medical View; 1996. p. 39–57. (in Japanese).
63. Adeghe J-HA, Cohen J, Sawers SR. Relationship between local and systemic autoantibodies to sperm, and evaluation of immunobead test for sperm surface antibodies. Acta Eur Fertil. 1986;17:99–105.
64. Clarke GN, Elliot PJ, Smaila C. Detection of sperm antibodies in semen using the immunobead test: a survey of 813 consecutive patients. Am J Reprod Immunol. 1985;7:118–23.
65. Centola GM, Andolina E, Deutsch A. Comparison of the immunobead binding test (IBT) and immunospheres (IS) assay for detecting serum antisperm antibodies. Am J Reprod Immunol. 1997;37:300–3.

66. Alexander N, Ackerman S, Windt ML. Immunology. In: Acosta A, Swanson R, Ackerman S, et al., editors. Human spermatozoa in assisted reproduction. Baltimore: Williams and Wilkins; 1990. p. 208–22.
67. Said TM, Agarwal A. Tests for sperm antibodies. In: Krause WKH, Naz RK, editors. Immune infertility. Cham: Springer; 2017. p. 197–207.
68. Ke RW, Dockter ME, Majumdar G, Buster JE, Carson SA. Flow cytometry provides rapid and highly accurate detection of antisperm antibodies. Fertil Steril. 1995;63:902–6.
69. Haas GG Jr, D'Cruz OJ, DeBault LE. Comparison of the indirect immunobead, radiolabeled, and immunofluorescence assays for immunoglobulin G serum antibodies to human sperm. Fertil Steril. 1991;55:377–88.

Chapter 3
Causes of Immune Infertility in Women with Anti-sperm Antibody (ASA)

Hiroaki Shibahara

Abstract Some of the anti-sperm antibodies (ASAs) have been shown to be highly related with reproductive failures. In them, sperm-immobilizing antibodies have been shown to be well related with refractory infertility.

The presence of ASA, especially sperm-immobilizing antibodies, in the sera of infertile women has been shown to inhibit sperm migration in the female genital tract. Such antibodies can also exert inhibitory effects on various stages of sperm-egg interaction and subsequent embryo development in vitro. However, ASA might have little impact on recurrent pregnancy loss (RPL).

In this chapter, the topic for the causes of immune infertility in women with ASA is provided.

3.1 Introduction

Some of the anti-sperm antibodies (ASAs) have been shown to be highly related with reproductive failures as shown in Table 3.1. In them, sperm-immobilizing antibodies have been shown to be well related with refractory infertility [1, 2]. The incidence of infertile women having sperm-immobilizing antibodies in their sera has been shown to be approximately 3% when it is tested for female patients who first visit infertility outpatient clinics [3].

The presence of ASA, especially sperm-immobilizing antibodies, in the sera of infertile women has been shown to inhibit sperm migration in the female genital tract. Such antibodies can also exert inhibitory effects on various stages of sperm-egg interaction and subsequent embryo development in vitro. However, ASA might have little impact on recurrent pregnancy loss (RPL).

H. Shibahara (✉)
Department of Obstetrics and Gynecology, School of Medicine, Hyogo Medical University, Nishinomiya, Hyogo, Japan
e-mail: sibahara@hyo-med.ac.jp

Table 3.1 Reproductive failures by anti-sperm antibodies

1.	Infertility related with ASA	
	(a)	Inhibition of sperm migration
		Through cervical mucus
		From uterine cavity to Fallopian tube
	(b)	Inhibitory effect on fertilization process
		Sperm-zona pellucida binding
		Acrosome reaction
		Sperm penetration through the zona pellucida
		Sperm-egg fusion
		Post-fusion events
	(c)	Harmful effect on embryo development
2.	Recurrent pregnancy loss related with ASA	

Recently, the author provided the topics for the roles of ASA in women with reproductive failures [4]. According to the report, the causes of immune infertility in women with ASA will be discussed.

3.2 Inhibitory Effect on Sperm Transport by ASA

3.2.1 ASA in the Cervical Mucus

If any adverse effects of ASA on the process of fertilization are present, ASA should be present in the secretory fluids of the reproductive tracts [5]. In female guinea pigs, Ashitaka et al. find a strong sperm-immobilizing activity in the vaginal secretions immunized with homologous sperm or testis [6].

Similarly, there have been several reports demonstrating the existence of ASA in the cervical mucus (CM) in human (Table 3.2). Parish et al. discovered at least two kinds of cytotoxic ASA in cervical mucus (CM) from infertile women [7]. For the detection of sperm-immobilizing antibodies in CM, the ordinary tube assay method is unsuitable for the examination of the antibody in a small volume of specimens like CM. Therefore, Isojima et al. developed a microtechnique of SIT [8]. By using the micro-SIT technique, they showed the results for the detection of sperm-immobilizing antibodies in CM [5]. In 28 infertile women possessing sperm-immobilizing antibodies in their sera, 22 (78.5%) had the antibodies in their CM, indicating that circulating sperm-immobilizing antibodies in the sera might be transmitted into CM. In 654 infertile women who did not have the antibodies in their sera, 82 (12.5%) had the antibodies in their CM, suggesting that there might be local production of sperm-immobilizing antibodies [8]. There was a significant difference of the incidence of sperm-immobilizing antibodies between the two groups. In total, 104 out of 682 infertile women tested were positive for micro-SIT in CM. The incidence of sperm-immobilizing antibodies in CM was 15.2%.

Later, Shulman et al. reported 7 (11.7%) out of 60 infertile women had ASA in their CM by using a modified immunobead method [9]. Eggert-Kruse et al. found among 192 infertile patients in only 2% of CM samples significant ASA levels by

Table 3.2 Summary of the incidence of anti-sperm antibodies (ASA) in cervical mucus

Authors	Assay used	No. of patients tested	No. of positive patients	Incidence of ASA (%)	References
Parish WE	Cytotoxic Ab test	48	3	6.3	[7]
Isojima S	Micro-SIT	682	104	15.2	[8]
Shulman S	CM-IBT	60	7	11.7	[9]
Eggert-Kruse	Indirect MAR test	192	4	2.1	[10]
Domagala	Indirect IBT	48	2	4.6	[11]
Kamieniczna	Indirect IBT	155	5	3.2	[12]
Total		**1185**	**125**	**10.5**	

Ab antibody, *SIT* sperm-immobilization test, *CM* cervical mucus, *MAR* mixed antiglobulin reaction, *IBT* immunobead test

using the indirect mixed antiglobulin reaction (MAR) test [10]. Among 48 patients of Domagala et al., there were only 2 CM samples (4.6%), which yielded positive results in the indirect immunobead test (IBT) [11]. Among 155 infertile women, Kamieniczna et al. demonstrated ASA in 3.2% of CM samples by means of the IBT [12]. Taken together, the incidence of ASA in CM was 10.5% from these six reports (Table 3.2).

3.2.2 *Inhibitory Effect on Sperm Transport by ASA in the Cervical Mucus*

ASA could cause infertility in female at different stages of reproduction including inhibitory effect on sperm transport in CM. Koyama et al. reported that the incidence of poor post-coital test (PCT) results was significantly higher in infertile women with sperm-immobilizing antibodies in their sera than in those without the antibodies [13]. Other investigators also arrived at similar conclusions [14–16]. These evidences indicate the inhibitory effect of sperm-immobilizing antibodies on sperm penetration through the CM.

In 1974, Isojima et al. developed a method to estimate quantitatively the sperm-immobilizing antibody, SI_{50}, in the sera of infertile women [17]. Later, Koyama et al. also described the usefulness of the evaluation of SI_{50} [18]. They reported that 3 (8%) of 37 patients with low sperm-immobilizing antibody levels in their sera spontaneously conceived. This result suggested that in such a small number of patients with sperm-immobilizing antibodies, sperm penetration through CM does not occur because of the lack of transfer of either serum antibodies or complement to the CM.

We investigated whether SI_{50} titer in infertile women with sperm-immobilizing antibodies correlates with the results of PCT [19]. Twenty-four (77.4%) of 31 women with sperm-immobilizing antibodies and 28 (20.4%) of 137 women

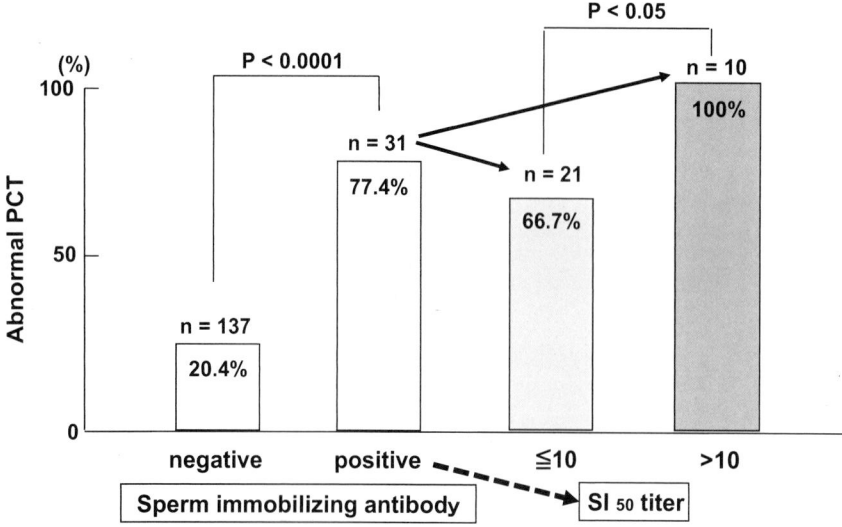

Fig. 3.1 Inhibitory effect on sperm migration in cervical mucus by sperm-immobilizing antibodies using post-coital test (PCT) [4]. Whether SI_{50} titer in infertile women with sperm-immobilizing antibodies correlates with the results of PCT was investigated. Twenty-four (77.4%) of 31 women with sperm-immobilizing antibodies and 28 (20.4%) of 137 women without the antibodies had abnormal PCT results ($P < 0.0001$). When the 31 patients with sperm-immobilizing antibodies were divided into 2 groups according to the SI_{50} titers, the abnormal result of PCT was obtained in all 10 patients with high (>10) SI_{50} titers, while that was 14 (66.7%) in 21 patients with low (≤ 10) SI_{50} titers ($P < 0.05$). These findings suggested that the SI_{50} titer in the serum can predict inhibitory effects on sperm migration through CM in infertile women with sperm-immobilizing antibodies

without the antibodies had abnormal PCT results ($P < 0.0001$) (Fig. 3.1). When the 31 infertile women with sperm-immobilizing antibodies were divided into 2 groups according to the SI_{50} titers, the abnormal result of PCT was obtained in all 10 women with high (>10) SI_{50} titers, while that was 14 (66.7%) in 21 women with low (≤ 10) SI_{50} titers ($P < 0.05$) (Fig. 3.1). These findings suggested that the SI_{50} titer in the serum can predict inhibitory effects on sperm migration through CM in infertile women with sperm-immobilizing antibodies [19].

3.2.3 ASA and Complement in the Peritoneal Fluid

It has not been shown whether sperm migration from the uterine cavity through the fallopian tubes was impaired by ASA. To test the possible impairment of sperm migration in the fallopian tubes, we performed laparoscopic examinations and investigated the presence of motile sperm in the peritoneal fluid following IUI [20], named peritoneal sperm recovery test (PSRT) [21].

In this experiment, both SI_{50} titer and complement activities ($C'H_{50}$) in peritoneal fluid were compared with those in patients' sera. A similar amount of sperm-

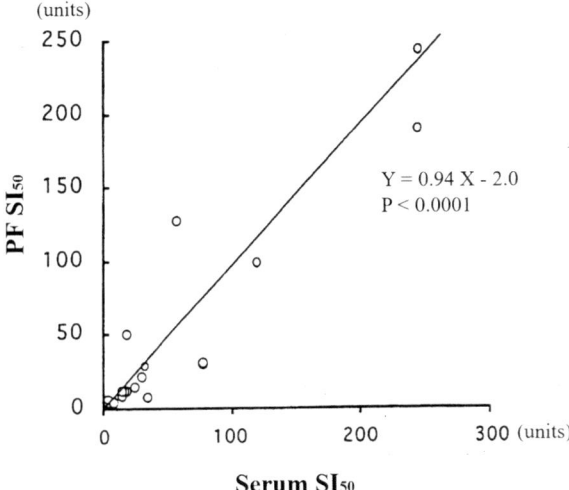

Fig. 3.2 Comparison of sperm-immobilizing antibody titers (SI_{50}) in patients' sera and peritoneal fluid (PF) [20]. To test the possible impairment of sperm migration in the fallopian tubes, we performed laparoscopic examinations and investigated the presence of motile sperm in the peritoneal fluid following IUI. SI_{50} titer in peritoneal fluid was compared with that in patients' sera. A similar amount of sperm-immobilizing antibodies was present in the peritoneal fluid and the sera

immobilizing antibodies was present in the peritoneal fluid and the sera (Fig. 3.2). Although the complement activities in the peritoneal fluid were less than those in sera, the former were still found to be sufficient to immobilize sperm in vivo [20].

3.2.4 Inhibitory Effect on Sperm Transport by ASA from Uterine Cavity Through Fallopian Tube

Patients with ASA, especially sperm-immobilizing antibodies, are less likely to conceive even with intrauterine insemination (IUI). Sperm-immobilizing antibodies not only interfere with sperm migration in CM as mentioned above but also interfere with various stages of the fertilization process.

Following unsuccessful attempts of several cycles of IUI treatments, the PSRT by the laparoscopic examinations was performed after IUI in 239 infertile women with confirmed tubal patency and normal semen characteristics in their husbands. Twenty-seven women had sperm-immobilizing antibodies in their sera, and 212 did not have the antibodies. The peritoneal fluid was aspirated from the Douglas pouch and centrifuged immediately. Then the sediment was microscopically examined for the presence of motile sperm.

As a result, sperm recovery in the peritoneal fluid was not observed in 24 (88.9%) of 27 patients with sperm-immobilizing antibodies, while it was not observed in 140 (66.0%) of 212 patients without the antibodies. There was a significant

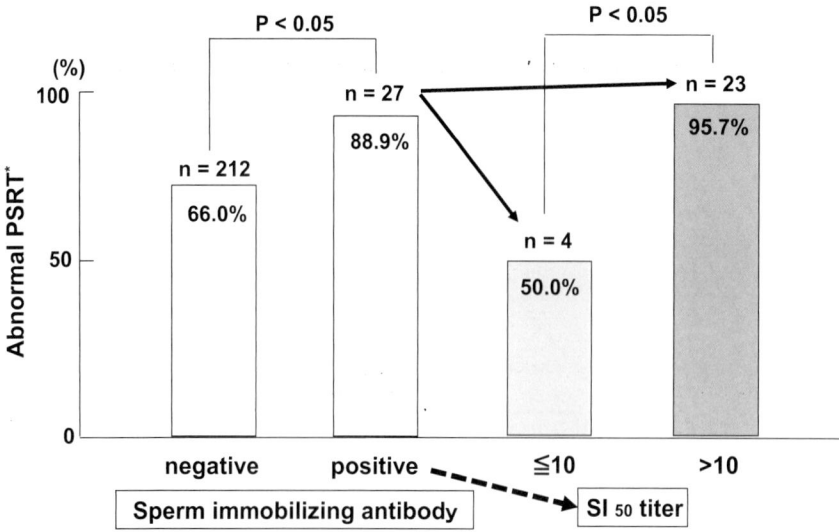

Fig. 3.3 Inhibitory effects on sperm migration from uterine cavity through the fallopian tube by sperm-immobilizing antibodies using peritoneal sperm recovery test (PSRT*) [4]. Sperm recovery in the peritoneal fluid using PSRT was not observed in 24 (88.9%) of 27 patients with sperm-immobilizing antibodies, while it was not observed in 140 (66.0%) of 212 patients without the antibodies. There was a significant difference between the two groups ($P < 0.05$). When the 27 patients with sperm-immobilizing antibodies were divided into 2 groups according to the SI_{50} titers, the abnormal result of PSRT was obtained in 22 (95.7%) of 23 patients with high (>10) SI_{50} titers, while that was 2 (50.0%) in 4 patients with low ($\leqq 10$) SI_{50} titers ($P < 0.05$). These findings suggested that the SI_{50} titer in the serum can predict inhibitory effects on sperm migration from uterine cavity through the fallopian tube in infertile women with sperm-immobilizing antibodies

difference between the two groups ($P < 0.05$) (Fig. 3.3) [20]. When the 27 patients with sperm-immobilizing antibodies were divided into 2 groups according to the SI_{50} titers, the abnormal result of PSRT was obtained in 22 (95.7%) of 23 patients with high (>10) SI_{50} titers, while that was 2 (50.0%) in 4 patients with low ($\leqq 10$) SI_{50} titers ($P < 0.05$) (Fig. 3.3). These findings suggested that the SI_{50} titer in the serum can predict inhibitory effects on sperm migration from uterine cavity through the fallopian tube in infertile women with sperm-immobilizing antibodies.

These results clearly suggested that the complement-dependent sperm-immobilizing antibodies could interfere with sperm migration in the female genital tract at the level of the fallopian tubes.

3.3   Inhibitory Effect of ASA on Sperm-Egg Interaction

3.3.1   ASA in Follicular Fluid

It has been shown that the concentrations of proteins and immunoglobulins in follicular fluid (FF) are lower or equal as compared with blood serum. The interrelationship between the protein fractions, however, is similar to that in serum [22].

Except for a study written by Vujisic et al. [23], the positive relationship of the presence of ASA between FF and serum was dominantly reported. Kohl et al. tested follicular fluid for ASA in 38 women by using an ELISA. Positive results were found only in women with antibodies circulating in serum ($r = 0.88$, $P < 0.001$) [24]. Nip et al. described a higher prevalence of ASA in infertile women and a relation between the concentration found in serum and FF [25]. Marin-Briggiler et al. demonstrated the presence of ASAs in the FF, which were able to induce the AR in capacitated human donor sperm [26]. They also showed that these ASAs were also able to block the zona binding of sperm [27].

We have also reported the effects of in vivo exposure to eggs with sperm-immobilizing antibodies in the FF on subsequent fertilization and embryo development in vitro [28]. In this study, patients' sera and their FF were collected in 15 IVF-ET or ICSI-ET treatment cycles from 11 infertile women with sperm-immobilizing antibodies in their sera. The SI_{50} titers in the sera and the FF were evaluated. There was a significant correlation in the SI_{50} titers between the patients' sera and their FF ($P < 0.0001$) (Fig. 3.4). Therefore, we concluded that thorough washing of oocytes collected in the culture medium should be carried out more carefully than usual to prohibit contamination of the antibodies in the fertilization medium in IVF to obtain a better IVF result. However, if fertilization is inhibited by

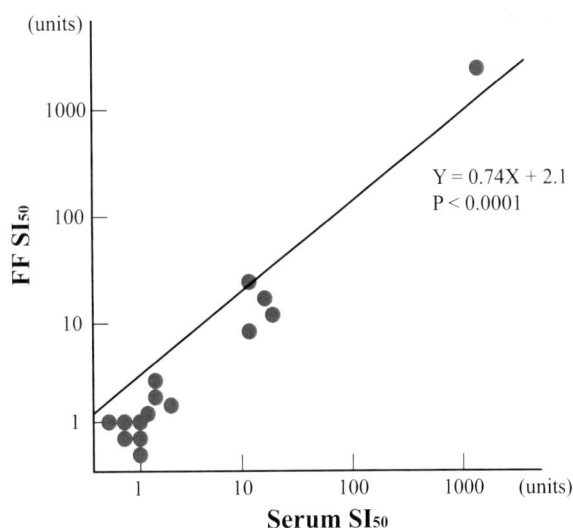

Fig. 3.4 Comparison of sperm-immobilizing antibody titers (SI_{50}) in patients' sera and follicular fluid (FF) [28]. Patients' sera and their FF were collected in 15 IVF-ET or ICSI-ET treatment cycles from 11 infertile women with sperm-immobilizing antibodies in their sera. The SI_{50} titers in the sera and the FF were evaluated. There was a significant correlation in the SI_{50} titers between the patients' sera and their FF ($P < 0.0001$)

$Y = 0.74X + 2.1$
$P < 0.0001$

the existence of a higher number of sperm-immobilizing antibodies, ICSI should be theoretically carried out.

3.3.2 Inhibitory Effect on Sperm-Egg Interaction by ASA in Follicular Fluid

3.3.2.1 Fertilization Failure by ASA in Clinical Practice

ASAs are presented in the FF, and they have been reported to show inhibitory effects on sperm-egg interaction. We retrospectively investigated the clinical effects of sperm-immobilizing antibodies in the sera of infertile women on fertilization rates [29].

Before the introduction of SIT as a routine test for female infertility, 85 oocytes were collected in 9 IVF cycles from 4 infertile women who were afterward found having had sperm-immobilizing antibodies in their sera, and the oocytes were inseminated with swim-up sperm in a medium containing the patient's serum. As controls, 50 oocytes were collected in 5 IVF cycles from 5 infertile women possessing the antibodies in their sera, and the oocytes were inseminated with swim-up sperm in a medium supplemented with human serum albumin (HSA). In the former group, 41 (48.2%) of 85 oocytes were fertilized. In the latter group, 43 (86.0%) of 50 oocytes were fertilized. There was a significant difference between the two groups ($P < 0.0001$) (Fig. 3.5). These findings indicated that the sperm-immobilizing antibodies in the sera of infertile women might cause low fertilization rate [29]. However, after thoroughly washing the collected eggs in culture medium without the patient's serum before IVF, there was no difference in the fertilization rate in the patients with high ($\geqq 10$) and low (<10) SI_{50} titers in their follicular fluid ($P = 0.62$) [28]. This finding is the rationale that ICSI is not necessary for the treatment of infertile women with ASA.

3.3.2.2 Inhibitory Effect of ASA on Sperm-Zona Pellucida Tight Binding

There are various stages of the normal fertilization process including sperm capacitation [30], sperm-zona pellucida binding [31], acrosome reaction [32], sperm penetration through zona pellucida [33], sperm-egg fusion [34], and post-fusion events. Several diagnostic tests were developed to predict the fertilization potential of sperm, including the hemizona assay (HZA) [35] for sperm-zona pellucida tight binding, zona pellucida penetration assay [33] for sperm penetration through zona pellucida, and zona-free hamster egg-sperm penetration assay [34] for sperm-egg fusion. Sperm-immobilizing antibodies can exert inhibitory effects on sperm-egg interaction at the level of the acrosome reaction [36] and zona pellucida recognition and penetration [37–43].

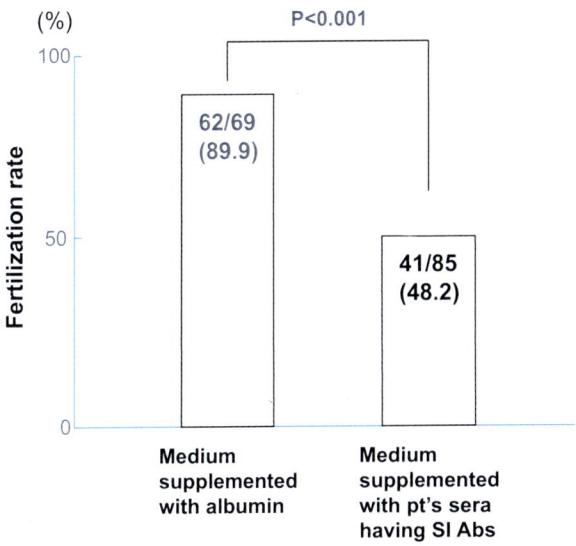

Fig. 3.5 Inhibition of IVF in the medium supplemented with sera from infertile women having sperm-immobilizing antibodies (SI Abs) [29]. Before the introduction of SIT as a routine test for female infertility, 85 oocytes were collected in 9 IVF cycles from 4 infertile women who were afterward found having had sperm-immobilizing antibodies in their sera, and the oocytes were inseminated with swim-up sperm in a medium containing the patient's serum. As controls, 50 oocytes were collected in 5 IVF cycles from 5 infertile women possessing the antibodies in their sera, and the oocytes were inseminated with swim-up sperm in a medium supplemented with human serum albumin (HSA). In the former group, 41 (48.2%) of 85 oocytes were fertilized. In the latter group, 43 (86.0%) of 50 oocytes were fertilized. There was a significant difference between the two groups ($P < 0.0001$)

3.3.2.2.1 Lessons from the Study Using Patient's Serum Containing Sperm-Immobilizing Antibodies

To prove if serum samples containing sperm-immobilizing antibodies exert blocking effect on sperm-egg interaction, we utilized the HZA to study the inhibitory effects of the sera from infertile women with and without sperm-immobilizing antibodies on sperm-zona pellucida tight binding [42]. The lower the hemizona index (HZI) is calculated, the stronger the sperm-zona pellucida tight binding is inhibited (see Chap. 2, Fig. 2.7).

Serum samples from 23 infertile patients with sperm-immobilizing antibodies and 1 pregnant patient were collected to screen sera to determine whether they contained factors to inhibit sperm-zona pellucida tight binding. The HZA showed that all 23 serum samples inhibited sperm-zona pellucida tight binding. The HZI ranged from 3 to 53 with a mean of 18.1 ± 12.0 (mean $\pm$ SD), which indicated a significant inhibitory effect on sperm-zona pellucida tight binding when compared to a normal HZI of 100. We offered that physicians should expect patients with low HZI to have more problems conceiving than those with normal HZI.

Then, we re-evaluated the in vitro effects of sperm-immobilizing antibodies on sperm-zona pellucida tight binding [43]. To prove if sperm-immobilizing antibodies themselves exert blocking effect on sperm-egg interaction, we utilized the HZA to study the inhibitory effects of the sera from infertile women with and without sperm-immobilizing antibodies on sperm-zona pellucida tight binding. When the HZI was 50% or less, the sperm binding to zona pellucida was considered to be significantly inhibited in the test serum as compared in the control serum [40, 44]. These results were compared with those of monoclonal sperm-immobilizing antibodies.

As a result, 23 (96%) of 24 patients' sera with sperm-immobilizing antibodies showed significant inhibitory effect on sperm-zona pellucida binding, whereas none of 16 patients' sera without sperm-immobilizing antibodies exhibited any inhibitory effect. However, there was no correlation between the SI_{50} titers of sperm-immobilizing antibodies and the HZI.

3.3.2.2.2 Lessons from the Study Using Sperm-Immobilizing Monoclonal Antibodies

To prove why the SI_{50} titers of sperm-immobilizing antibodies did not correlate with the HZI, several monoclonal sperm-immobilizing antibodies established in our laboratory were used for the HZA. Among four sperm-immobilizing monoclonal antibodies (2H12 [45], 2C6 [46], H6-3C4 [47], En46 [48]) tested, only one mono-clonal antibody (2C6) showed a significant inhibitory effect on the sperm-zona tight binding. A human monoclonal antibody (H6-3C4) derived from an infertile woman with sperm-immobilizing antibodies, whose serum showed an inhibitory effect on HZA, did not inhibit the HZA [43].

Later, we tested other sperm-immobilizing monoclonal antibodies to study if they show inhibitory effects on sperm-zona pellucida tight binding. The results are summarized in Table 3.3. Sperm-immobilizing monoclonal antibodies that belonged

Table 3.3 Classification of anti-human sperm-immobilizing monoclonal antibodies (mAb) [4]

Group	mAb	Ig class	Immunogen	SI_{50} titer	SAT	HZI
A	1A11	Mouse M	Sperm fraction	1000	1:20	33
	2C6	Mouse M	HSP	200	1:20	9
	1G12	Mouse M	Sperm fraction	100	1:20	7
B	2H12	Mouse M	JEG-3	15000<	1:1	111
	2E5	Mouse G_3	Whole sperm	3000	1:20	83
	YTH-1	Mouse M	Placenta	3000	1:1	103
	YTH-3	Mouse M	Placenta	1000	1:80	117
	2B6	Mouse G_3	Whole sperm	300	1:80	114

Group A: mAbs that have both sperm-immobilizing activities and inhibitory effect on sperm-zona tight binding
Group B: mAbs that have only sperm-immobilizing activities
HSP human seminal plasma, *JEG-3* human choriocarcinoma cell line, *SAT* sperm-agglutination test, *HZI* hemizona index

to Group A (1A11 [49], 2C6 [46], 1G12 [50]) have both sperm-immobilizing activities and also inhibitory effect on sperm-zona pellucida tight binding. On the other hand, sperm-immobilizing monoclonal antibodies that belonged to Group B (2H12 [45], 2E5 [49], YTH-1, YTH-3 [49], 2B6 [49]) have only sperm-immobilizing activities.

Therefore, it was concluded that there are at least two kinds of sperm-immobilizing antibodies, one that causes sperm-immobilization in the presence of complement as well as directly blocking sperm-zona pellucida tight binding and another which only causes sperm-immobilization. And the vast majority of sperm-immobilizing antibodies reduce binding of sperm on zona pellucida even without the presence of complement.

We speculate that there are at least two mechanisms in which ASA inhibits sperm-zona pellucida tight binding. One is that ASA directly binds to the ligand for zona pellucida in the sperm membrane (Fig. 3.6a). The other is that ASA binds to the corresponding antigen near the ligand for zona pellucida in the sperm membrane, resulting in steric hindrance (Fig. 3.6b). Both of them interfere the binding between the ligand for zona pellucida in the sperm membrane and the sperm receptor in zona pellucida.

a. **ASA directly bind to the ligand in the sperm membrane for zona pellucida (ZP)**

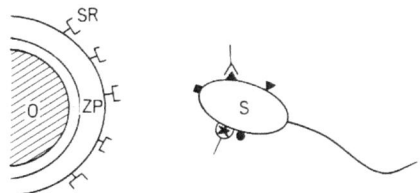

b. **ASA bind to the corresponding antigen near the ligand in the sperm membrane for ZP, resulting a steric hindrance**

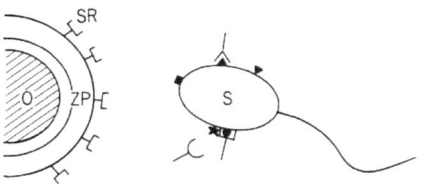

ZP : Zona pellucida
O : Ovum
SR : Sperm receptor in ZP
S : Sperm
★ : Ligand for ZP
● ⎫
■ ⎬ Sperm antigens
▲ ⎭ in membrane
▼ ⎫ ASA against the
Ⴤ ⎬ ligand for ZP
Ⴤ ⎫ ASA against other
Y ⎬ antigens in membrane

Fig. 3.6 Variation of the inhibitory effect on sperm-zona tight binding by anti-sperm antibodies (ASAs) [4]. We speculate that there are at least two mechanisms in which ASA inhibits sperm-zona pellucida (ZP) tight binding. (**a**) One is that ASA directly binds to the ligand (★) for ZP in the sperm membrane. (**b**) The other is that ASA binds to the corresponding antigen (●) near the ligand (★) for ZP in the sperm membrane, resulting in steric hindrance. Both of them interfere the binding between the ligand for ZP in the sperm membrane and the sperm receptor in ZP

3.3.2.3   Each ASA Blocks Fertilization at Specific Stage

Previous reports concerning the blocking effects of ASA on fertilization have involved only a single diagnostic test of sperm function. As there are various stages of the normal fertilization process, it might be suggested that the use of the HZA alone may not be sufficient to detect the blocking effects of ASA on fertilization. Therefore, we investigated by using several diagnostic tests for predicting the fertilization potential of sperm, including the HZA [33], zona pellucida penetration assay (ZPA) [34], and zona-free hamster egg-sperm penetration assay (SPA) [35], to study the diversity of the blocking effects of ASA on fertilization in humans and mice [51].

Use of these three tests allowed us to differentiate whether the antibody blocked sperm-zona pellucida tight binding and/or sperm penetration into the ooplasm. The ZPA was also used to study the effects of four monoclonal antibodies (2C6 [46], 1G12 [50], 3B10 [52], H6-3C4 [47]) on human sperm penetration into the zona pellucida. Seven monoclonal antibodies (Vx4, Vx5, Vx8, Vx10, Vx13, Vx16, Vx18) [53] against murine sperm were tested for their inhibitory effects on IVF and HZA in mice. Of 22 patient sera with sperm-immobilizing antibodies, 21 (95.5%) inhibited HZA attachment and penetration, whereas this did not occur in any of 13 patient sera without these antibodies. However, 19 of 22 (86.4%) patient sera with sperm-immobilizing antibodies and 8 of 13 (61.5%) patient sera without these antibodies inhibited the SPA. Two (2C6, 1G12) of four monoclonal antibodies against human sperm showed strong inhibitory effects in all the assays (HZA, ZPA, and SPA). One monoclonal antibody (3B10) did not inhibit HZA but blocked ZPA and SPA. Another monoclonal antibody (H6-3C4) seemed to have no inhibitory effects on fertilization. Two (Vx5 and Vx8) of the seven monoclonal antibodies against murine sperm inhibited IVF in mice but did not block mouse HZA.

These findings suggest that ASAs block fertilization at specific stages. Some of them may inhibit sperm capacitation and thus prevent all processes of fertilization that follow. Some other antibodies may not affect capacitation and sperm binding to zona pellucida but inhibit the acrosome reaction, followed by the blocking of sperm penetration through zona pellucida and ooplasm [51].

3.4   Impairment of Early Embryo Development by ASA

3.4.1   Early Studies for the Impairment of Early Embryo Development by ASA in Rodent

As the sperm surface antigens were proved to represent in the developing embryos, the embryo development could be impaired by ASA [54, 55]. Koyama et al. [55] have shown that the sperm-specific antigens were transferred to the fertilized eggs in rats by using immunofluorescent staining. Such antigens were located on the cell membrane in the various developmental stages from one cell to blastocyst. They found that ASA against rat sperm impaired the in vitro development of fertilized eggs in the presence of complement.

3.4.2 Sperm-Immobilizing Antibodies in the Sera of Infertile Women Cause Poor Embryo Quality

In human, we found that the sperm-immobilizing antibodies in the sera of infertile women cause poor embryo quality following fertilization in vitro [29]. As mentioned above, the sperm-immobilizing antibody in the sera of infertile women caused low fertilization rate in vitro. Following fertilization in a medium containing the patient's serum with sperm-immobilizing antibody, an evaluation of embryo cleavage was performed on the second day or the third day after oocyte retrieval according to Veeck's classification [56]. The grade 1 and grade 2 embryos were considered good quality. Using this classification, 16 (39.0%) of 41 embryos incubated in the medium containing the patient's serum were good quality, while 34 (79.1%) of 43 embryos incubated in the medium supplemented with HSA were good quality. There was a significant difference between the two groups ($P = 0.0003$). These findings indicated that sperm-immobilizing antibodies in the sera of infertile women could cause poor embryo quality in vitro.

3.4.3 Further Analysis of the Inhibitory Effect on Embryo Development by ASA Using Mice

To investigate if patients' sera with anti-human sperm antibodies show inhibitory effects on IVF and embryo development, we established sperm-immobilization test (SIT) in mice to evade the ethical problems [57]. Patients' sera were collected from 8 infertile women having sperm-immobilizing antibodies and 17 infertile women without the antibodies. Male ICR mice and female F1 mice (BALB/c X C57BL/6J) were used. In mouse IVF, pre-incubated sperm were cultured in the medium containing patient's serum with or without sperm-immobilizing antibodies or bovine serum albumin (BSA) as a control. The fertilization rates and the incidences of blastocyst formation were compared.

Five (62.5%) of 8 serum samples with sperm-immobilizing antibodies and 9 (52.9%) of 17 serum samples without the antibodies showed sperm-immobilizing activities in mice. There was no significant difference between the two groups. Five sera with sperm-immobilizing activities in human and mice and five sera without sperm-immobilizing activities in human or mice were used for further experiments. The fertilization rates in BSA, patient's serum with sperm-immobilizing antibodies, and that without the antibodies were 82.5% (746/904), 43.6% (508/1165), and

64.5% (669/1037), respectively. There were significant differences between the groups. The incidences of blastocyst formation were 59.9% (447/746), 31.7% (161/508), and 47.7% (319/669), respectively. There were also significant differences between the groups.

Some of the patient's serum with and without sperm-immobilizing antibodies could immobilize sperm with complement. However, as compared with control, sera with sperm-immobilizing activities against human and mouse sperm significantly blocked IVF and inhibited embryo development in mice.

In conclusion, it was again suggested that SIT in the sera of infertile women should be performed at least before proceeding IVF. The manipulation of gametes and embryos from patients having sperm-immobilizing antibodies should be carefully carried out especially to avoid contaminating patient's serum and FF in the culture medium in order to overcome the immunological causes of female infertility by ASA, and satisfactory results under suitable conditions for gametes and embryos have been obtained [58–62].

3.5   ASA and Recurrent Pregnancy Loss (RPL)

3.5.1   The Relationship Between ASA and RPL Is Controversial

The relationship between ASA and recurrent pregnancy loss (RPL) was controversially reported. Increased miscarriage rates in women with ASA were demonstrated by some authors [63–67]. In contrast, lack of association between ASA and RPL was reported [68, 69].

Haas GG et al. reported that no pregnancies were subsequently gestated to term in women who were ASA-positive unless they were inoculated with their husband's leukocytes as treatment for an immune basis (not related to ASA) for their recurrent abortions [63]. In this paper, they used a radiolabeled antiglobulin assay (RAA), a modified enzyme-linked immunosorbent assay (ELISA), a tray agglutination test (TAT), and SIT to detect ASA for 173 women with a history of 3 or more recurrent consecutive abortions.

Mathur S et al. investigated the types for human leukocyte antigens (HLA) of A and B loci and screened for the presence of cytotoxic and hemagglutinating antibodies to sperm in 26 couples with 2 or more early spontaneous abortions and 53 couples with recently proven fertility (control group) [64]. They suggested that the presence of ASA is associated with early pregnancy wastage. Histocompatibility antigens B7 and B35 may play a role through their association with ASA and early spontaneous abortions.

Witkin SS et al. examined the relationship between ASA, conception, and miscarriage in 109 infertile couples [65]. ASAs present on the surface of husbands' ejaculated sperm and antibodies in husbands' or wives' sera that reacted with a purified population of the husbands' motile sperm were detected by an ELISA. They concluded that ASAs in female sera and on ejaculated sperm were associated with a failure to conceive and first-trimester miscarriage. Later, they also reported a statistically significant association ($P < 0.005$) between the presence of IgG tail-directed ASA and a history of unexplained RPL [66]. ASA may be a marker for defective immunosuppression in women with RPL. Alternatively, exposure of sperm-sensitized pregnant women to sperm may activate the maternal immune system to respond to paternal antigens present on the embryo.

Naz RK found that early cleavage of zygotes was abnormal after oocytes were fertilized with sperm cells from 1 of 64 infertile couples studied in their human IVF and embryo development procedure [67]. The abnormal cleavage, which did not affect fertilization, was not due to polyspermy or parthenogenetic activation of oocytes, but to ASA present on the sperm cells, which reacted with the acrosomal (strongly) and tail (weakly) regions of human sperm. The antibodies recognized a double band comprising two glycoproteins of 14 ± 3 and 18 ± 3 kDa and a protein band of 22 ± 3 kDa, and antisera were raised in rabbits against the double band and the protein. Passive transfer of affinity-purified human or rabbit immunoglobulins directed against the double band, but not those directed against the protein, caused a significant inhibition of early cleavage of oocytes without affecting pronuclear formation in mice. It was suggested that the sperm surface antigens of 14 ± 3 and 18 ± 3 kDa may provide an extra-nuclear signal to oocytes to divide in mice and humans.

In contrast, lack of association between ASA and RPL was reported [68, 69]. Both of them indicated that women with a history of RPL have a significantly lower incidence of circulating ASA.

3.5.2 Higher Titer of ASA Might Impair Embryo Development and the Very Early Stage of Pregnancy

In our previous study, the effects of sperm-immobilizing antibodies on pregnancy outcome were examined in infertile women treated with IVF-ET [62]. One-hundred forty three ET cycles in 58 infertile women with sperm-immobilizing antibodies and 363 ET cycles in patients with tubal infertility as control were compared. Diagnosis of chemical pregnancy was done when the urinary hCG level had risen over 50 IU/L but a gestational sac could not be demonstrated later. Thirty-three (23.1%) of 143 cycles in the patients with sperm-immobilizing antibodies and 56 (15.4%) of 363 cycles in the control patients were diagnosed as pregnancy. There was a significant difference between the two groups ($P < 0.05$). In the patients with sperm-immobilizing antibodies, 12 (36.4%) were chemical pregnancies, 5 (15.2%) were clinical abortions, and 16 (48.5%) had deliveries. In the control group, 18 (32.1%) were chemical pregnancies, 10 (17.9%) were clinical abortions including ectopic pregnancies, and 28 (50.0%) had deliveries. There were no significant differences in each category. When the SI_{50} titers at the time of conception were considered, chemical pregnancy rates were 22.2% (4/18) in patients with SI_{50} titers below 10 units, but those in patients with SI_{50} titers above 10 were 50.0% (5/10) and above 100 were 60.0% (3/5), respectively ($P > 0.05$). In four of five patients who had both chemical and clinical pregnancies, the SI_{50} titers at the time of conception were higher in the chemical pregnancy cycles than in the clinical pregnancy cycles. Though the pregnancy rates were significantly higher in the patients with sperm-immobilizing antibodies as compared to those with tubal infertility, chemical pregnancy rates were also higher in the patients with higher sperm-immobilizing antibody titers. These results might suggest that sperm-immobilizing antibodies may cause the damage of early development of human embryos in vivo in the patients with a high titer of the antibodies.

In our later study that demonstrated the adverse effects of sperm-immobilizing antibodies in the sera of infertile women on fertilization and embryo quality, there were no patients who aborted after successfully establishing pregnancies following IVF-ET, although the number of patients was low [29].

These findings might suggest that higher titer of sperm-immobilizing antibodies could impair embryo development and the very early stage of pregnancy; however, the antibodies might have little impact on RPL.

References

1. Isojima S, Li TS, Ashitaka Y. Immunologic analysis of sperm immobilizing factor found in sera of women with unexplained sterility. Am J Obstet Gynecol. 1968;101:677–83.
2. Isojima S, Tsuchiya K, Koyama K, Tanaka C, Naka O, Adachi H. Further studies on sperm immobilizing antibody found in sera of unexplained cases of sterility in women. Am J Obstet Gynecol. 1972;112:199–207.
3. Isojima S. Recent advances in reproductive immunology. Asia-Oceania J Obstet Gynaecol. 1983;9:15–26.
4. Shibahara H, Wakimoto Y, Fukui A, Hasegawa A. Anti-sperm antibodies and reproductive failures. Am J Reprod Immunol. 2021;85:e13337.
5. Isojima S, Koyama K. Techniques for sperm immobilization test. Arch Androl. 1989;23:185–99.
6. Ashitaka Y, Isojima S, Ukita H. Mechanism of experimental sterility induced in guinea pigs by injections of homologous testis and sperm. II. Relationship between sterility and a sperm-immobilizing antibody. Fertil Steril. 1964;15:213–21.
7. Parish WE, Ward A. Studies of cervical mucus and serum from infertile women. J Obstet Gynecol British Common. 1968;75:1089–100.
8. Isojima S, Koyama K. Microtechnique of sperm immobilization test. In: Immunology of reproduction (Proc 3rd Int Symp). Sofia: Bulgarian Academy of Sciences; 1975. p. 215–9.
9. Shulman S, Hu C. A study of the detection of sperm antibody in cervical mucus with a modified immunobead method. Fertil Steril. 1992;58:387–91.
10. Eggert-Kruse W, Bockem-Hellwig S, Doll A, Rohr G, Tilgen W, Runnebaum B. Antisperm antibodies in cervical mucus in an unselected subfertile population. Hum Reprod. 1993;8:1025–31.
11. Domagala A, Kasprzak M, Kurpisz M. Immunological characteristics of cervical mucus in infertile women. Zentralbl Gynakol. 1997;119:616–20.

12. Kamieniczna M, Domagala A, Kurpisz M. The frequency of antisperm antibodies in infertile couples-a polish pilot study. Med Sci Monit. 2003;9:CR142–9.
13. Koyama K, Ikuma K, Kubota K, Isojima S. Effects of antisperm antibodies on sperm migration through cervical mucus. Excertta Med Int Congr Ser. 1980;512:705–8.
14. Cantuaria AA. Sperm immobilizing antibodies in the serum and cervicovaginal secretions of infertile and normal women. Br J Obstet Gynaecol. 1977;84:865–8.
15. Chen C, Jones WR. Application of a sperm micro-immobilization test to cervical mucus in the investigation of immunologic infertility. Fertil Steril. 1981;35:542–5.
16. Jager S, Kremer J, De Wilde-janssen IW. Are sperm immobilizing antibodies in cervical mucus an explanation for a poor post-coital test? Am J Reprod Immunol. 1984;5:56–60.
17. Isojima S, Koyama K. Quantitative estimation of sperm immobilizing antibody in the sera of women with sterility of unknown etiology: the 50% sperm immobilization unit (SI_{50}). Excerpta Med Int Congr Ser. 1974;370:10–5.
18. Koyama K, Kubota K, Ikuma K, Shigeta M, Isojima S. Correlation between quantitative antibody titers of sperm immobilizing antibodies and pregnancy rates by treatments. Fertil Steril. 1988;33:201–6.
19. Shibahara H, Shiraishi Y, Hirano Y, Kasumi H, Koyama K, Suzuki M. Relationship between level of serum sperm immobilizing antibody and its inhibitory effect on sperm migration through cervical mucus in immunologically infertile women. Am J Reprod Immunol. 2007;57:142–6.
20. Shibahara H, Shigeta M, Toji H, Koyama K. Sperm immobilizing antibodies interfere with sperm migration from the uterine cavity through the fallopian tubes. Am J Reprod Immunol. 1995;34:120–4.
21. Templeton AA, Mortimer D. Laparoscopic sperm recovery in infertile women. Br J Obstet Gynaecol. 1980;87:1128–31.
22. Munuce MJ, Quintero I, Caille AM, Ghersevich S, Berta CL. Comparative concentrations of steroid hormones and proteins in human peri-ovulatory peritoneal and follicular fluids. Reprod Biomed Online. 2006;13:202–7.
23. Vujisic S, Lepej SZ, Jerkovic L, Emedi I, Sokolic B. Antisperm antibodies in semen, sera and follicular fluids of infertile patients: relation to reproductive outcome after in vitro fertilization. Am J Reprod Immunol. 2005;54:13–20.
24. Kohl B, Kohl H, Krause W, Deichert U. The clinical significance of antisperm antibodies in infertile couples. Hum Reprod. 1992;7:1384–7.
25. Nip MM, Taylor PV, Rutherford AJ, Hanecock KW. Autoantibodies and antisperm antibodies in sera and follicular fluids of infertile patients; relation to reproductive outcome after in-vitro fertilization. Hum Reprod. 1995;10:2564–9.
26. Marin-Briggiler CI, Vazquez-Levin MH, Gonzalez-Echeverria F, Blaquier JA, Miranda PV, Tezon JG. Effect of antisperm antibodies present in human follicular fluid upon the acrosome reaction and sperm-zona pellucida interaction. Am J Reprod Immunol. 2003;50:209–19.
27. Vazquez-Levin MH, Marin-Briggiler CI, Veaute C. Antisperm antibodies: invaluable tools toward the identification of sperm proteins involved in fertilization. Am J Reprod Immunol. 2014;72:206–18.
28. Shibahara H, Hirano Y, Shiraishi Y, Shimada K, Kikuchi K, Suzuki T, Takamizawa S, Suzuki M. Effects of in vivo exposure to eggs with sperm-immobilizing antibodies in follicular fluid on subsequent fertilization and embryo development in vitro. Reprod Med Biol. 2006;5:137–43.
29. Taneichi A, Shibahara H, Hirano Y, Suzuki T, Obara H, Fujiwara H, Takamizawa S, Sato I. Sperm immobilizing antibodies in the sera of infertile women cause low fertilization rates and poor embryo quality in vitro. Am J Reprod Immunol. 2002;47:46–51.
30. Benoff S, Cooper GW, Hurley I, Mandel FS, Rosenfeld DL. Antisperm antibody binding to human sperm inhibits capacitation induced changes in the levels of plasma membrane sterols. Am J Reprod Immunol. 1993;30:113–30.

31. Coddington CC, Alexander NJ, Fulgham D, Mahony M, Johnson D, Hodgen GD. Hemizona assay (HZA) demonstrates effects of characterized mouse antihuman sperm antibodies on sperm zona binding. Andrologia. 1992;24:271–7.
32. Zouari R, De Almeida M, Feneux D. Effect of sperm-associated antibodies on the dynamics of sperm movement and on the acrosome reaction of human spermatozoa. J Reprod Immunol. 1992;22:59–72.
33. Burkman LJ, Coddington CC, Franken DR, Kruger TF, Rosenwaks Z, Hodgen GD. The hemizona assay (HZA)- development of a diagnostic test for the binding of human spermatozoa to the human hemizona pellucida to predict fertilization potential. Fertil Steril. 1988;49:688–97.
34. Yanagimachi R, Lopata A, Odom CB, Bronson RA, Mahi CA. Penetration of biologic characteristics of zona pellucida in highly concentrated salt solution the use of salt stored eggs for assessing the fertilizing capacity of spermatozoa. Fertil Steril. 1979;35:562–74.
35. Yanagimachi R, Yanagimachi H, Rogers BJ. The use of zona-free animal ova as a test-system for the assessment of the fertilizing capacity of human spermatozoa. Biol Reprod. 1976;15:471–6.
36. Bandoh R, Yamano S, Kamada M, Daitoh T, Aono T. Effect of sperm-immobilizing antibodies on the acrosome reaction of human spermatozoa. Fertil Steril. 1992;57:387–92.
37. Bronson RA, Cooper GW, Rosenfeld DL. Sperm-specific isoantibodies and autoantibodies inhibit the binding of human sperm to the human zona pellucida. Fertil Steril. 1982;38:724–9.
38. Alexander NJ. Antibodies to human spermatozoa impede sperm penetration of cervical mucus or hamster eggs. Fertil Steril. 1984;41:433–9.
39. Kamada M, Daitoh T, Hasebe H, Irahara M, Yamano S, Mori T. Blocking of human fertilization in vitro by sera with sperm-immobilizing antibodies. Am J Obstet Gynecol. 1985;153:328–31.
40. Mahony MC, Blackmore PF, Bronson RA, Alexander NJ. Inhibition of human sperm-zona pellucida tight binding in the presence of antisperm antibody positive polyclonal patient sera. J Reprod Immunol. 1991;19:287–301.
41. Tsukui S, Noda Y, Fukuda A, Matsumoto H, Tatsumi K, Mori T. Blocking effect of sperm immobilizing antibodies on sperm penetration of human zonae pellucidae. J In Vitro Fert Embryo Transf. 1988;5:123–8.
42. Shibahara H, Shigeta M, Koyama K, Burkman LJ, Alexander NJ, Isojima S. Inhibition by the hemizona assay (HZA). Acta Obstet Gynaecol Jpn. 1991;43:237–8.
43. Shibahara H, Burkman LJ, Isojima S, Alexander NJ. Effects of sperm-immobilizing antibodies on sperm-zona pellucida tight binding. Fertil Steril. 1993;60:533–9.
44. Mahony MC, Fulgham DL, Blackmore PF, Alexander NJ. Evaluation of human sperm-zona pellucida tight binding by presence of monoclonal antibodies to sperm antigens. J Reprod Immunol. 1991;19:269–85.
45. Tsuji Y, Fukuda H, Iuchi A, Ishizuka I, Isojima S. Sperm immobilizing antibodies react to 3-0-sulfated galactose residue of seminolipid on human sperm. J Reprod Immunol. 1992;22:225–36.
46. Kameda K, Takada Y, Hasegawa A, Tsuji Y, Koyama K, Isojima S. Sperm immobilizing and fertilization-blocking monoclonal antibody 2C6 to human seminal plasma antigen and characterization of the antigen epitope corresponding to the monoclonal antibody. J Reprod Immunol. 1991;20:27–41.
47. Isojima S, Kameda K, Tsuji Y, Shigeta M, Ikeda Y, Koyama K. Establishment and characterization of a human hybridoma secreting monoclonal antibody with high titers of sperm immobilizing and agglutinating activities against human seminal plasma. J Reprod Immunol. 1987;10:67–78.
48. Komori S, Yamasaki N, Shigeta M, Isojima S, Watanabe T. Production of heavy-chain class-switch variants of human monoclonal antibody by recombinant DNA technology. Clin Exp Immunol. 1988;71:508–16.
49. Kameda K, Tsuji Y, Koyama K, Isojima S. Comparative studies of the antigens recognized by sperm-immobilizing monoclonal antibodies. Biol Reprod. 1992;46:349–57.

50. Komori S, Kameda K, Sakata K, Hasegawa A, Toji H, Tsuji Y, Shibahara H, Koyama K, Isojima S. Characterization of fertilization-blocking monoclonal antibody 1G12 with human sperm-immobilizing activity. Clin Exp Immunol. 1997;109:547–54.
51. Shibahara H, Shigeta M, Inoue M, Hasegawa A, Koyama K, Alexander NJ, Isojima S. Diversity of the blocking effects of antisperm antibodies on fertilization in human and mouse. Hum Reprod. 1996;11:2595–9.
52. Toji HA. Fertilization-blocking antibody which recognizes an antigenic epitope for lysozyme on human spermatozoa (in Japanese). Adv Obstet Gynecol. 1991;43:519–527.
53. Ben KL, Hamilton MS, Alexander NJ. Vasectomy-induced autoimmunity: monoclonal antibodies affect sperm function and in vitro fertilization. J Reprod Immunol. 1988;13:73–84.
54. Menge AC. Effect of isoimmunization and isoantisera against seminal antigens on fertility processes in female rabbits. Biol Reprod. 1971;4:137–44.
55. Koyama K, Hasegawa A, Isojima S. Effects of antisperm antibody on the rat embryos. Gamete Res. 1984;10:143–52.
56. Veeck LL. Atlas of the human oocytes and early conceptus, vol. 2. Baltimore: Williams & Wilkins; 1991.
57. Taneichi A, Shibahara H, Takahashi K, Sasaki S, Kikuchi K, Sato I, Yoshizawa M. Effects of sera from infertile women with sperm immobilizing antibodies on fertilization and embryo development in vitro in mice. Am J Reprod Immunol. 2003;50:146–51.
58. Yovich J, Stranger JD, Kay D, Boettcher B. In vitro fertilization of oocytes from women with serum antisperm antibodies. Lancet. 1984;1:369–70.
59. Sugimoto Y, Shigeta M, Ikeda Y, Funauchi H, Taira S, Koyama K, Isojima S. Successful application of in vitro fertilization and embryo replacement for the treatment of infertile women with sperm immobilizing antibody. Acta Obstet Gynaecol Jpn. 1986;38:1135–6.
60. Kobayashi S, Bessho T, Shigeta M, Koyama K, Isojima S. Correlation between quantitative antibody titers of sperm immobilizing antibodies and pregnancy rates by treatments. Fertil Steril. 1990;54:1107–13.
61. Daitoh T, Kamada M, Yamano S, Maruyama S, Kobayashi T, Maegawa M, Aono T. High implantation rate and consequently high pregnancy rate by in vitro fertilization-embryo transfer in infertile women with antisperm antibody. Fertil Steril. 1995;63:87–91.
62. Shibahara H, Mitsuo M, Ikeda Y, Shigeta M, Koyama K. Effects of sperm immobilizing antibodies on pregnancy outcome in infertile women treated with IVF-ET. Am J Reprod Immunol. 1996;36:96–100.
63. Haas GG, Kubota K, Quebbeman JF, Jijon A, Menge AC, Beer AE. Circulating antisperm antibodies in recurrently aborting women. Fertil Steril. 1986;45:209–15.
64. Mathur S, Neff MR, Williamson HO, Genco PV, Rust PF, Glassman AB. Sperm antibodies and human leukocyte antigens in couples with early spontaneous abortions. Int J Fertil. 1987;32:59–65.
65. Witkin SS, David SS. Effect of sperm antibodies on pregnancy outcome in a subfertile population. Am J Obstet Gynecol. 1988;158:59–62.
66. Witkin SS, Chaudhry A. Association between recurrent spontaneous abortions and circulating IgG antibodies to sperm tails in women. J Reprod Immunol. 1989;15:151–8.
67. Naz RK. Effects of antisperm antibodies on early cleavage of fertilized ova. Biol Reprod. 1992;46:130–9.
68. Maier DB, Parke A. Subclinical autoimmunity in recurrent aborters. Fertil Steril. 1989;51:280–5.
69. Clarke GN, Gordon Baker HW. Lack of association between sperm antibodies and recurrent spontaneous abortion. Fertil Steril. 1993;59:463–4.

Chapter 4
Production of Anti-sperm Antibody (ASA) in Women

Hiroaki Shibahara

Abstract The factors that affect production of anti-sperm antibodies (ASAs) in some women are not fully understood. Moreover, the reasons why majority of women do not develop ASA on exposure to sperm are not clear.

By examining the frequencies of HLA-DQA1, HLA-DQB1, and HLA-DRB1 genes, the high frequency of HLA-DRB1*0901 and DQB1*0303 genes in the Japanese population may account for higher frequency of sperm-immobilizing antibody.

Successful heterotransplantation with human peripheral blood lymphocytes (Hu-PBLs) from infertile women with sperm-immobilizing antibodies into severe combined immunodeficient (SCID) mice indicates that the immune response resulting in the production of the antibodies in SCID mice might require the ongoing presence of the eliciting human sperm antigens.

There is a possibility that ASAs are produced in infertile women treated with artificial insemination; however, it is still controversial. There was a significant difference of the incidence in sperm-immobilizing antibodies between the women treated by ICSI because of a severe male factor and those whose husbands had a normal sperm count. Therefore, the production of sperm-immobilizing antibodies is likely to occur in women with particular HLA haplotypes after repeated exposure to a large enough amount of sperm.

Characterization of sperm-immobilizing antibodies may help in the identification of sperm-specific antigens that can be used as candidate antigens for the development of sperm-based contraceptive vaccines.

H. Shibahara (✉)
Department of Obstetrics and Gynecology, School of Medicine, Hyogo Medical University, Nishinomiya, Hyogo, Japan
e-mail: sibahara@hyo-med.ac.jp

H. Shibahara, A. Hasegawa (eds.), *Gamete Immunology*,
https://doi.org/10.1007/978-981-16-9625-1_4

4.1 Introduction

Anti-sperm antibodies (ASAs) are produced in both women and men. Females do not generally produce antibodies against sperm; however, some infertile women have been found to possess ASA which may contribute to their infertility. Moreover, the reasons why most women do not develop an immune response following exposure to sperm are not clear yet.

Several assay methods were developed to detect ASA. Among them, the sperm-immobilization test (SIT), which detects sperm-immobilizing antibodies, has been shown to be the most reliable assay for detecting ASA closely related to female infertility [1, 2]. This observation was confirmed by other studies [3–6]. The possible roles of sperm-immobilizing antibodies in infertility are well known. However, the factors that affect the production of sperm-immobilizing antibodies in infertile women are not fully understood.

4.2 Which Women Produce Anti-sperm Antibodies (ASAs)?

It has been demonstrated that susceptibility to various immune disorders including autoimmune diseases has a genetic background. In our group, Tsuji et al. [7] analyzed HLA genotypes of 38 infertile women possessing sperm-immobilizing antibodies to examine whether susceptibility to antibody production is influenced by HLA. DNA samples of infertile patients possessing sperm-immobilizing anti-bodies were obtained from peripheral white blood cells. The typing of HLA-DR and HLA-DQ was performed by polymerase chain reaction-sequence-specific oligonu-cleotide probes (PCR-SSOP) and PCR-restriction fragment length polymorphism (RFLP), respectively. By examining the frequencies of HLA-DQA1, HLA-DQB1, and HLA-DRB1 genes in comparison to the normal Japanese population, we found that infertile women with sperm-immobilizing antibodies had a significantly higher frequency of genes encoding HLA-DRB1*0901 (26.3% vs 13.6%, $P < 0.005$) and HLA-DQB1*303 (26.3% vs 14.8%, $P < 0.01$) (Table 4.1).

HLA-DRB1*0901 and HLA-DQB1*0303 are very rare among Caucasians but characteristically high among Japanese. In conclusion, the high frequency of

Table 4.1 Gene frequencies with significantly higher incidences in infertile women possessing sperm-immobilizing antibodies compared with that of normal Japanese controls

HLA gene type	Patients' gene frequency (%) ($n = 38$)	Japanese controls' gene frequency (%) ($n = 493$)	P-value
HLA-DRB1*0901	26.3	13.6	< 0.005
HLA-DQB1*0303	26.3	14.8	< 0.01

HLA-DRB1*0901 and DQB1*0303 genes in the Japanese population may account for higher frequency of sperm-immobilizing antibody in Japanese compared to other ethnic groups.

Other immune diseases such as juvenile-onset myasthenia gravis [8] and systemic lupus erythematosus with antiphospholipid syndrome [9] are reported to be linked to these genotypes. These immune diseases also occur at a higher frequency in the Japanese population.

4.3 Why Do Some Women Produce Anti-sperm Antibodies (ASAs)?

4.3.1 Immunization of Females with Sperm or Testis

Some studies had indicated that homologous or heterologous immunization of females with sperm or testis could induce sperm antibody activity and infertility in animals [10, 11]. In human, Baskin [12] reported on a study of 20 fertile women immunized three times intramuscularly at weekly intervals, with their partner's whole ejaculate. All but one of the women showed sperm-immobilizing activity in their serum by 1 week after the last injection which persisted for up to 1 year. One woman became pregnant after 12 months when the sperm-immobilizing activity was no longer detectable in her serum. These results demonstrated that women could be immunized to develop sperm-immobilizing activity, resulting in reduced fecundity.

4.3.2 Reconstruction of Severe Combined Immunodeficient (SCID) Mice with Human Peripheral Blood Lymphocytes (Hu-PBLs) from Infertile Women with Sperm-Immobilizing Antibodies

In the present age, it is difficult to analyze the mechanism of the ASA production by direct immunization with sperm in humans for the ethical reasons. By using severe combined immunodeficient (SCID) mice, an in vitro model of the human immune system [13], that were heterotransplanted with human peripheral blood lymphocytes (Hu-PBLs) from infertile women with sperm-immobilizing antibodies, we previously reported that human sperm could be responsible for the immune response and for the production of sperm-immobilizing antibodies in infertile women [14].

In brief, eight infertile women with sperm-immobilizing antibodies in their sera served as blood donors. The Hu-PBLs were isolated and injected intraperitoneally into female SCID mice. Fresh ejaculated human sperm were washed and used for immunization. As shown in Table 4.2, three mice of the first group were engrafted with Hu-PBLs without any treatment. In the second group, three mice were immunized with washed human sperm on the same day and one day after the engrafting of

Table 4.2 Production of anti-sperm antibodies in SCID mice reconstituted with human peripheral lymphocytes (Hu-PBL) from infertile women with sperm-immobilizing antibodies

Group	Pretreatments (day −1 and day 0)	Reconstitution (day 0)	Immunization (day 0 and day 1)	Human IgG[a] (mg/mL)	SIV[b]	SpermCheck assay[c] (%)
I	None	Hu-PBL	None	0.18–15.0	1.2	12
II	None	Hu-PBL	Human sperm	0.27–3.60	7.0	8
III	Anti-asialo GM1 and irradiation	Hu-PBL	Human sperm	0.50–22.0	13.3	80

[a] Human IgG was measured by ELISA
[b] The value was expressed as sperm-immobilizing test value by sperm immobilization test. SIV of 2 or more was judged as positive
[c] The maximum number of immunolatex beads bound to motile sperm was shown. Value of 20 or more was judged as positive

Hu-PBLs. In the third group, two mice were prepared with a single dose of anti-asialo GM1 antibodies for depleting murine NK activity one day before the injection of Hu-PBLs, followed by X-ray irradiation with a dose of 3 Gy before injection of Hu-PBLs, and then immunized twice with washed human sperm. As a control, four mice were engrafted with Hu-PBLs from four different women without sperm-immobilizing antibodies in their sera following the same protocol used in the third group.

After the injection of Hu-PBLs, mice were bled every 2 weeks. Human serum IgG levels were measured by the enzyme-linked immunosorbent assay (ELISA). ASAs were measured by both the complement-dependent SIT [1] and the SpermCheck assay [15] that could detect ASA bound to sperm surface. As shown in Table 4.2, human IgG appeared in every mouse serum and rapidly increased in concentration reaching peak levels between 2 and 6 weeks after transfer of Hu-PBLs in SCID mice. Sperm-immobilizing antibodies were detected in the second and the third groups. In the third group, SIV and the number of immunolatex beads bound to motile sperm were higher than those in the second group, indicating that pretreatments of the SCID mice with anti-asialo GM1 antibodies and irradiation for depleting NK activities were a useful procedure for reconstituting them with Hu-PBLs. The SIV were lower than those in the patients' sera. Reconstruction of SCID mice with Hu-PBLs from infertile women without the antibodies failed to produce ASA.

In conclusion, we have successfully heterotransplanted Hu-PBLs from infertile women with sperm-immobilizing antibodies, especially following pretreatment of SCID mice with anti-asialo GM1 antibodies and irradiation to deplete NK cells. These findings might indicate that the immune response resulting in the production of sperm-immobilizing antibodies in SCID mice reconstituted with Hu-PBLs from infertile women with the antibodies might require the ongoing presence of the eliciting human sperm antigens. The ASAs in infertile women seem to be produced by an antigen-driven mechanism.

4.3.3   Anti-sperm Antibody Production and Artificial Insemination

Artificial insemination such as intrauterine insemination (IUI) has been a popular treatment for couples with infertility before they proceed to assisted reproductive technology (ART). Its indications are unexplained infertility, mild male factor infertility, inadequate secretion of the cervical mucus, and immunological infertility. However, the washing technique used to prepare a population of sperm with enhanced motility for IUI will also remove other components of semen. Seminal plasma is known to contain immunosuppressive factors, which have been postulated to prevent an immune response in most women following exposure to an antigenic load (i.e., sperm) delivered during intercourse [16]. It may be hypothesized that artificial insemination will increase the risk of a woman developing ASA. However, the clinical findings have been controversial so far.

4.3.3.1   Reports that Do Not Support the Hypothesis

To test the hypothesis that IUI with washed sperm induces ASA formation, Horvath et al. [17] measured serum ASA levels by the immunobead test (IBT) in infertile women. ASA levels were measured before therapy (baseline) and then serially during subsequent cycles, for a maximum of 6 cycles. Twenty-eight patients underwent IUI, and an additional 25 patients who did not undergo IUI were selected as a control. ASA levels were compared between the two groups. Of the 53 enrolled patients, 18 completed 6 treatment cycles, and 35 achieved pregnancy before 6 cycles. Forty-five patients (85%) had less than 10% immunobead binding, six (11%) had binding between 10% and 25% (mean 16%, range 14% to 20%), and two (4%) had binding greater than 25% (28% and 42%, respectively). Mean binding was similar (less than 10%) in the IUI and the control groups. Eighteen patients conceived in the IUI group and 17 in the control group. Of patients who conceived, all but one had less than 10% immunobead binding at the time of conception (mean 1.6 months). In patients who did not conceive, there was no difference in immunobead binding between the control and the IUI groups after 6 months of therapy. They concluded that they did not support the hypothesis that serum ASA levels, as detected by immunobead binding, will increase by undergoing IUI over a prolonged treatment period.

Similarly, Goldberg JM et al. [18] designed to confirm or refute the theoretical concern that IUI may induce ASA in infertile women. Serum and cervical mucus were obtained at the first, fourth, and sixth IUIs. The serum was screened by the IBT for IgG and IgA. If screening results were positive (greater than 10% binding), ASAs were titered by the microimmobilization and microagglutination tests. The IBT was performed on the cervical mucus after liquefaction with bromelin. Ninety-three patients were followed up prospectively. Of these, 40 completed 6 IUI cycles, and the remaining 53 completed 4 cycles. Low transient ASA levels were detected in

10.8% of the patients and would not be expected to affect the prognosis for fertility. They also concluded that IUI does not induce significant ASA production in women.

Moretti-Rojas I et al. [19] investigated the possible development of serum ASA in women receiving repeated IUI. Patients acted as its own control and were evaluated before and after various (1–15) IUI cycles using three different assays for ASA. It was found that only 2 (4.9%) out of 41 women developed ASA. They concluded that exposure of the upper reproductive tract to washed sperm during repeated IUI with partners' sperm does not significantly stimulate the appearance of serum ASA.

From these three reports, the clinical importance of the development of ASA by IUI is not entirely clear, although IUI might predispose small number of women to develop evidence of immunity to sperm.

4.3.3.2 Reports that Support the Hypothesis

After intercourse, sperm can be found in the upper female genital tract and in the peritoneal fluid. It is thus not surprising that ASAs are produced in the female. It has been shown that after coitus, the maximum number of motile sperm found in the ampulla of the fallopian tubes and/or in the peritoneal fluid is less than 200 and, after classical intracervical insemination, about 1100 [20, 21]. IUI will result in a far higher number of sperm in the female genital tract than following normal intercourse. It is conceivable that this can induce an immune response.

Friedman et al. [22] indicated that IUI with capacitated spermatozoa did induce serum ASA in the women inseminated. They carried out a prospective study to assess whether sperm processing and IUI may lead to the production of cervical mucus or circulating ASA in the female partner. Fifty-one infertile women were studied prospectively to assess whether IUI led to the development of ASA in cervical mucus and/or serum. All women were tested for the presence of ASA in cervical mucus and serum before and after IUI. Each woman underwent between one and nine cycles of IUI (mean, 4.0 cycles). Five women (9.8%) developed serum ASA after IUI treatment. Three of 49 women (6.1%) developed cervical mucus ASA, and three other women demonstrated disappearance of antibodies following IUI treatment. IUI did not increase mucus or serum antibody titers in women who presented with ASA. The number of IUI cycles did not correlate with the development of ASA. They concluded that IUI increases the risk of the female partner developing systemic ASA, but the clinical significance of this finding is unclear.

There have been reported several methods for artificial insemination, such as IUI, intra-tubal insemination (ITI) [23–25], direct intraperitoneal insemination (DIPI) [26], direct intrafollicular insemination (DIFI) [27], and fallopian tube sperm perfusion (FSP) [28, 29].

Kahn et al. [29] have investigated whether the FSP method will result in the formation of serum ASA in the female. They showed that some infertile women without ASA detected by IBT or tray agglutination test (Friberg) developed serum ASA after treatments with FSP, in which a large volume (4 mL) of insemination containing $10^7–10^8$ sperm were used for IUI. A total of 184 treatment cycles were given to 128 women. Prior to treatment, 11 (8.6%) women had a positive tray agglutination test and/or a positive IBT. After completing one to four treatment cycles, another six (4.7%) women had developed serum ASA. The antibodies induced by the treatment were of isotype IgM and directed against the tail tip of the sperm. Moreover, 2 of 11 infertile women with ASA diagnosed prior to FSP showed an increase in antibody titer during the FSP treatment. There was no statistically significant difference in the pregnancy rate between the women with ASA and the women without. They concluded that the small risk of developing ASA is not a contraindication for treating infertile couples with FSP.

It has been shown by Livi et al. that some infertile women can conceive after receiving direct intraperitoneal insemination (DIPI) treatment, while others produced ASA after receiving a second insemination procedure without becoming pregnant [30]. To evaluate the occurrence of ASA in women, with no prior sensitization, they tested the serum ASA with the SIT and the IBT for 112 couples undergoing DIPI. A serum sample was taken from each of the 112 patients immediately before the first DIPI. Another sample was taken from 58 patients who underwent a second insemination procedure. In 16 of the 58 patients, the IBT results were positive for one or more immunoglobulin classes. Five patients showed positive SIT. In 7 out of these 16 subjects (12%), the antibodies were bound to the head and to the shaft of the sperm tail. Five of the six patients submitted to a third DIPI procedure showed unchanged SIT values and IBT binding percentages. In one subject, SIT (6 months after the third insemination) became negative. They concluded that the antibody production observed in the study might be either a transient response to massive antigen stimulation or the first step toward systemic immunity.

From these three reports, IUI, FSP, and DIPI might have a small risk of developing ASA. However, treatment for infertile couples by these artificial inseminations seems to have no contraindication.

4.3.4 Anti-sperm Antibody Production and the Partners' Sperm Count

We also found that the production of sperm-immobilizing antibodies in infertile women is associated with their husbands' sperm count [31]. In this study, the sperm-immobilization test (SIT) was performed on 221 infertile women whose husbands had normal semen characteristics according to the criteria by WHO; 160 patients were treated by intracytoplasmic sperm injection (ICSI) because of poor semen characteristics, and 1013 virgin female children acted as the controls. A significant difference of the incidence in SIT was observed between the virgin female children and the women whose husbands had a normal sperm count ($P < 0.0001$). There was also a significant difference of the incidence in SIT between the women treated by ICSI because of a severe male factor and those whose husbands had a normal sperm count ($P < 0.05$) (Tables 4.3 and 4.4).

In this study, the relationship between the SI_{50} titers in the sera of seven infertile women and their husbands' total sperm count was investigated. The mean $\pm$ SD of the SI_{50} titers, the total sperm count, and the total motile sperm count was 71.9 $\pm$ 119.9,

Table 4.3 Incidence of sperm-immobilizing antibodies in infertile women. Comparison between patients whose husbands are with severe oligozoospermia or azoospermia and those with normozoospermia

Women whose husbands are with	N	Sperm-immobilizing antibodies	
		Positive	Negative
Severe oligozoospermia or azoospermia	160	0 (0)[*]	160
Normozoospermia	221	7 (3.12)[*]	214
Total	381	7 (1.84)	374

[*]$P < 0.05$

Table 4.4 Incidence of sperm-immobilizing antibodies in infertile women. Comparison between patients with and without past *Chlamydia trachomatis* (Ct) infection

Anti-Ct antibodies	N	Sperm-immobilizing antibodies	
		Positive (%)	Negative
Positive	78	5 (6.4)[*]	73
Negative	195	3 (1.5)[*]	192
Total	273	8 (2.9)	265

[*]$P < 0.05$

$5.9 \pm 2.0 \times 10^8$, and $3.7 \pm 1.3 \times 10^8$, respectively. There were no significant differences between the SI_{50} titers and the total sperm count or the total motile sperm count. It has been shown that undulations of SI_{50} titers in infertile women with the antibodies exist [32]. Moreover, the semen characteristics are not always reproducible. Taken this into account, there seems to be no relationship between the SI_{50} titers and the sperm counts. Further studies are required to prove why some women produce high titers of the antibodies while others produce low titers.

In conclusion, these results indicate that the production of sperm-immobilizing antibodies in women begins after they have been exposed to a large enough amount of sperm. However, the precise amount of sperm required to produce the antibodies was not clarified.

Taken together, the production of sperm-immobilizing antibodies is likely to be triggered in women with specific types of HLA after being exposed to a large enough amount of sperm. These findings might contribute to the development of immunocontraceptives based on sperm-specific antigens, which can induce sperm-immobilizing antibodies at a sufficient level to block sperm penetration through the female reproductive tract and thus inhibit fertilization.

4.4   What Are the Target Antigens of Sperm-Immobilizing Antibodies?

Biochemical and molecular analysis of the target antigens of sperm-immobilizing antibodies would contribute to elucidation of the immunological mechanism for infertility in women, as well as contribute to a better understanding of the physiological functions of sperm surface antigens.

To identify human sperm antigens recognized by the sera from infertile women having sperm-immobilizing antibodies, two-dimensional gel electrophoresis was employed to separate Percoll-purified human sperm proteins using isoelectric focusing (IEF), followed by polyacrylamide gel electrophoresis (PAGE) [33]. Sperm proteins were transferred to the nitrocellulose membranes and immunoblotted with seven sera from infertile women with high titers of sperm-immobilizing antibodies and six sera from those without SI antibodies [34]. The blots were compared to the 2D composite image of human sperm proteins [sperm protein encyclopedia] and sperm surface index [33], and the sperm surface proteins recognized by infertile sera were identified. Fifty-two human sperm surface proteins reacted with 7 sera from infertile women with sperm-immobilizing antibodies, while 35 of these were reactive with the 6 sperm-immobilizing antibody-negative control sera. The average numbers of protein spots that reacted with test and control sera were 24.6 and 15.0, respectively. A subset of sperm surface proteins which were unique to the sperm-immobilizing antibodies were identified by the following criteria: the sperm protein spots which were highly reactive with the infertile sera containing sperm-immobilizing antibodies but not reactive with any of the sperm-immobilizing antibody-negative infertile sera. As shown in Fig. 4.1, the coordinates of four prominent immunoreactive sperm surface proteins which were unique to the sperm-immobilizing antibodies were identified on the basis of their high reactivity with the sera from infertile women containing sperm-immobilizing antibodies but lack of reactivity with any of the sperm-immobilizing antibody-negative infertile sera.

These sperm surface antigens might be considered as the ideal targets in developing immunocontraceptives because the patients with high SI_{50} titers are refractory to infertility treatments except for IVF-ET. However, it is reported that most of the sperm-immobilizing antibodies in sera of infertile women as well as monoclonal antibodies might be generated to carbohydrate structures of the sperm-coating antigens or sperm membrane proteins [35, 36]. In addition, it was found that mAb H6-3C4 reacts with the N-acetyllactosamine structure in glycolipids from different origins, although the naturally occurring antigen on spermatozoa was not determined [37].

Later, Hasegawa et al. [38] analyzed the epitope for human sperm monoclonal antibodies. Human monoclonal antibody, mAb H6-3C4, that possesses strong sperm-immobilizing activity, was obtained from a human-mouse heterohybridoma that was generated using peripheral B cells of an infertile patient bearing sperm-immobilizing antibodies by Isojima et al. [39]. mAb H6-3C4 has been suggested by several research groups to react with a carbohydrate moiety of male reproductive tract CD52 (mrt-CD52) [40, 41]. CD52 was identified as an antigen for a monoclonal antibody called campath-1 that was produced using a human spleen cell antigen [42, 43]. Human CD52 is expressed in virtually all lymphocytes, mature sperm, and seminal plasma [44]. The sperm maturation-associated CD52 is produced and secreted by epithelial cells of the cauda epididymis. Mature sperm acquire the molecules from the epididymal fluid secreted by the epithelium during their passage, though extra CD52 can be shed into seminal plasma [41, 45]. The core protein of lymphocyte and male reproductive tract CD52 (mrt-CD52) has a common small

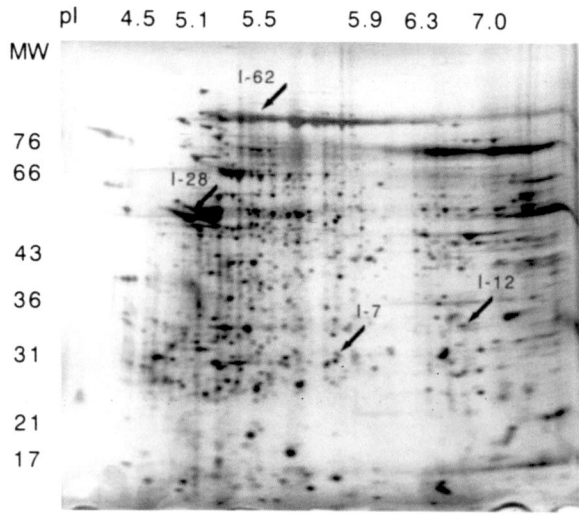

Fig. 4.1 The coordinates of prominent immunoreactive sperm proteins corresponding to sperm-immobilizing antibodies. A subset of sperm surface proteins which were unique to the sperm-immobilizing antibodies were identified by the following criteria: the sperm protein spots which were highly reactive with the infertile sera containing sperm-immobilizing antibodies but not reactive with any of the sperm-immobilizing antibody-negative infertile sera. The coordinates of four prominent immunoreactive sperm proteins are as follows: (MW, pI) = (32.6, 5.9), (35.6, 6.6), (58.6, 4.9), (93.8, 5.5)

peptide comprising 12 amino acids encoded by a single copy gene located on chromosome 1 [46] and a glycosylphosphatidylinositol (GPI) anchor attached to Ser at the COOH-terminus [46, 47]. The 12-amino acid peptide backbone carries a multi-branched N-linked carbohydrate attached to the Asn3 position, having distinct structural differences between lymphocytes and sperm [47, 48]. Male reproductive tract CD52 is characterized by extremely heterogeneous carbohydrate molecules compared to lymphocyte CD52.

Hasegawa et al. [38] analyzed the epitope on mrt-CD52 for mAb H6-3C4 and found that it was polymorphic in Western blot analysis and disappeared after enzymatic removal of the N-linked carbohydrate moiety. Two other monoclonal antibodies (1G12 [49], campath-1 [42]) with sperm-immobilizing activity recognized mrt-CD52 in a polymorphic manner similar to mAb H6-3C4. Further analysis showed that 1G12 recognized a structure formed by the peptide and/or a glycosylphosphatidylinositol (GPI) anchor portion as does campath-1. Results of a lectin binding assay suggested the presence of O-linked carbohydrates on mrt-CD52. These results also indicated that the peptide portion of CD52 could serve as an epitope for sperm-immobilizing antibodies. It was concluded that the epitope of mAb H6-3C4 is similar to, but distinct from, those of 1G12 and campath-1 and that mrt-CD52 contains different antigenic epitopes (Fig. 4.2).

These results suggested that SI Abs in some infertile women target a sperm-specific carbohydrate antigen on CD52 molecules which are secreted from the

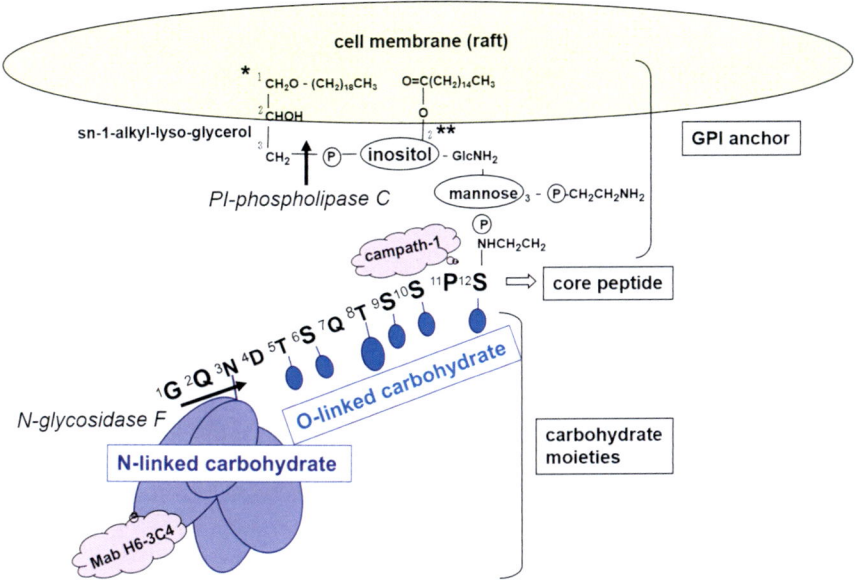

Fig. 4.2 Hypothetical structure of male reproductive tract CD52 (mrt-CD52). The core peptide of mrt-CD52 is composed of 12 amino acid residues, N-linked and O-linked carbohydrates, and a GPI anchor portion inserted in the plasma membrane. Both mAbs, H6-3C4 and Campath-1, are reactive to the mrt-CD52; however, the epitopes are different. H6-3C4 recognizes the mrt-specific N-linked carbohydrate

epididymis and coat the sperm surface. Thereafter, Koyama et al. [50] have found the mrt-CD52 molecule acts to suppress complement-dependent hemolytic activity through the classical pathway. CD52 possibly regulates C1q binding to the antigen-antibody complex, suggesting that CD52 protects sperm from the cytotoxic effects of complement activation in the female genital tract.

4.5   Risk Factors for the Production of Anti-sperm Antibodies (ASAs)

4.5.1   Inflammatory Diseases of the Genital Tract

Among the risk factors for ASA production, inflammatory diseases of the genital tract are believed to play an important role [51]. *Chlamydia trachomatis* (*C. trachomatis*) infection is one of the most common sexually transmitted diseases. There are some reports suggesting that human sperm have antigens that cross-react immunologically with certain microbial antigens, such as *C. trachomatis* [52–54]. However, this is still controversial.

We performed a retrospective study to investigate the correlation between anti-chlamydial antibodies and sperm-immobilizing antibodies in infertile women [55]. The presence of sperm-immobilizing antibodies was examined by the sperm-immobilization test using sera from 273 infertile women. Anti-chlamydial antibodies (IgG and IgA) were examined to prove past *C. trachomatis* infection by ELISA using the same sera from infertile women. As shown in Table 4.4, the overall incidence of sperm-immobilizing antibodies was 2.9% (8/273) in infertile women. The incidences of sperm-immobilizing antibodies were 6.4% (5/78) in cases with past *C. trachomatis* infection and 1.5% (3/195) in cases without past *C. trachomatis* infection ($P = 0.031$). A significantly higher incidence of sperm-immobilizing antibodies was noted in infertile women with past *C. trachomatis* infection compared with that of those without past *C. trachomatis* infection. This was the first demonstration that *C. trachomatis* infection could play a role in the production of sperm-immobilizing antibodies in infertile women.

In this study, there were no significant differences in anti-*C. trachomatis* antibody titers (IgA, IgG) between infertile women with and without sperm-immobilizing antibodies. There was also no significant difference in SI_{50} titers between infertile women with and without past *C. trachomatis* infection. These findings might suggest that *C. trachomatis* infection is one of the triggers for the production of sperm-immobilizing antibodies in women.

4.5.2   Unprotected Sexual Activity During Menses

Under physiological conditions, the cervical epithelium and endometrium function as an integral part of the innate and adaptive immune systems by recognizing and protecting against sperm challenges, to ensure female reproductive health [56].

Defense responses of mucosal immunity in the female genital tract regulated by sex hormones would be impaired during menses. Wang et al. [57] investigated to clarify whether unprotected sexual activity during menses will increase the risk for ASA production in female.

Twenty-seven infertile women who had a usual practice of vaginal intercourse during menses were included, while 30 age-matched infertile women without this practice were set as controls. Indirect immunobead test (I-IBT) was used to measure the ASA levels. No case was revealed to develop significant ASA level, in which 50% or more motile sperm were attached to the immunobeads. Six of 27 women (22.2%) in the case group and 1 of 30 women (3.3%) in the control group were detected to be ASA subpositive, in which 10–50% of motile sperm were attached to the immunobeads. There was a significant difference found in the subpositive incidence between these two groups ($P < 0.05$). Among these six subpositive cases, three became pregnant in the subsequent 2-year follow-up after condom therapy for 6 months. They concluded that sperm exposure during menses is a risk factor for ASA production in female. Although a precise causal linkage between

ASA and infertility in these women cannot be drawn, the potential disadvantages of sexual activity during menses should still be given importance.

References

1. Isojima S, Li TS, Ashitaka Y. Immunologic analysis of sperm-immobilizing factor found in sera of women with unexplained sterility. Am J Obstet Gynecol. 1968;101:677–83.
2. Isojima S, Tsuchiya K, Koyama K, Tanaka C, Naka O, Adachi H. Further studies on sperm-immobilizing antibody found in sera of unexplained cases of sterility in women. Am J Obstet Gynecol. 1972;112:199–207.
3. Ansbacher R, Keung-Yeung K, Behrman SJ. Clinical significance of sperm antibodies in infertile couples. Fertil Steril. 1973;24:305–8.
4. Petrunia DM, Taylor PJ, Watson JI. A comparison of methods of screening for sperm antibodies in the serum of women with otherwise unexplained infertility. Fertil Steril. 1976;27:655–61.
5. Jones WR, Ing RM, Kaye MD. A comparison of screening tests for anti-sperm activity in the serum of infertile women. J Reprod Fertil. 1973;32:357–64.
6. Cantuaria AA. Sperm immobilizing antibodies in the serum and cervicovaginal secretions in infertile and normal women. Br J Obstet Gynaecol. 1977;84:865–8.
7. Tsuji Y, Mitsuo M, Yasunami R, Sakata K, Shibahara H, Koyama K. HLA-DR and HLA-DQ gene typing of infertile women possessing sperm-immobilizing antibody. J Reprod Immunol. 2000;46:31–8.
8. Matsuki K, Juji T, Tokunaga K, Takamizawa M, Maeda H, Soda M, Nomura Y, Sagawa M. HLA antigens in Japanese patients with myasthenia gravis. J Clin Invest. 1990;86:392–9.
9. Matsushita S, Fujisao S, Nishimura Y. Molecular mechanisms underlying HLA-DR-associated susceptibility to autoimmunity. Int J Cardiol. 1996;54:S81–90.
10. Katsh S. In vitro demonstration of uterine anaphylaxis in guinea pigs sensitized with homologous testis or sperm. Nature. 1957;180(4594):1047–8.
11. Isojima S, Graham RM, Graham JB. Sterility in female guinea pigs induced by injection with testis. Science. 1959;129(3340):44.
12. Baskin MJ. Temporary sterilization by the injection of human spermatozoa. Am J Obstet Gynecol. 1932;24:892–7.
13. Mosier DE, Gulizia RJ, Baird SM, Wilson DB. Transfer of a functional human immune system to mice with severe combined immunodeficiency. Nature. 1988;335(6187):256–9.
14. Shibahara H, Tsuji Y, Koyama K. Production of human sperm immobilizing antibodies in severe combined immunodeficient (SCID) mice reconstituted with human peripheral blood lymphocytes from infertile women. Am J Reprod Immunol. 1996;35:57–60.
15. McClure RD, Tom RA, Watkins M, Murthy S. SpermCheck: a simplified screening assay for immunological infertility. Fertil Steril. 1989;52:650–4.
16. Alexander NJ, Anderson DJ. Immunology of semen. Fertil Steril. 1987;47:192–205.
17. Horvath PM, Beck M, Bohrer MK, Shelden RM, Kemmann E. A prospective study on the lack of development of antisperm antibodies in women undergoing intrauterine insemination. Am J Obstet Gynecol. 1989;160:631–7.
18. Goldberg JM, Haering PL, Friedman CI, Dodds WG, Kim MH. Antisperm antibodies in women undergoing intrauterine insemination. Am J Obstet Gynecol. 1990;163(1 Pt 1):65–8.
19. Moretti-Rojas I, Rojas FJ, Leisure M, Stone SC, Asch RH. Intrauterine inseminations with washed human spermatozoa does not induce formation of antisperm antibodies. Fertil Steril. 1990;53:180–2.
20. Ahlgren M. Sperm transport and survival in the human fallopian tube. Gynecol Invest. 1975;6:206–14.

21. Mortimer D, Templeton AA. Sperm transport in human female reproductive tract in relation to semen analysis characteristics and time of ovulation. J Reprod Fertil. 1982;64:401–8.
22. Friedman AJ, Juneau-Norcross M, Sedensky B. Antisperm antibody production following intrauterine insemination. Hum Reprod. 1991;6:1125–8.
23. Jansen RPS, Anderson JC. Catheterisation of the fallopian tubes from the vagina. Lancet. 1987;2:309–10.
24. Berger GS. Intratubal insemination. Fertil Steril. 1987;48:328–30.
25. Shibahara H, Hayashi T, Yamada Y, Shiotani T, Ikuma K. Ultrasound-guided intratubal insemination as a treatment for the infertile patients who were refractory to intrauterine insemination. Jpn J Fertil Steril. 1994;39:198–203. [in Japanese].
26. Forrler A, Dellenbach P, Nisand I, Moreau L, Cranz C, Clavert A, Rumpler Y. Direct intraperitoneal insemination in unexplained and cervical infertility. Lancet. 1986;1:916–7.
27. Lucena E, Ruiz JA, Mendoza JC, Lucena A, Lucena C, Arango A. Direct intrafollicular insemination: a case report. J Reprod Med. 1991;36:525–6.
28. Kahn JA, von During V, Sunde A, Sørdal T, Molne K. Fallopian tube sperm perfusion: first clinical experience. Hum Reprod. 1992;7(Suppl 1):19–24.
29. Kahn JA, Sunde A, von During V, Sørdal T, Remen A, Lippe B, Siegel J, Molne K. Formation of antisperm antibodies in women treated with fallopian tube sperm perfusion. Hum Reprod. 1993;8:1414–9.
30. Livi C, Coccia E, Versari L, Pratesi S, Buzzoni P. Does intraperitoneal insemination in the absence of prior sensitization carry with it a risk of subsequent immunity to sperm? Fertil Steril. 1990;53:137–42.
31. Shibahara H, Kikuchi K, Shiraishi Y, Suzuki M, Shigeta M, Koyama K. Infertile women without sensitization to an appropriate amount of sperm do not produce sperm-immobilizing antibodies in their sera. Reprod Med Biol. 2003;2:105–8.
32. Koyama K, Kubota K, Ikuma K, Shigeta M, Isojima S. Application of the quantitative sperm immobilization test for follow-up study of sperm-immobilizing antibody in the sera of sterile women. Int J Fertil. 1988;33:201–6.
33. Naaby-Hansen S, Flickinger CJ, Herr JC. Two-dimensional gel electrophoretic analysis of vectorially labeled surface proteins of human spermatozoa. Biol Reprod. 1997;56:771–87.
34. Shibahara H, Sato I, Shetty J, Naaby-Hansen S, Herr JC, Wakimoto E, Koyama K. Two-dimensional electrophoretic analysis of sperm antigens recognized by sperm immobilizing antibodies detected in infertile women. J Reprod Immunol. 2002;53:1–12.
35. Koyama K, Kubota K, Kameda K, Shigeta M, Nakamura N, Isojima S. Recognition of carbohydrate antigen epitopes by sperm immobilizing antibodies in sera of infertile women. Fertil Steril. 1991;56:954–9.
36. Kameda K, Tsuji Y, Koyama K, Isojima S. Comparative studies of the antigens recognized by sperm-immobilizing monoclonal antibodies. Biol Reprod. 1992;46:349–57.
37. Tsuji Y, Clausen H, Nudelman E, Kaizu T, Hakomori S, Isojima S. Human sperm carbohydrate antigens defined by an antisperm human monoclonal antibody derived from an infertile woman bearing antisperm antibodies in her serum. J Exp Med. 1988;168:343–56.
38. Hasegawa A, Fu Y, Tsubamoto H, Tsuji Y, Sawai H, Komori S, Koyama K. Epitope analysis for human sperm-immobilizing monoclonal antibodies, MAb H6-3C4, 1G12 and campath-1. Mol Hum Reprod. 2003;9:337–43.
39. Isojima S, Kameda K, Tsuji Y, Shigeta M, Ikeda Y, Koyama K. Establishment and characterization of a human hybridoma secreting monoclonal antibody with high titers of sperm immobilizing and agglutinating activities against human seminal plasma. J Reprod Immunol. 1987;10:67–78.
40. Diekman AB, Norton EJ, Klotz KL, Westbrook VA, Shibahara H, Naaby-Hansen S, Flickinger CJ, Herr JC. N-linked glycan of a sperm CD52 glycoform associated with human infertility. FASEB J. 1999;13:1303–13.
41. Kirchhoff C, Schroter S. New insights into the origin, structure and role of CD52: a major component of the mammalian sperm glycocalyx. Cells Tissues Organs. 2001;168:93–104.

42. Hale G, Xia MQ, Tighe HP, Dyer MJ, Waldmann H. The CAMPATH-1 antigen (CDw52). Tissue Antigens. 1990;35:118–27.
43. Valentin H, Gelin C, Coulombel L, Zoccola D, Morizet J, Bernard A. The distribution of the CDW52 molecule on blood cells and characterization of its involvement in T cell activation. Transplantation. 1992;54:97–104.
44. Hale G, Rye PD, Warford A, Lauder I, Brito-Babapulle A. The glycosylphosphatidylinositol-anchored lymphocyte antigen CDw52 is associated with the mrt maturation of human spermatozoa. J Reprod Immunol. 1993;23:189–205.
45. Kirchhoff C. Molecular characterization of male reproductive tract proteins. Rev Reprod. 1998;3:86–95.
46. Xia MQ, Tone M, Packman L, Hale G, Waldmann H. Characterization of the CAMPATH-1 (CDw52) antigen: biochemical analysis and cDNA cloning reveal an unusually small peptide backbone. Eur J Immunol. 1991;21:1677–84.
47. Schroter S, Derr P, Conradt HS, Nimtz M, Hale G, Kirchhoff C. Male-specific modification of human CD52. J Biol Chem. 1999;274:29862–73.
48. Treumann A, Lifely R, Schneider P. Primary structure of CD52. J Biol Chem. 1995;270:6088–99.
49. Komori S, Kameda K, Sakata K, Hasegawa A, Toji H, Tsuji Y, Shibahara H, Koyama K, Isojima S. Characterization of fertilization-blocking monoclonal antibody 1G12 with human sperm-immobilizing activity. Clin Exp Immunol. 1997;109:547–54.
50. Koyama K, Hasegawa A, Komori S. Functional aspects of CD52 in reproduction. J Reprod Immunol. 2009;83:56–9.
51. Witkin SS, Toth A. Relationship between genital tract infections, sperm antibodies in seminal fluid, and infertility. Fertil Steril. 1983;40:805–8.
52. Blum M, Pery J, Blum I. Antisperm antibodies in young oral contraceptive users. Adv Contracept. 1989;5:41–6.
53. Cunningham DS, Fulgham DL, Rayl DL, Hansen KA, Alexander NJ. Antisperm antibodies to sperm surface antigens in women with genital tract infection. Am J Obstet Gynecol. 1991;164:791–6.
54. Dimitrova D, Kalaydjiev S, Hristov L, Nikolov K, Boyadjiev T, Nakov L. Antichlamydial and antisperm antibodies in patients with chlamydial infections. Am J Reprod Immunol. 2004;52:330–6.
55. Hirano Y, Shibahara H, Koriyama J, Tokunaga M, Shimada K, Suzuki M. Incidence of sperm-immobilizing antibodies in infertile women with past chlamydia trachomatis infection. Am J Reprod Immunol. 2010;65:127–32.
56. Wira CR, Fahey JV, Sentman CL, Pioli PA, Shen L. Innate and adaptive immunity in female genital tract: cellular responses and interactions. Immunol Rev. 2005;206:306–35.
57. Wang Y-X, Zhu W-J, Jiang H. Sperm exposure during menses is a risk factor for developing antisperm antibody (ASA) in female. Arch Gynecol Obstet. 2013;288:1145–8.

Chapter 5
Non-assisted Reproductive Technology (ART) Treatments for Infertile Women with Anti-sperm Antibody (ASA)

Hiroaki Shibahara

Abstract Before the introduction of in vitro fertilization and embryo transfer (IVF-ET), there are essentially two strategies for the treatments of infertile women with anti-sperm antibodies (ASAs). One is to reduce ASA production, and the other is to increase opportunities that sperm and egg encounter each other.

The principle of the constant use of condom could reduce ASA production by avoiding the sperm sensitization in women with ASA. However, most of the previous studies concluded that the results of condom therapy in terms of pregnancies were disappointing.

The potential adverse effects and lack of effectiveness in many cases have decreased the enthusiasm for the use of oral corticosteroids for the treatment of infertile women with ASA. Meanwhile, there was a report that hydrocortisone for local immunosuppression becomes a valuable method of therapy in cervical immunological infertility caused predominantly by anti-sperm sIgA.

The immunosuppressive effects of danazol have been reported in patients with autoimmune diseases such as idiopathic thrombocytopenic purpura and others. From our clinical experience using danazol as a therapy for infertile women with sperm-immobilizing antibodies, further clinical trials for the treatment of such patients with ASA by using danazol might be necessary to clarify its usefulness.

Recently, the usefulness of cyclosporine A therapy for pregnancy outcome in women with recurrent pregnancy loss was suggested; it might encourage the clinical trial for the treatment of infertile women with ASA by using cyclosporine A if necessary, because the low-dose cyclosporine A therapy has no inhibitory effect on ovulation and also no side effects on both the mother and the fetus.

Intrauterine insemination (IUI) can overcome problems related with sperm passage through the cervical mucus, and it has been shown to be useful for the treatment of a part of infertile women with ASA. However, when the titer of sperm-immobilizing antibodies is higher, sperm inserted in uterine cavity by IUI procedure

H. Shibahara (✉)
Department of Obstetrics and Gynecology, School of Medicine, Hyogo Medical University, Nishinomiya, Hyogo, Japan
e-mail: sibahara@hyo-med.ac.jp

H. Shibahara, A. Hasegawa (eds.), *Gamete Immunology*,
https://doi.org/10.1007/978-981-16-9625-1_5

would be impaired by the antibodies secreted in uterine cavity, in fallopian tube, and even in follicular fluid. Therefore, the ultrasound-guided intra-tubal insemination (ITI) therapy for infertile women with sperm-immobilizing antibodies might be effective as a substitute for IUI.

5.1   Introduction

Women with anti-sperm antibodies (ASAs), especially sperm-immobilizing antibodies, are refractory to traditional treatments for infertility except for in vitro fertilization and embryo transfer (IVF-ET). Before the introduction of IVF-ET, there are essentially two strategies for the treatments of infertile women with ASA, especially sperm-immobilizing antibodies. One is to reduce ASA production, and the other is to increase opportunities that sperm and egg encounter each other. For the former, constant use of condom or abstinence to avoid the sperm sensitization and immunosuppressive therapy such as corticosteroid therapy to suppress the antibody production were once tried. However, they are no longer used due to no efficacy and possible side effects, respectively [1]. For the latter, intrauterine insemination (IUI) and IVF-ET have been successfully applied to eliminate the impairment of sperm passage through the female reproductive tract by ASA [2, 3].

Here, our clinical experiences using unique therapy for infertile women with sperm-immobilizing antibodies are also introduced. One is administration of danazol that have been used for the treatment of endometriosis and also for auto-immune diseases such as idiopathic thrombocytopenic purpura (ITP) as it has an immunosuppressive effect. The other is the ultrasound-guided intra-tubal insemination (ITI) therapy as a substitute for IUI.

5.2   Condom Therapy

Condoms have been widely used for the purpose of contraception and also for avoiding infection of sexually transmitted illness such as human immunodeficiency virus (HIV). Besides, condom therapy has been historically used in the treatment of infertility to counteract the supposed harmful effects of ASA. The principle of the therapy was that the constant use of condom could reduce ASA production by avoiding the sperm sensitization in women with ASA. Such a constant condom therapy as described by Dukes and Franklin [4] for periods of up to 6 and 12 months has been widely used to counteract the supposed harmful effect of ASA detected by sperm-agglutination test when infertility problems are present.

In the earlier study, Schwimmer et al. [5] reported that circulating sperm-agglutinating antibodies in women fell to undetectable levels after 2–12 months of condom therapy. Careful timing of ovulation is essential because of the increase in antibodies after reexposure. However, they concluded that the results of condom therapy in terms of pregnancies were disappointing.

Isojima et al. [6] examined the condom applications for seven cases of infertile women with sperm-immobilizing antibodies and found that the sperm-immobilization value (SIV) remained more or less constant and was unaffected by the condom therapy.

Jones [7] reported that occlusion (condom) therapy in patients with ASA, detected by the three most commonly used tests for humoral anti-sperm activity which are microagglutination, macroagglutination, and immobilization (complement-dependent cytotoxicity), has proved unsuccessful when compared with the natural reproductive patterns in untreated patients. Similarly, Kremer et al. [8] reported that condom therapy in three couples with ASA, for at least 6 months, had no result.

Interestingly, Franken et al. [9] experienced a case of infertile woman with ASA who was treated by 7 months of condom therapy. At the start of therapy, female titer was 1:256; after a month, the female's sperm antibody titer was 1:64; and by 7 months, the titer was in the range of 1:8. The first insemination attempt had been unsuccessful, and after this single introduction of sperm into the woman's reproduction tract, ASA titer again rose to 1:256.

In the early 1980s, Greentree [10] insisted that it is possible that pregnancies occur in marriages that are barren due to the presence of ASA, but it often occurs when these antibodies disappear spontaneously. A programmed schedule of the gradual decrease of coitus with condoms after a period of normal sexual intercourse has theoretic immunologic merit in bringing about normal pregnancies in previously infertile couples. The author has sounded the alarm that, from an immunologic point of view, this approach, although it may work, is illogical and concluded that the lengthy continuous condom therapy to improve fertility when harmful ASAs are present would be put to rest once and for all. Shushan A et al. [11] also recognized that condom therapy for the treatment of infertile women with ASA has been widely applied, but with controversial results.

As for sperm-immobilizing antibodies, Koyama et al. [12] reported that when the quantitative antibody titers, SI_{50}, were followed over 3 years in sterile women with the sperm-immobilizing antibody, the antibody titers were found to be unstable and undulated over a period of several months. The undulation of SI_{50} units seemed to be unrelated to the separation from sperm contact by condom therapy (Fig. 5.1). They concluded that the therapy consisting of use of a condom seemed to be ineffective in decreasing the antibody titers.

In the era of assisted reproductive technology (ART) since the first successful application of IVF-ET in 1978 by Steptoe and Edwards [13], it has been shown that infertile women with sperm-immobilizing antibodies have higher successful pregnancies than those with other indications [2, 3, 14, 15]. Therefore, the use of condom therapy came into oblivion. When least expected, Wang et al. [16] recently reported that among six subpositive (10–50% of immunobead bound) cases with ASA detected by indirect immunobead test (I-IBT), three became pregnant in the subsequent 2-year follow-up after condom therapy for 6 months. However, there has been no report that supported their results so far.

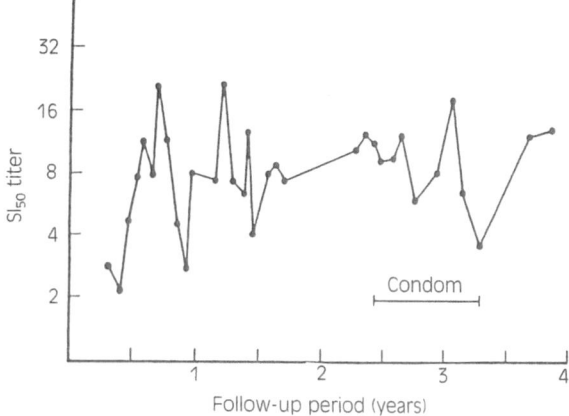

Fig. 5.1 Undulation of SI$_{50}$ units and the condom therapy [12]. When the quantitative antibody titers, SI$_{50}$, were followed over 3 years in sterile women with the sperm-immobilizing antibody, the antibody titers were found to be unstable and undulated over a period of several months. The undulation of SI$_{50}$ units seemed to be unrelated to the separation from sperm contact by condom therapy. The therapy consisting of use of a condom seemed to be ineffective in decreasing the antibody titers

5.3   Corticosteroid Therapy

5.3.1   Systemic Corticosteroid Therapy

There have been few suggestions for therapeutic interventions for infertile women with ASA when compared with infertile men with ASA. Haas GG et al. [17] first reported the usefulness of systemic corticosteroid treatment for an infertile woman with ASA. Corticosteroids were administered according to a modification of the regimen of Shulman [18]. An infertile woman, who had unexplained infertility for 2 years and whose plasma initially showed increased IgG anti-sperm activity on two occasions, was treated with methylprednisolone. She was begun on a 10-day regimen of methylprednisolone, 32 mg by mouth three times a day, beginning on day 21 of her cycle. The dosage was gradually decreased over the next 3 days. Thus, she was treated with corticosteroids for 13 days. She also was suggested to use condoms or abstinence during steroid therapy. She was then encouraged to have sexual intercourse at least every 2 or 3 days until ovulation occurred, as manifested by a rise in basal body temperature (BBT). An additional plasma sample was obtained at the time of ovulation for testing (21 days after start of corticosteroid therapy).

Corticosteroid therapy was associated with a decrease in IgG anti-sperm activity to values well within the normal range when undiluted plasma was used as the antibody source. At this time, sexual intercourse was encouraged. Ovulation subsequently occurred, as manifested by a rise in the BBT. When the BBT reached a sustained elevation, the human chorionic gonadotropin result in her serum was positive, and the pregnancy has progressed to term. However, they also reported two women with similar IgG anti-sperm activities, who had been infertile for 2 years and 5 years, had spontaneous remission of their IgG ASA without corticosteroid therapies. In both cases, these patients conceived within 1 month after the negative test result. Therefore, their report seems very interesting for showing the possible effectiveness of corticosteroid therapy; however, further analysis must be required to have precise evaluation for the treatment.

Later, Alexander et al. [19] reported the pregnancy rates in patients treated for ASA with corticosteroid therapy. In this study, five women with ASA were given 60 mg of prednisone daily for 1 week. Before the treatments, all of them exhibited poor post-coital tests. Although the data for women and men were analyzed together, the corticosteroid therapy caused significant reduction in levels of sperm-agglutinating antibodies ($P < 0.001$) and also those of sperm-immobilizing antibodies ($P < 0.04$). Of the five treated women, four conceived. They mentioned that the choice of immunosuppressive medicines for treatment of infertility remains empirical. In this study, they used 60 mg of prednisone; however, they suggested that whether higher levels, longer durations of treatment, or more repetitions of treatment would be better is not known. Their conclusion was that significantly more pregnancies occurred in couples in which one partner had ASA and was treated with 60 mg of prednisone per day for at least 7 days than in the untreated group.

Following these reports, there have been no additional papers showing the effectiveness of oral corticosteroid therapy. One of the reasons might be that the ART has been widely accepted for the treatment of infertile women with ASA. As shown in the section of the condom therapy, it has been shown that infertile women with sperm-immobilizing antibodies have higher successful pregnancies than those with other indications [2, 3, 14, 15]. The other reason is that the corticosteroid therapy could cause minor problems, such as mood changes, fluid retention, and dyspepsia, and/or serious complications, such as gastrointestinal bleeding, aseptic necrosis of the hip joint [20], and bone loss [21]. The potential adverse effects and lack of effectiveness in many cases have decreased the enthusiasm for the use of oral corticosteroids for the treatment of infertile women with ASA.

5.3.2 Local Corticosteroid Therapy

Zdenka [22] reported the usefulness of local corticosteroid treatment for infertile women with sperm-agglutinating antibodies in the cervical mucus. Cervical mucus sperm-agglutinating antibodies of secretory immunoglobulin A (sIgA) or IgG detected by the tray agglutination test (TAT) and the indirect mixed antiglobulin reaction (MAR) test were found in 234 unexplained infertile women.

Hydrocortisone was applied to the ectocervix. Hydrocortisone (Spofa) at a dose of 100 mg was incorporated into 20 mL of water-soluble resin gel (Carbopol 940, Goodrich) and filled a 20 mL polyethylene sterile syringe, where 1 mL represents 5 mg of hydrocortisone. The specification for one syringe is Carbopol 940, 0.24 g; hydrocortisone, 0.1 g; triethanolamine, 0.24 g; and distilled water up

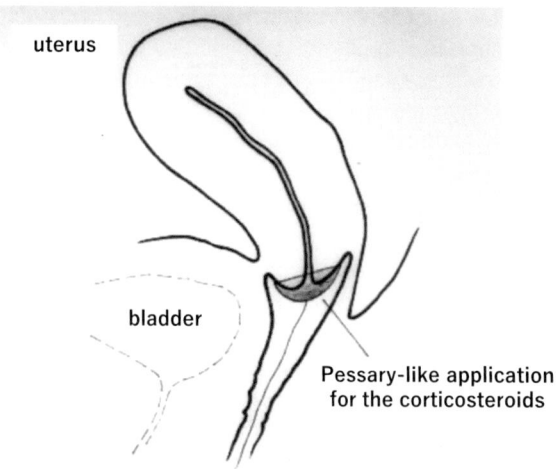

Fig. 5.2 Local corticosteroid therapy [22]. Zdenka reported the usefulness of local corticosteroid treatment for infertile women with sperm-agglutinating antibodies in the cervical mucus. Hydrocortisone was applied to the ectocervix. Hydrocortisone (Spofa) at a dose of 100 mg was incorporated into 20 mL of water-soluble resin gel (Carbopol 940, Goodrich) and filled a 20 mL polyethylene sterile syringe, where 1 ml represents 5 mg of hydrocortisone. The specification for one syringe is Carbopol 940, 0.24 g; hydrocortisone, 0.1 g; triethanolamine, 0.24 g; and distilled water up to 20.0 mL. All patients were instructed on how to use a pessary-like applicator for the corticosteroids

to 20.0 mL. All patients were instructed on how to use a pessary-like applicator for the corticosteroids (Fig. 5.2) [23]. Gel is applied to the ectocervix daily for 4 h, except during the bleeding period. Only coitus with a condom is allowed during the treatment. Hydrocortisone was used at a daily dose of 20 mg for the first 4 days, 10 mg for the next 7 days, and then 5 mg until menstruation each month. The surface of the external cervical os must be wiped off slightly before aspiration of the mucus with a Fortuna syringe (Optima Denmark) from the cervical canal directly into glass capillaries. The Kremer test, tray agglutination test (TAT) and indirect mixed antiglobulin reaction (MAR) tests, colposcopy, and hormonal and oncologic cytology were repeated every period during the treatment and for at least two cycles after. Local sperm-agglutinating antibodies were monitored every cycle, and hydrocortisone therapy was continued until the absence of these antibodies, but to a maximum of 5 months.

Sperm-agglutinating antibodies of sIgA disappeared totally in 102 patients (91 resulting in deliveries of healthy babies, 3 in ectopic pregnancies, and 8 in spontaneous miscarriages). A decrease of sperm-agglutinating antibodies in ovulatory mucus during hydrocortisone application without pregnancy was registered in 60 infertile women. This group was referred to treatment by IVF-ET. No hydrocortisone effect on immunocompetent cells producing anti-sperm IgG alone or combined with sIgA was seen in 72 patients. Vaginal mycosis, a side effect of hydrocortisone treatment, was seen in seven cases.

The author concluded that hydrocortisone for local immunosuppression becomes a valuable method of therapy in cervical immunological infertility caused predominantly by anti-sperm sIgA. So far, there have been no other papers that reported the experiences of this treatment.

5.4 Danazol Therapy

Infertile women with sperm-immobilizing antibodies are less likely to conceive than those without the antibodies. The patients with lower SI_{50} titers (<10) could possibly conceive with the treatment of IUI; however, those with higher SI_{50} titers ($\geqq 10$) have almost never been able to conceive by such therapies. Therefore, the latter patients are able to conceive only by IVF-ET.

The immunosuppressive effects of danazol, an attenuated androgen developed for the treatment of endometriosis, have been reported in patients with autoimmune diseases such as idiopathic thrombocytopenic purpura (ITP) [24], systemic lupus erythematosus (SLE) [25, 26], and autoimmune hemolytic anemia (AIHA) [27], paroxysmal nocturnal hemoglobinuria [28], myelodysplastic syndrome [29], and others, which are generally much less serious than with glucocorticoids. Therefore, we evaluated the effect of the danazol therapy on infertile women associated with endometriosis and also sperm-immobilizing antibodies by following the SI_{50} titers examined by the quantitative sperm-immobilization test every 1–2 months [30].

In this retrospective study, four infertile women with sperm-immobilizing antibodies were treated by danazol (400 mg/day) for 4–6 months because they were diagnosed with endometriosis or adenomyosis by the transvaginal ultrasonography or under the diagnostic laparoscopy following several cycles of unsuccessful IUI. As a control, three infertile women with sperm-immobilizing antibodies were treated by gonadotropin-releasing hormone (GnRH) agonist, buserelin acetate (900 μg/day), for 3–6 months because of endometriosis or adenomyosis.

Although the SI_{50} units were not affected by the danazol therapy in the patients with higher titers ($\geqq 10$) of sperm-immobilizing antibodies, it declined 1 month after the therapy in some patients with lower SI_{50} titers (<10) (Fig. 5.3). However, it lasted only 1 month after finishing the administration of danazol. No pregnancy was obtained during the period of this study. In all patients who were treated by buserelin acetate, the changes of the SI_{50} units were within the natural undulation range, indicating buserelin acetate had no effect on the production of sperm-immobilizing antibodies as we expected.

One infertile man with sperm-immobilizing antibodies, who was refractory to glucocorticoid's treatment, was also treated with danazol. The patient's SI_{50} units lowered slightly; however, the sperm-fertilizing ability tested by the zona-free hamster penetration test did not improve as a result of the danazol therapy.

In the recent review article by Letchumanan P et al. [31], danazol is a useful drug in the treatment of SLE patients, especially in patients with refractory thrombocytopenia, AIHA, and premenstrual flares, and in some mild nonhematologic

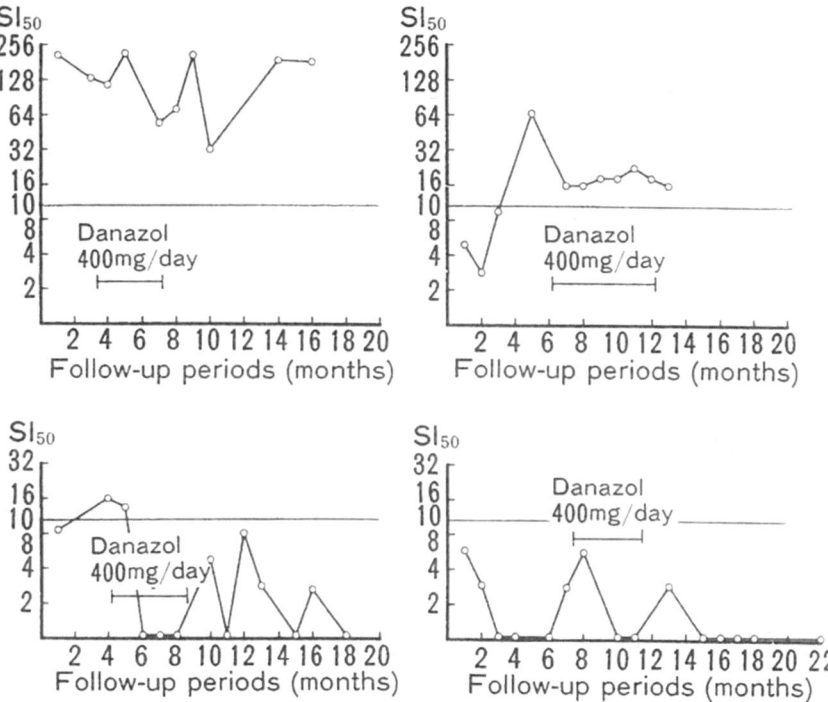

Fig. 5.3 The danazol therapy declined 1 month after the therapy in some patients with lower SI_{50} titers (<10) [30]. The effect of the danazol therapy on infertile women associated with endometriosis and also sperm-immobilizing antibodies was evaluated by following the SI_{50} titers examined every 1–2 months. Four infertile women with sperm-immobilizing antibodies were treated by danazol (400 mg/day) for 4–6 months. Although the SI_{50} units were not affected by the danazol therapy in the patients with higher titers (≥ 10) of sperm-immobilizing antibodies, it declined 1 month after the therapy in some patients with lower SI_{50} titers (<10). However, it lasted only 1 month after finishing the administration of danazol. No pregnancy was obtained during the period of this study

manifestations of SLE. They concluded that it appears to be relatively well tolerated, safe, and efficacious. So far, our experience for the evaluation of the effect of danazol on the treatment of ASA in infertile women with endometriosis has been the only one; further clinical trials for the treatment of such patients with ASA by using danazol might be necessary to clarify its usefulness.

5.5 Cyclosporine A Therapy

As an alternative treatment of corticosteroid therapy, Bouloux PM et al. [32] reported the usefulness of cyclosporine A, an immunosuppressant widely used after organ transplantation and to treat autoimmune diseases, in male autoimmune infertility. In brief, nine infertile men with ASA were treated with cyclosporine A for 6 months with a dose of 5–10 mg/kg/day. Seminal plasma and serum ASA fell in three subjects on treatment, and sperm count and motility increased substantially in one patient. Three successful pregnancies (33.3%) occurred in the study group: one on treatment, one in the first cycle of IUI with the husband's sperm after treatment (twin infants), and one 3 months after cessation of treatment (twin infants). Successful conceptions with cyclosporine A were unrelated to falls in ASA titer. As this study did not have placebo controls, definite conclusion could not be drawn. So far, this is the only report of the cyclosporine A therapy for infertile men with ASA. As for infertile women with ASA, there has been no report demonstrating its effectiveness.

Recently, Azizi R et al. [33] reported that cyclosporine A improved pregnancy outcomes in women with unexplained recurrent pregnancy loss (RPL). Sixty-six patients in the treatment group were treated with oral low-dose cyclosporine A 100 mg/day for 30 days. Twenty patients in the control group were treated with progesterone. Both groups started treatment as soon as the pregnancy test was positive. The live birth rate was significantly higher in the study group (41/66, 62.1%) than in the control group (6/20, 30.0%) ($P < 0.001$). There were no obvious side effect and adverse consequence in the pregnant women. No IUGR or birth defect was observed in fetus in this study. They concluded that the mechanism of cyclosporine A therapy may be related to immune regulation which may favor the success of pregnancy in women with unexplained RPL.

This observation might encourage the clinical trial for the treatment of infertile women with ASA by using cyclosporine A if necessary, because the low-dose cyclosporine A therapy has no inhibitory effect on ovulation and also no side effects on both the mother and the fetus.

5.6 Artificial Insemination Therapy

5.6.1 Intrauterine Insemination (IUI) Therapy

Intrauterine insemination (IUI) can overcome problems related with sperm passage through the cervical mucus. IUI has been shown to be useful for the treatment of a part of infertile women with ASA. It depends on what kind of ASA is tested for the diagnosis of immunological infertility in each clinic for infertile couples.

Check et al. [34] reported that the pregnancy rates after IUI were identical to women who did not have ASA, if the male partner did not have ASA or male factor infertility. In two other studies, IUI treatment did not increase the pregnancy rates per couple or per cycle in infertile women with ASA [35, 36]. The low pregnancy rates (13%) may occur when women with ASA are treated with IUI; however, the use of clomiphene citrate or gonadotropins, along with IUI, results in cumulative pregnancy rates after three to four cycles of therapy of 30% and 39%, respectively [35].

For the treatment of infertile women with sperm-immobilizing antibodies, IUI has been selected to overcome the factors that inhibit conception such as the impairment of sperm passage through cervical mucus because of the antibodies

[37, 38]. Kobayashi et al. [39] reported that IUI for patients with higher SI_{50} titers ($\geqq 10$ units) is invalid even repeated IUI, as described below, was applied. They suggested that when the titer of sperm-immobilizing antibodies is higher, sperm inserted in uterine cavity by IUI procedure would be impaired by the antibodies secreted in uterine cavity [40], in fallopian tube [41], and even in follicular fluid [42].

There are some infertile women with sperm-agglutinating activity without the presence of sperm-immobilizing antibodies in their sera. In our group, Koriyama et al. [43] have investigated if the results of sperm-agglutination test based on Franklin et al. [4, 44] affect infertility treatments. Infertile women who visited our department were divided into the three groups:

Group I: Patients with sperm-immobilizing antibodies ($SI_{50} \geq 1$) ($n = 20$)

Group II: Patients with sperm-agglutinating activity but not with sperm-immobilizing antibody ($n = 25$)

Group III: Patients without sperm-agglutinating activity and without sperm-immobilizing antibody ($n = 205$)

There were no significant differences of the history of past pregnancy, age of the couples, and infertility periods between the three groups. The results of post-coital test (PCT) performed in some of the patients were compared. The positive rates of PCT were 45.5% (5/11) in Group I, 66.7% (8/12) in Group II, and 84.7% (105/124) in Group III, respectively (Fig. 5.4). There was a significant difference only between

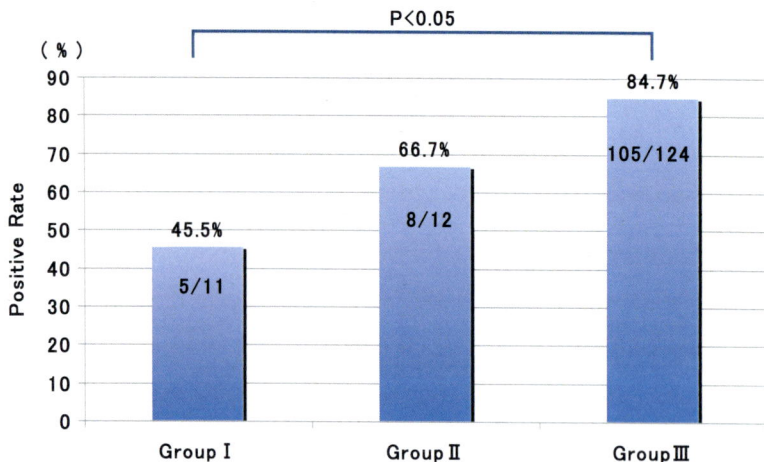

Fig. 5.4 Comparison of post-coital test (PCT) results between the three groups [43]. Infertile women who visited our department were divided into three groups: Group I, patients with sperm-immobilizing antibodies ($SI_{50} \geq 1$); Group II, patients with sperm-agglutinating activity but not with sperm-immobilizing antibody; and Group III, patients without sperm-agglutinating activity and without sperm-immobilizing antibody. The positive rates of PCT were 45.5% (5/11) in Group I, 66.7% (8/12) in Group II, and 84.7% (105/124) in Group III, respectively. There was a significant difference only between Group I and Group III ($P < 0.05$). The judgment of PCT was followed by the criteria in the WHO manual (1999). The result was considered normal when at least one progressively motile sperm was detected in the cervical mucus

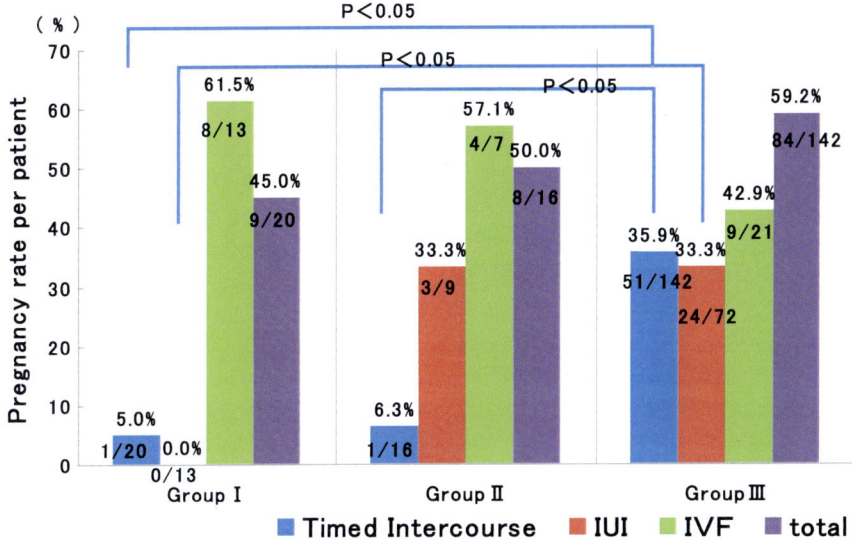

Fig. 5.5 Comparison of pregnancy rates between the three groups. Pregnancy rates per patients were 45.0% (9/20) in Group I, 50.0% (8/16) in Group II, and 59.2% (84/142) in Group III, respectively. There were no significant differences between the three groups. However, the pregnancy rates per patient by timed intercourse were 5.0% (1/20) in Group I, 6.3% (1/16) in Group II, and 35.9% (51/142) in Group III, respectively. There were significant differences between Group I and Group III ($P < 0.05$) and Group II and Group III ($P < 0.05$). The pregnancy rates per patient by IUI were 0% (0/13) in Group I, 33.3% (3/9) in Group II, and 33.3% (24/729) in Group III, respectively. There were also significant differences between Group I and Group III ($P < 0.05$) and Group II and Group III ($P < 0.05$). These findings suggested that in patients with sperm-agglutination, the pregnancy rate by timed intercourse was significantly lower than that in controls. In contrast, favorable pregnancy results were obtained by both IUI and IVF

Group I and Group III ($P < 0.05$). In total, pregnancy rates per patients who came to our hospital for the infertility treatments were 45.0% (9/20) in Group I, 50.0% (8/16) in Group II, and 59.2% (84/142) in Group III, respectively. There were no significant differences between the three groups. However, as shown in Fig. 5.5, the pregnancy rates per patient by timed intercourse were 5.0% (1/20) in Group I, 6.3% (1/16) in Group II, and 35.9% (51/142) in Group III, respectively. There were significant differences between Group I and Group III ($P < 0.05$) and Group II and Group III ($P < 0.05$). The pregnancy rates per patient by IUI were 0% (0/13) in Group I, 33.3% (3/9) in Group II, and 33.3% (24/729) in Group III, respectively. There were also significant differences between Group I and Group III ($P < 0.05$) and Group II and Group III ($P < 0.05$).

These findings suggested that in patients with sperm-agglutination, the pregnancy rate by timed intercourse was significantly lower than that in controls. In contrast, favorable pregnancy results were obtained by both IUI and IVF. They concluded that evaluation of the existence of sperm-agglutination seems to be useful for decision-making in infertile women without the presence of sperm-immobilizing antibody regarding whether they have the possibility of conception by timed intercourse.

5.6.2 Repeated IUI Therapy

As a method to improve the pregnancy rates in infertile women with sperm-immobilizing antibodies, it was attempted that IUI was repeated 3 h after the first IUI just before ovulation as judged by luteinizing hormone surge. This technique has been called as repeated IUI, and it was performed to neutralize ASA in the uterine cavity with repeated introduction of husband's semen; however, the results were not satisfactory [1, 39].

These results strongly suggested that the quantitative estimation of sperm-immobilizing antibodies is important for selecting the appropriate method of treatment for infertile women with sperm-immobilizing antibodies. Furthermore, as the antibody titers, SI_{50} units, are not constant but fluctuated over a period of several months, it is necessary to estimate SI_{50} titers periodically by the quantitative SIT.

5.6.3 Intra-tubal Insemination (ITI) Therapy

As the successful pregnancy rates by the use of IUI are not always satisfactory, especially for infertile women with ASA, we applied ultrasound-guided intra-tubal insemination (ITI) [45–47] therapy for infertile women with sperm-immobilizing antibodies as a substitute for IUI [48].

Between May 1999 and July 2007, sperm-immobilization test (SIT) was performed in 1333 infertile women. In them, 28 women were tested positive, giving the incidence of 2.1%. For women whose SI_{50} titers were lower or middle level, they were initially treated with repeated IUI as described above, and if it was not successful after several attempts, ITI or IVF-ET was selected according to the patients' choice. After determining that the patient's tubes appeared normal by hysterosalpingography (HSG), ITI was performed using the Labotect tubal cannulation set (Zinnanti Surgical Instruments, Chatsworth, CA) for transcervical tubal cannulation [46]. This procedure could be performed with ultrasound guidance or tactile guidance [49]. ITI was performed with a 0.2 mL of washed and concentrated husband's sperm.

Table 5.1 Comparison of the pregnancy rates (PR) by intra-tubal insemination (ITI) therapy for infertile women with various indications

Indications	No. of cases	No. of cycles	No. of patients conceived	PR (%) Per case	Per cycle
Disturbance of sperm transport between uterus and tube	5	17	2	40.0	11.8
Male factor (severe cases are not included)	9	25	1	11.1	4.0
Unexplained	3	12	0	0	0
Sperm-immobilizing antibodies	7	25	3	42.9	12.0
Total	24	79	6	25.0	7.6

After placement of the Labotect bivalve vaginal speculum, the cervix was grasped posteriorly with the Labotect tenaculum. The tenaculum was attached to the speculum with enough tension to result in some uterine descent and theoretically straighten the uterine canal. The 20-cm-long, flexible, metal ball-tipped cannula was advanced until it was tactilely determined to be located in the uterine cornua. The 40-cm-long specimen catheter (external diameter, 3 French with a 2 French tapered tip) was then placed through the guide cannula and advanced 4 cm into the tube, again under ultrasound guidance or tactile guidance. After the insemination, the specimen catheter was withdrawn and inspected for evidence of blood or kinking.

The pregnancy rates by treatments were compared between repeated IUI for 68 cycles in 12 women and ITI for 25 cycles in 7 women. There was no pregnancy by the treatment with repeated IUI. Meanwhile, there were three successful pregnancies by the treatment with ITI. There was a significant difference of the pregnancy rates per patient between the two treatment groups (0% vs 42.9%, $P < 0.05$). The SI_{50} titers in infertile women who conceived by ITI were below 10 units, which means relatively lower sperm-immobilizing antibody titers.

We compared the pregnancy rates by ITI therapy with other indications, such as disturbance of sperm transport between uterus and tube diagnosed under laparoscopy, male factor infertility (severe cases are not included), and unexplained infertility. As shown in Table 5.1, ITI therapy was performed in 24 infertile women for 79 cycles in total. The pregnancy rates per case and per cycle were 25.0% and 7.6%, respectively. The pregnancy rates per patient and also per cycle in infertile women with sperm-immobilizing antibodies were the best in all the indications even the numbers of cases were not enough.

In conclusion, ITI therapy could be one of the choices of the treatment strategy for infertile women with sperm-immobilizing antibodies, especially those with relatively lower SI_{50} units (Figs. 5.6 and 5.7).

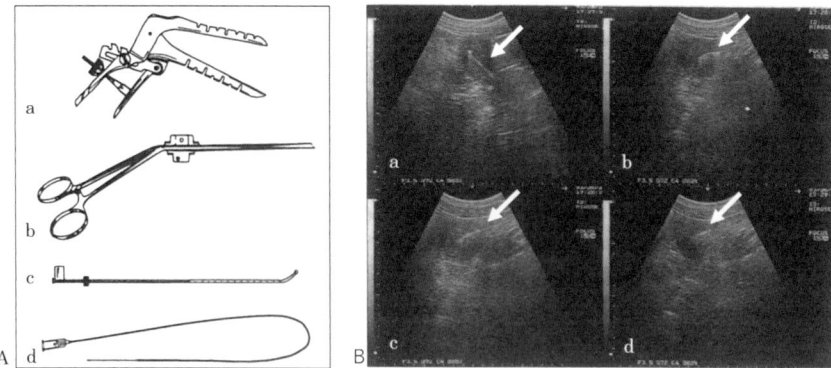

Fig. 5.6 Procedures of the intra-tubal insemination (ITI) therapy by using the fallopian tube catheter set. (**a**) Instruments for ITI using the Labotect tubal cannulation set. After placement of the bivalve vaginal speculum (a), the cervix was grasped posteriorly with the tenaculum (b). The tenaculum was attached to the speculum with enough tension to result in some uterine descent and theoretically straighten the uterine canal. The 20-cm-long, flexible, metal ball-tipped cannula (c) was advanced until it was tactilely determined to be located in the uterine cornua. The 40-cm-long specimen catheter (external diameter, 3 French with a 2 French tapered tip) (d) was then placed through the guide cannula. (**b**) Transabdominal ultrasound-guided procedure for ITI. Under ultrasound guidance or tactile guidance, the specimen catheter was inserted into the uterine cavity (a). The catheter was advanced until it was tactilely determined to be located in the uterine cornua (b) and further advanced 4 cm into the tube (c), again under ultrasound guidance or tactile guidance. ITI was performed with a 0.2 mL of washed and concentrated husband's sperm (d). After the insemination, the specimen catheter was withdrawn and inspected for evidence of blood or kinking

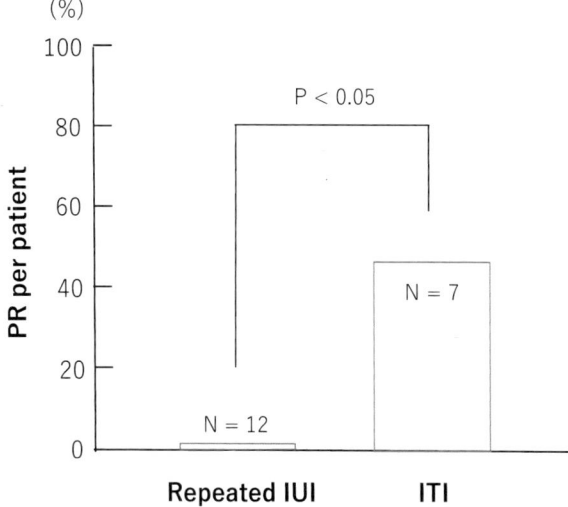

Fig. 5.7 Comparison of the pregnancy rates between repeated IUI and ITI in infertile women with sperm-immobilizing antibodies. The pregnancy rates by treatments were compared between repeated IUI for 68 cycles in 12 women and ITI for 25 cycles in 7 women. All the women had sperm-immobilizing antibodies in their sera. There was no pregnancy by the treatment with repeated IUI. Meanwhile, there were three successful pregnancies by the treatment with ITI. There was a significant difference of the pregnancy rates per patient between the two treatment groups (0% vs 42.9%, $P < 0.05$). The SI_{50} titers in infertile women who conceived by ITI were below 10 units, which means relatively lower sperm-immobilizing antibody titers

References

1. Koyama K. Gamete immunology: infertility and contraception. Reprod Immunol Biol. 2009;24: 1–17.
2. Sugimoto Y, Hasegawa A, Yokoyama K, Ikeda Y, Bessho T, Shigeta M, Ikuma K, Taira S, Koyama K, Isojima S. Successful application of in vitro fertilization and embryo replacement in the treatment of infertile women with sperm immobilizing antibody. Acta Obstet Gynaecol Jpn. 1987;38:1135–6.
3. Kobayashi S, Bessho T, Shigeta M, Koyama K, Isojima S. Correlation between quantitative antibody titers of sperm immobilizing antibodies and pregnancy rates by treatments. Fertil Steril. 1990;54:1107–13.
4. Dukes CD, Franklin RR. Sperm agglutinins and human infertility: female. Fertil Steril. 1968;19: 263–7.
5. Schwimmer WB, Ustay KA, Behrman SJ. An evaluation of immunologic factors of infertility. Fertil Steril. 1967;18:167–80.
6. Isojima S, Tsuchiya K, Koyama K, Tanaka C, Naka O, Adachi H. Further studies on sperm-immobilizing antibody found in sera of unexplained cases of sterility in women. Am J Obstet Gynecol. 1972;112:199–207.
7. Jones WR. The use of antibodies developed by infertile women to identify relevant antigens. Acta Endocrinol Suppl (Copenh). 1975;194:376–404.
8. Kremer J, Jager S, Kuiken J. Treatment of infertility caused by antisperm antibodies. Int J Fertil. 1978;23:270–6.
9. Franken DR, Slabber CF. Experimental findings with sperm antibodies: condom therapy (a case report). Andrologia. 1979;11:413–6.
10. Greentree LB. Antisperm antibodies in infertility: the role of condom therapy. Fertil Steril. 1982;37:451–2.
11. Shushan A, Schenker JG. Immunological factors in infertility. Am J Reprod Immunol. 1992;28: 285–7.
12. Koyama K, Kubota K, Ikuma K, Shigeta M, Isojima S. Application of the quantitative sperm immobilization test for follow-up study of sperm-immobilizing antibody in the sera of sterile women. Int J Fertil. 1988;33:201–6.
13. Steptoe PC, Edwards RG. Birth after the reimplantation of a human embryo. Lancet. 1978;2 (8085):366.
14. Daitoh T, Kamada M, Yamano S, Murayama S, Kobayashi T, Maegawa M, Aono T. High implantation rate and consequently high pregnancy rate by in vitro fertilization-embryo transfer treatment in infertile women with antisperm antibody. Fertil Steril. 1995;63:87–91.
15. Shibahara H, Mitsuo M, Ikeda Y, Shigeta M, Koyama K. Effects of sperm immobilizing antibodies on pregnancy outcome in infertile women treated with IVF-ET. Am J Reprod Immunol. 1996;36:96–100.
16. Wang YX, Zhu WJ, Jiang H. Sperm exposure during menses is a risk factor for developing antisperm antibody (ASA) in female. Arch Gynecol Obstet. 2013;288:1145–8.
17. Haas GG Jr, Cines DB, Schreiber AD. Immunologic infertility: identification of patients with antisperm antibody. N Engl J Med. 1980;303:722–7.
18. Shulman S. Treatment of immune male infertility with methylprednisolone. Lancet. 1976;2: 1243.
19. Alexander NJ, Sampson JH, Fulgham DL. Pregnancy rates in patients treated for antisperm antibodies with prednisone. Int J Fertil. 1983;28:63–7.
20. Hendry WF. Bilateral aseptic necrosis of femoral heads following intermittent high-dose steroid therapy. Fertil Steril. 1982;38:120.
21. Pearce G, Tabensky DA, Delmas PD, Gordon Baker HW, Seeman E. Corticosteroid-induced bone loss in men. J Clin Endocrinol Metab. 1998;83:801–6.
22. Ulčová-Gallová Z. Ten-year experience with antispermatozoal activity in ovulatory cervical mucus and local hydrocortisone treatment. Am J Reprod Immunol. 1997;38:231–4.

23. Ulčová-Gallová Z. Immunological and physicochemical properties of cervical ovulatory mucus. J Reprod Immunol. 2010;86:115–21.
24. Ahn YS, Harrington WJ, Simon SR, Mylvaganam R, Pall LM, So AG. Danazol for the treatment of idiopathic thrombocytopenic purpura. N Engl J Med. 1983;308:1369–99.
25. Morley KD, Parke A, Hughes GR. Systemic lupus erythematosus: two patients treated with danazol. Br Med J. 1982;284:1431–2.
26. Agnello V, Gell J, Gelfand J, Turksoy RU. Preliminary observations on danazol therapy of systemic lupus erythematosus: effect on DNA antibodies, thrombocytopenia and complement. J Rheumatol. 1983;10:682–7.
27. Ahn YS, Harrington WJ, Mylvaganam R, Cayub J, Pall LM. Danazol therapy for autoimmune hemolytic anemia. Ann Intern Med. 1985;102:298–301.
28. Gertz MA. Possible response of paroxysmal cold hemoglobinuria to danazol. Ann Intern Med. 1987;106:635.
29. Kornberg A, Yona R, Iuklea S, Kaufman S. Danazol and myelodysplastic syndromes. Ann Intern Med. 1986;104:445–6.
30. Shibahara H, Shigeta M, Kato H, Koyasu Y, Koyama K, Isojima S. Preliminary observation of the effect of danazol therapy on serum sperm immobilizing antibody titers (SI_{50} unit) in infertile women. Jpn J Fertil Steril. 1992;37:591–6. [in Japanese].
31. Letchumanan P, Thumboo J. Danazol in the treatment of systemic lupus erythematosus: a qualitative systematic review. Semin Arthritis Rheum. 2011;40:298–306.
32. Bouloux PM, Wass JA, Parslow JM, Hendry WF, Besser GM. Effect of cyclosporin A in male autoimmune infertility. Fertil Steril. 1986;46:81–5.
33. Azizi R, Ahmadi M, Danaii S, Abdollahi-Fard S, Mosapour P, Eghbal-Fard S, Dolati S, Kamrani A, Rahnama B, Mehdizadeh A, Jadidi-Niaragh F, Yousefi B, Yousefi M. Cyclosporine A improves pregnancy outcomes in women with recurrent pregnancy loss and elevated Th1/Th2 ratio. J Cell Physiol. 2019;234:19039–47.
34. Check JH, Bollendorf A, Katsoff D, Kozak J. The frequency of antisperm antibodies in the cervical mucus of women with poor postcoital tests and their effect on pregnancy rates. Am J Reprod Immunol. 1994;32:38–42.
35. Margalloth EJ, Sauter E, Bronson RA, Rosenfeld DL, Scholl GM, Cooper GW. Intrauterine insemination as treatment for antisperm antibodies in the female. Fertil Steril. 1988;50:441–6.
36. Gregoriou O, Vitoratos N, Papdias C, Konidaris S, Maragudakis A, Zcourlas PA. Intrauterine insemination as a treatment of infertility in women with antisperm antibodies. Int J Gynecol Obstet. 1991;35:151–6.
37. Koyama K, Ikuma K, Kubota K, Isojima S. Effects of antisperm antibodies on sperm migration through cervical mucus. Excerpta Med Int Congr Ser. 1980;512:705–8.
38. Shibahara H, Shiraishi Y, Hirano Y, Kasumi H, Koyama K, Suzuki M. Relationship between level of serum sperm immobilizing antibody and its inhibitory effect on sperm migration through cervical mucus in immunologically infertile women. Am J Reprod Immunol. 2007;57:142–6.
39. Kobayashi S, Bessho T, Shigeta M, Koyama K, Isojima S. Correlation between quantitative antibody titers of sperm immobilizing antibodies and pregnancy rates by treatment. Fertil Steril. 1990;54:1107–13.
40. Koyama K, Isojima S, Adachi H. Direct proof of transmission of antibodies against spermatozoa into the uterine cavity of the rat. Excerpta Medica Int Congr Ser. 1971;278:562–3.
41. Shibahara H, Shigeta M, Toji H, Koyama K. Sperm immobilizing antibodies interfere with sperm migration from the uterine cavity through the fallopian tubes. Am J Reprod Immunol. 1995;34:120–4.
42. Shibahara H, Hirano Y, Shiraishi Y, Shimada K, Kikuchi K, Suzuki T, Takamizawa S, Suzuki M. Effects of in vivo exposure to eggs with sperm-immobilizing antibodies in follicular fluid on subsequent fertilization and embryo development in vitro. Reprod Med Biol. 2006;19:137–43.
43. Koriyama J, Shibahara H, Shiraishi Y, Shimada K, Hirano Y, Suzuki T, Takamizawa S, Suzuki M. Diversity of the antisperm antibodies; a strategy for the treatment of infertile women with

sperm agglutinating activity without the presence of sperm immobilizing antibody in their sera. 6th European Congress of Reproductive Immunology, Moscow, WTC. (Abstract); 2008.

44. Franklin RR, Dukes CD. Antispermatozoal antibody and unexplained infertility. Am J Obstet Gynecol. 1964;89:6–9.
45. Jansen RP, Anderson JC. Catheterisation of the fallopian tubes from the vagina. Lancet. 1987;2: 309–10.
46. Lisse K, Sydow P. Fallopian tube catheterization and recanalization under ultrasonic observation: a simple technique to evaluate tubal patency and open proximally obstructed tubes. Fertil Steril. 1991;56:198–201.
47. Shibahara H, Hayashi T, Yamada Y, Shiotani T, Ikuma K. Ultrasound-guided intratubal insemination as a treatment for the infertile patients who were refractory to intrauterine insemination. Jpn J Fertil Steril. 1994;39:198–203. [in Japanese].
48. Asada K, Shibahara H, Shiraishi Y, Shimada K, Kikuchi K, Hirano Y, Suzuki T, Takamizawa S, Suzuki M. Effectiveness of the intra-tubal insemination for infertile women with sperm-immobilizing antibodies. Acta Obstet Gynaecol Jpn. 2008;60(Suppl):510 (S-230).
49. Pratt DE, Shangold G, Bieber E, Vignovic E, Barnes R, Schreiber J. Transvaginal intratubal insemination by tactile sensation: a preliminary report. Fertil Steril. 1991;56:984–6.

Chapter 6
Assisted Reproductive Technology (ART) as a Treatment for Infertile Women with Anti-sperm Antibody (ASA)

Hiroaki Shibahara

Abstract For infertile women with anti-sperm antibodies (ASAs), treatments such as condom therapy, abstinence, and steroid suppression therapy have been applied, and they resulted in a generally poor consequence. Some success has been achieved with intrauterine insemination (IUI), but not when serum ASA levels are high. There have been some reports that described unsuccessful application of IVF-ET for the treatment of infertile women with ASA. They presumed that these results underline the importance of using replacement serum in cases where the wife has significant sperm antibody levels in her serum.

After paying a special attention to remove the ASA adhered to cumulus cells around oocytes and avoiding the contamination of blood in the aspirated follicular fluid and moreover removing the antibodies in the follicular fluid by transferring oocytes several times to the new culture medium with replacement serum that was added to the culture medium, fertilization and cleavage rates of mature oocytes from the patients with ASA were not different from those in patients without the antibodies. Therefore, each infertile woman entering an IVF-ET program should have the ASA assay performed as a preliminary screening. It is also important that substitution of the patients' serum by replacement serum in the fertilization and embryo growth media may prove to be an effective means of improving IVF treatment for women with ASA.

For infertile women with sperm-immobilizing antibodies, it is important to assess the SI_{50} titers by the quantitative method to select appropriate treatments. A strategy for the treatment of infertile women with sperm-immobilizing antibodies was suggested. The strategy emphasizes the importance of assessing SI_{50} titers to select treatments for infertile women with sperm-immobilizing antibodies.

H. Shibahara (✉)
Department of Obstetrics and Gynecology, School of Medicine, Hyogo Medical University, Nishinomiya, Hyogo, Japan
e-mail: sibahara@hyo-med.ac.jp

H. Shibahara, A. Hasegawa (eds.), *Gamete Immunology*,
https://doi.org/10.1007/978-981-16-9625-1_6

6.1 Introduction

Anti-sperm antibodies (ASAs) in the sera of infertile women have been thought to account for the failure of conception. Treatments such as condom therapy, abstinence, and steroid suppression therapy have been applied with generally poor results as previously described in this text. Some success has been achieved with intrauterine insemination (IUI), but not when serum ASA levels are high [1]. It is therefore of interest to know whether oocytes from patients with ASA are capable of fertilization by using in vitro fertilization (IVF).

However, the presence of ASA, especially sperm-immobilizing antibodies, in the sera of infertile women has been shown to exert inhibitory effects on various stages of sperm-egg interaction and subsequent embryo development in vitro [2–10]. We have reported the effects of in vivo exposure to eggs with sperm-immobilizing antibodies in follicular fluid on subsequent fertilization and embryo development in vitro [9]. The SI_{50} titers in the sera and follicular fluid of the patients were significantly correlated ($P < 0.0001$). Indeed at the time, fertilization failure occurred even with IVF if the culture medium for insemination contained serum from infertile women with sperm-immobilizing antibodies [10]. To overcome the fertilization failure in IVF by some means, alternative serum or protein source should be used for culture medium. Intracytoplasmic sperm injection (ICSI) is not usually necessary when the husbands' sperm of the infertile women with ASA are normal and also without ASA bound to sperm.

In this chapter, the inhibitory effect by ASA on IVF and also the usefulness of IVF-embryo transfer (ET) without using serum containing ASA for the treatment of infertile women with ASA will be discussed.

6.2 Fertilization Failure or Low Fertilization Rate in In Vitro Fertilization (IVF) Using the Patient's Serum Containing Anti-sperm Antibodies (ASAs) in Culture Medium

There have been some reports that described unsuccessful application of IVF-ET for the treatment of infertile women with ASA.

Pexieder et al. [11] have reported by comparing the cumulus/corona coagulation, the absence of fertilization, and the presence of ASA evaluated using an enzyme-linked immunosorbent assay (ELISA)-based test [12] in the sera of 32 infertile patients undergoing IVF-ET. Observation of the morphology of fertilization in eight patients with ASA and seven patients without ASA has shown a significant association ($P < 0.025$) among the cumulus/corona coagulation, the absence of fertilization, and the presence of ASA. They concluded that cumulus/oocyte complex washing and enzymatic cumulus removal are considered as elective interventions in the case of ASA. Each patient entering an IVF-ET program should have the ASA assay performed as a preliminary screening.

Clarke et al. [13] reported the effect of ASA, detected by the immunobead test (IBT) [14], derived from the sera from infertile women on fertilization and embryo cleavage. In a group of patients with IBT-IgG and IBT-IgA sperm antibody titers of ≥ 10 in serum, a low fertilization rate (15%) was obtained when the wife's serum was used as serum supplement in the IVF culture medium. Where replacement (antibody-negative donor or cord) serum was used in the culture medium, a higher fertilization rate (69%) was obtained ($P < 0.01$). Six pregnancies were obtained in the antibody-positive group ($n = 20$), five of which occurred in patients with IBT-IgG and IBT-IgA titers <10, for a pregnancy rate of 5/9 in this subgroup. Four of these patients delivered (4/9). They presumed that these results underline the importance of using replacement serum in cases where the wife has significant sperm antibody levels in her serum.

Mandelbaum et al. [15] reported that ASAs appear to impair reproduction; however, their clinical significance in IVF-ET is unestablished. For examination of this question, the IBT was used to identify IgA, IgG, and IgM ASA in the serum and follicular fluid of patients undergoing IVF-ET. ASA binding to sperm tail tip did not predict the fertilization rate of uniformly inseminated mature oocytes. However, the fertilization rate in couples with ASA to sperm head (ASA-H) of at least one isotype in female serum ($n = 6$) was significantly less than in those without ASA-H ($n = 34$, 34% versus 74%, $P < 0.01$). Among these women, oocyte fertilization rates were 33% versus 71% ($P < 0.001$). Sixty percent of women whose ova did not fertilize ($n = 5$) had ASA-H in their serum versus 6% of those whose ova did ($n = 35$, $P < 0.05$). The presence of ASA-H in follicular fluid also correlated with fertilization. ASA-H in female serum reduced the zygote cleavage rate from 91–67% ($P = 0.51$). They concluded that the presence of ASA-H in female serum and follicular fluid is associated with reduced fertilization in IVF-ET.

We retrospectively investigated the clinical effects of sperm-immobilizing antibodies in the sera of infertile women on fertilization rates [10]. Before the introduction of sperm-immobilization test (SIT) [16, 17] as a routine test for female infertility, nine infertile women treated by IVF-ET were afterward found having had sperm-immobilizing antibodies in their sera. By using their own sera in the culture medium, 41 (48.2%) of 85 oocytes were fertilized. As controls, 5 infertile women with sperm-immobilizing antibodies were treated by using the culture medium supplemented with human serum albumin (HSA), and 43 (86.0%) of 50 oocytes were fertilized. There was a significant difference between the two groups ($P < 0.0001$). We also concluded that the sperm-immobilizing antibodies in the sera of infertile women caused low fertilization rate.

Taken together, each infertile woman entering an IVF-ET program should have the ASA assay performed as a preliminary screening. It is also important that substitution of the patients' serum by replacement serum in the fertilization and embryo growth media may prove to be an effective means of improving IVF treatment for women with ASA.

6.3   Successful Application of In Vitro Fertilization (IVF) Using Alternative Serum or Protein Source for Culture Medium for Infertile Women with Anti-sperm Antibodies (ASAs)

As described above, for infertile women with ASA, especially sperm-immobilizing antibodies, IUI is first applied, and if it is not successful after several attempts, IVF-ET is used. IVF-ET is an ideal method for the treatment of infertile women with sperm-immobilizing antibodies only if a special attention is paid to remove the antibodies adhered to cumulus cells around oocytes before insemination. The contamination of blood in the aspirated follicular fluid should be avoided as much as possible, and the antibodies in the follicular fluid are removed by transferring oocytes several times to the new culture medium. The patient's serum should never be added to the culture medium [18].

Yovich et al. [19] first reported the successful results of IVF of oocytes from five infertile women with ASA in their serum. The antibodies examined were sperm-agglutinating antibodies as assessed by the gelatin agglutination test (GAT) [20] and tray slide agglutination test (TSAT) [21] and sperm-immobilizing antibodies as assessed by the SIT. In the two patients who conceived, serum collected at the time of ovum aspiration was also assessed for ASA. The remaining three women were known to have raised antibody levels during the preceding 2 months. Donor, rather than maternal, serum was used in the culture medium because these patients possessed serum ASA. Twenty oocytes were collected from 5 patients, and 15 of these were fertilized in the presence of donor serum. All embryos developed to morphologically normal two-cell and four-cell embryos. All five women proceeded to ET, and the two women who became pregnant had serum GAT titers of 40 and SIT titers of 16. They concluded that the high rate of fertilization suggests that oocytes previously bathed in ASA are not permanently affected by this exposure.

Ackermann et al. [22] reported a case of woman who had ASA diagnosed using serum anti-sperm agglutinins [20] (titer range, 1:64 to 1:256) and immobilizins [23] (titer range, 1:16 to 1:256). The patient had a result of repeated poor post-coital tests, and she was treated by 6-month condom therapy followed by five attempts at IUI. Then she was successfully treated with IVF-ET using the culture medium supplemented with donor serum.

Clarke et al. [24] reported a preliminary study by analyzing IVF-ET results for three women with ASA in their plasma and follicular fluid. They suggested that substitution of the patients' serum by replacement serum in the fertilization and embryo growth media may prove to be an effective means of improving IVF-ET treatment for women with ASA.

From our group, Sugimoto et al. [25] reported high fertilization and pregnancy rates by IVF-ET in infertile women with sperm-immobilizing antibodies. They paid a special attention to remove the antibodies adhered to cumulus cells around oocytes, and the contamination of blood in the aspirated follicular fluid was avoided. Moreover, the antibodies in the follicular fluid were removed by transferring oocytes several times to the new culture medium. Human cord serum was added to the culture medium. By these procedures, fertilization and cleavage rates of mature oocytes from the patients with sperm-immobilizing antibodies were not different from those in patients without the antibodies. In brief, 5 of 16 patients who were treated with IVF-ET had sperm-immobilizing antibodies in their sera. The antibody titers, SI_{50} units, determined by the quantitative sperm-immobilization test [26, 27] were ranged from 14.5 to 243 units, and all patients had IUI therapy repeatedly without successful pregnancy. For the patients with sperm-immobilizing antibodies, culture medium was supplemented with human cord serum to avoid the inhibitory effect on fertilization by the antibodies. For the patients with other indications for IVF-ET, their own sera were used in the culture medium. The fertilization rates in patients with and without sperm-immobilizing antibodies were 88.5% (23/26) and 81.5% (22/27), respectively. There was no significant difference between the two groups ($P > 0.05$); however, the fertilization rate was better in women with sperm-immobilizing antibodies. Three of 5 patients with the antibodies and 1 of 11 patients without the antibodies conceived. They concluded that IVF-ET could be a hopeful method for the treatment of infertile women with sperm-immobilizing antibodies.

Later, Kobayashi et al. [1] additionally reported over the paper by Sugimoto et al. [25] that of the 36 infertile women with sperm-immobilizing antibodies treated with IVF-ET, 14 (38.9%) successfully conceived. Patients with high SI_{50} titers ($10 \leq$ units) did not conceive by ordinary or repeated IUI with husband's semen except when treated with IVF-ET. Those with relatively low SI_{50} titers (<10 units) could conceive either by repeated or ordinary IUI, though the success rates were lower than by IVF-ET.

Koyama et al. [18] reported the average SI_{50} titers in infertile women with sperm-immobilizing antibodies at the time of pregnancy by various treatments. As shown in Table 6.1, the number of pregnancies and average SI_{50} units were 3 women and 2.9 units by timed intercourse, 12 women and 10.8 units by IUI, and 30 women and 66.0 units by IVF-ET, respectively. SI_{50} titers of three patients who became pregnant by timed intercourse were much lower than those of patients who became pregnant by IUI or IVF-ET treatment. SI_{50} titers were much lower in the IUI group than in the IVF-ET group. They also reported the correlation between SI_{50} titers and pregnancy rates in patients treated by IUI and IVF-ET as shown in Table 6.2. When SI_{50} titers were constantly lower than 10 units or in the medium range, IUI was effective in some cases; however, when the titer was constantly higher than 10 units, IUI was no longer effective, and the pregnancy was obtained only by IVF-ET. They concluded

Table 6.1 Average SI50 titers in infertile women with sperm-immobilizing antibodies at the time of pregnancy by various treatments

Treatment	No. of pregnancies	SI_{50} units
Timed intercourse	3	2.9
IUI	12	10.8
IVF-ET	30	66.0

Adapted from ref. [12]

Table 6.2 Correlation between SI_{50} titers and pregnancy rates in patients treated by IUI and IVF-ET

SI_{50} units	Low (<10)	Medium (around 10)	High (>10)
	No. of pregnancy/patients treated (%)		
IUI	7/25 (28.0)	5/39 (12.8)	0/29 (0)
IVF-ET	3/11 (27.3)	17/31 (54.1)	10/27 (37.0)

Adapted from ref. [12]

that it is important to assess the SI_{50} titers by the quantitative method to select appropriate treatments for infertile women with sperm-immobilizing antibodies.

Daitoh et al. [28] investigated the effect of anti-sperm immunity on IVF-ET treatment of women with sperm-immobilizing antibodies. Eighteen women with sperm-immobilizing antibodies and 122 infertile patients with non-immune etiology as controls were selected. Infertile couples due to a male factor and with unknown etiology were excluded. Fertilization rate and cleavage, implantation rate per embryo transferred, and pregnancy rate in both groups were compared. Although the culture medium was supplemented with donor serum in patients with sperm-immobilizing antibodies, the fertilization rate in this group (61.3%) was significantly lower than that in the control group (76.8%) ($P < 0.05$). They did not mention the reason why the fertilization rate in patients with the antibodies was lower than expected even if alternative sera were used.

To answer this question, we investigated if exposing retrieved eggs to a high amount of sperm-immobilizing antibodies in the follicular fluid in vivo affected subsequent fertilization and embryo development in vitro, even if they were washed with an antibody-free culture medium [9]. Patients' sera and their follicular fluid were collected in 15 IVF-ET or ICSI-ET treatment cycles from 11 infertile women with sperm-immobilizing antibodies in their sera. The fertilization rate, good-quality embryo rate, and implantation rate by IVF-ET were compared between patients having higher (>10 units) SI_{50} titers and lower (<10 units) SI_{50} titers in their follicular fluid. As mentioned above, there was a significant correlation in the SI_{50} titers between the patients' sera and their follicular fluid ($P < 0.0001$). Thorough washing of the collected eggs in culture medium without the patient's serum before IVF was performed. Consequently, the fertilization rates in the patients with high (>10 units) and low (<10 units) SI_{50} titers in their follicular fluid were 77.8% (28/36) and 82.8% (24/29), respectively. There was no significant difference between the two groups ($P = 0.62$). Therefore, high SI_{50} titers did not show stronger inhibitory effect on fertilization if enough care was taken for manipulation of the gametes. ICSI did not seem to be necessary in patients having the antibodies if their sera were not supplemented in the culture media.

6.4   Changes of Sperm-Immobilizing Antibody Titers During Pregnancy

In our group, Kobayashi et al. [1] reported that SI_{50} units, sperm-immobilizing antibody titers, tended to decline as pregnancy proceeded in the follow-up studies of the infertile women who conceived successfully. The SI_{50} units of sperm-immobilizing antibody in infertile women's sera were titrated by quantitative sperm-immobilization test (SIT) [26, 27] and were followed every 1 or 2 months. In nine pregnancies of eight patients who successfully conceived, the SI_{50} titers were examined through the gestational periods (Fig. 6.1). Interestingly, the SI_{50} units gradually decreased during pregnancy. They speculated some possible reasons: (1) dilution effect of the antibodies during pregnancy, (2) adsorption effect of antibodies to the placenta, (3) avoiding antigenic stimulation by blockade of sperm migration into the uterine or peritoneal cavity, and (4) suppression of the production of antibodies by the large amount of steroid hormone excreted in pregnancy. In some patients, the SI_{50} titers increased again during or after puerperal periods.

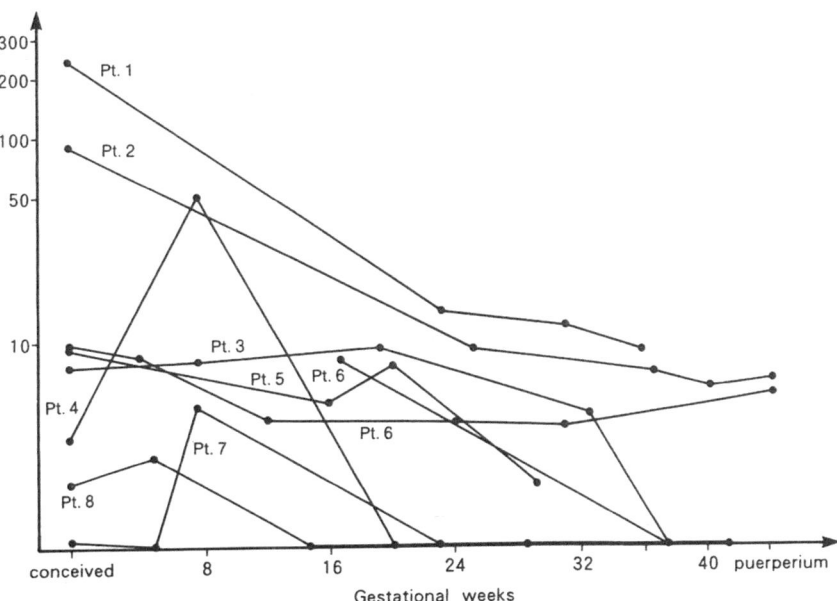

Fig. 6.1 Changes of SI_{50} units during pregnancy [1]. Sperm-immobilizing antibody titers, SI_{50} units, tended to decline as pregnancy proceeded in the follow-up studies of the infertile women who conceived successfully. The SI_{50} units of sperm-immobilizing antibody in their sera were titrated by quantitative sperm-immobilization test and were followed every 1 or 2 months. In nine pregnancies of eight patients who successfully conceived, the SI_{50} titers were examined through the gestational periods. The SI_{50} units gradually decreased during pregnancy. In some patients, the SI_{50} titers increased again during or after puerperal periods

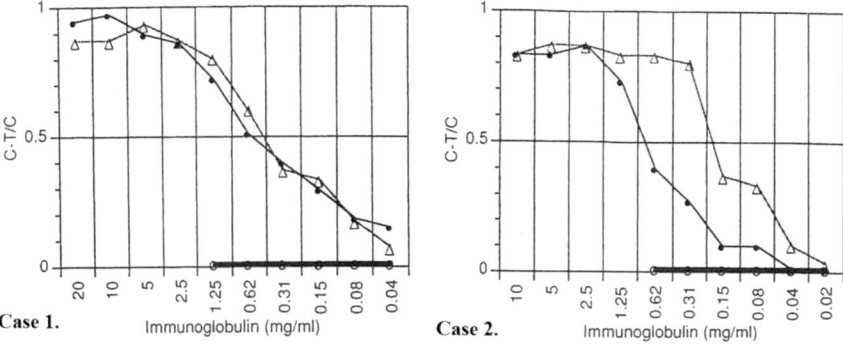

Case 1. Immunoglobulin (mg/ml)   **Case 2.** Immunoglobulin (mg/ml)

● : Patient's serum,   △ : Umbilical cord serum,   ○ : Placental extract

Fig. 6.2 Sperm-immobilizing activities and immunoglobulin levels in patient's serum, umbilical cord serum, and placental extracts. From two women with sperm-immobilizing antibodies who conceived, sera from the patients and the umbilical cord and also about 20 g of placenta were collected after successful delivery. Both the sera and the placental extracts were examined for sperm-immobilizing activities. The percentage of motile sperm in each diluted test serum or placental extract (T%) and that in the control serum (C%) were counted, and the antibody activity for sperm-immobilization was calculated by the formula C − T/C. The vertical axis represents that the sperm-immobilizing activity is higher when C − T/C value comes close to 1. The horizontal axis represents the concentration of IgG in each sample. Sera from the patient and umbilical cord showed enough sperm-immobilizing activities in both cases. However, no sperm-immobilizing activities were detected in the placental extracts even if high concentrations of immunoglobulin were confirmed. The adsorption effect of the antibodies to the placenta might not be the reasons for a decrease of sperm-immobilizing antibody titers during pregnancy

Later, our group investigated the reason why sperm-immobilizing antibody titers, SI_{50} units, decreased during pregnancy. Taya et al. [29] investigated if antigenic cross-reactivity of carbohydrate epitopes both on human sperm and placental cells, that was suggested previously [30, 31], was one of the reasons for a decrease of sperm-immobilizing antibody titers during pregnancy. From two women with sperm-immobilizing antibodies who conceived by treatments, sera from the patients and the umbilical cord and also about 20 g of placenta were collected after successful delivery. Both the sera and the placental extracts were examined for sperm-immobilizing activities. As shown in Fig. 6.2, sera from the patient and umbilical cord showed enough sperm-immobilizing activities in both cases. However, no sperm-immobilizing activities were detected in the placental extracts even if high concentrations of immunoglobulin were confirmed. These findings suggested that the fact that the sperm-immobilizing activities decrease during pregnancy course might not be caused by the adsorption effect of the antibodies to the placenta. Other mechanisms such as dilution effect might be suggested.

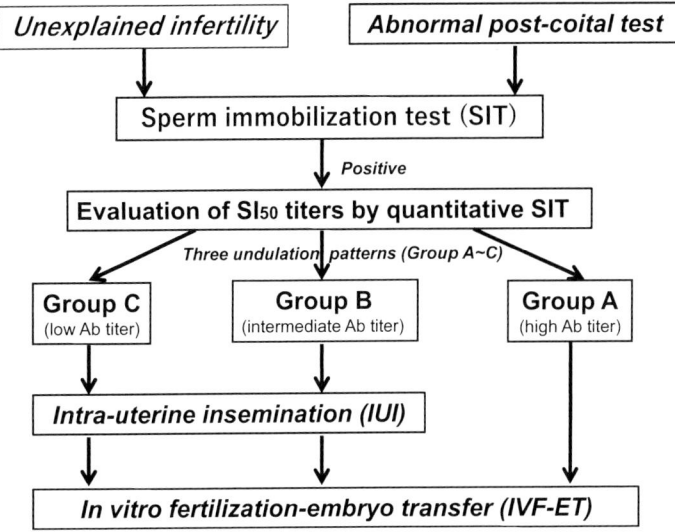

Fig. 6.3 Strategy for the treatment of infertile women with sperm-immobilizing antibodies [35]. It is important to assess the SI_{50} titers to select treatments for infertile women with sperm-immobilizing antibodies. A strategy for the treatments of infertile women with sperm-immobilizing antibodies is shown according to the undulation patterns of individual patient's SI_{50} titers. Patients with sperm-immobilizing antibodies are divided into three groups according to their follow-up SI_{50} titers: Group A, the patients with continuously high SI_{50} titers (>10 units) are recommended to be treated by IVF-ET; Group B, the patients have intermediate SI_{50} titer patterns around 10; and Group C, the patients with continuously low SI_{50} titers (<10 units) are recommended to be treated by repeated or ordinary IUI. If they are not able to be conceived by several cycles of IUI, then the treatment by IVF-ET should be considered

6.5   Strategy for the Treatment of Infertile Women with Sperm-Immobilizing Antibodies

It is well known that infertile women having sperm-immobilizing antibodies are refractory to the conventional treatments with timed intercourse or IUI because the antibodies secreted in the female genital tract inhibit sperm migration within cervical mucus [32, 33] and the fallopian tubes [34]. To overcome this cause of infertility in vivo, IVF-ET has been applied, and satisfactory outcomes resulting from suitable culture conditions for gametes and embryo have been obtained [1, 8, 9, 25].

Kobayashi et al. [1] proposed a strategy for the treatment of infertile women with sperm-immobilizing antibodies according to the undulation patterns of individual patient's SI_{50} titers. They divided patients with sperm-immobilizing antibodies into three groups according to their follow-up SI_{50} titers. Group A, which consisted of patients with continuously high SI_{50} titers (>10 units), did not conceive by ordinary or repeated IUI; however, a satisfactory pregnancy rate was obtained by IVF-ET. Group B, in which the patients had intermediate SI_{50} titer patterns around 10, showed rare rates of success with IUI. In Group C, which consisted of patients with continuously low SI_{50} titers (<10 units), conception by repeated or ordinary IUI was achieved, although the success rates were lower than by IVF-ET.

Taken together, a strategy for the treatment of infertile women with sperm-immobilizing antibodies was suggested in Fig. 6.2 [35]. This strategy emphasizes the importance of assessing SI_{50} titers to select treatments for infertile women with sperm-immobilizing antibodies.

References

1. Kobayashi S, Bessho T, Shigeta M, Koyama K, Isojima S. Correlation between quantitative antibody titers of sperm immobilizing antibodies and pregnancy rates by treatments. Fertil Steril. 1990;54:1107–13.
2. Kamada M, Daitoh T, Hasebe H, Irahara M, Yamano S, Mori T. Blocking of human fertilization in vitro by sera with sperm-immobilizing antibodies. Am J Obstet Gynecol. 1985;153:328–31.
3. Tsukui S, Noda Y, Fukuda A, Matsumoto H, Tatsumi K, Mori T. Blocking effect of sperm immobilizing antibodies on sperm penetration of human zonae pellucidae. J In Vitro Fert Embryo Transf. 1988;5:123–8.
4. Shibahara H, Shigeta M, Koyama K, Burkman LJ, Alexander NJ, Isojima S. Inhibition of sperm-zona pellucida tight binding by sperm immobilizing antibodies as assessed by the hemizona assay (HZA). Acta Obstet Gynaecol Jpn. 1991;43:237–8.
5. Bandoh R, Yamano S, Kamada M, Daitoh T, Aono T. Effect of sperm-immobilizing antibodies on the acrosome reaction of human spermatozoa. Fertil Steril. 1992;57:387–92.
6. Shibahara H, Burkman LJ, Isojima S, Alexander NJ. Effects of sperm-immobilizing antibodies on sperm-zona pellucida tight binding. Fertil Steril. 1993;60:533–9.
7. Shibahara H, Shigeta M, Inoue M, Hasegawa A, Koyama K, Alexander NJ, Isojima S. Diversity of the blocking effects of antisperm antibodies on fertilization in human and mouse. Hum Reprod. 1996;11:2595–9.
8. Shibahara H, Mitsuo M, Ikeda Y, Shigeta M, Koyama K. Effects of sperm immobilizing antibodies on pregnancy outcome in infertile women treated with IVF-ET. Am J Reprod Immunol. 1996;36:96–100.
9. Shibahara H, Hirano Y, Shiraishi Y, Shimada K, Kikuchi K, Suzuki T, Takamizawa S, Suzuki M. Effects of in vivo exposure to eggs with sperm-immobilizing antibodies in follicular fluid on subsequent fertilization and embryo development in vitro. Reprod Med Biol. 2006;5:137–43.
10. Taneichi A, Shibahara H, Hirano Y, Suzuki T, Obara H, Fujiwara H, Takamizawa S, Sato I. Sperm immobilizing antibodies in the sera of infertile women cause low fertilization rates and poor embryo quality in vitro. Am J Reprod Immunol. 2002;47:46–51.
11. Pexieder T, Boillat E, Janecek P. Antisperm antibodies and in vitro fertilization failure. J In Vitro Fertil Embryo Transf. 1985;2:229–32.
12. Ackerman SB, Wortham JW, Swanson RJ. An indirect enzyme-linked immunosorbent assay (ELISA) for the detection and quantitation of antisperm antibodies. Am J Reprod Immunol. 1981;1:199–205.
13. Clarke GN, Lopata A, Johnston WIH. Effects of sperm antibodies in females on human in vitro fertilization. Fertil Steril. 1986;46:435–41.
14. Bronson RA, Cooper GW, Rosenfeld DL. Sperm-specific isoantibodies and autoantibodies inhibit the binding of human sperm to the human zona pellucida. Fertil Steril. 1982;38:724–9.
15. Mandelbaum SL, Diamond MP, DeCherney AH. Relationship of antisperm antibodies to oocyte fertilization in in vitro fertilization-embryo transfer. Fertil Steril. 1987;47:644–51.
16. Isojima S, Li TS, Ashitaka Y. Immunologic analysis of sperm-immobilizing factor found in sera of women with unexplained sterility. Am J Obstet Gynecol. 1968;101:677–83.

17. Isojima S, Tsuchiya K, Koyama K, Tanaka C, Naka O, Adachi H. Further studies on sperm-immobilizing antibody found in sera of unexplained cases of sterility in women. Am J Obstet Gynecol. 1972;112:199–207.
18. Koyama K. Gamete immunology: infertility and contraception. Reprod Immunol Biol. 2009;24: 1–17.
19. Yovich JL, Stanger JD, Kay D, Boettcher B. In-vitro fertilisation of oocytes from women with serum antisperm antibodies. Lancet. 1984;1(8373):369–70.
20. Kibrick S, Belding DL, Merrill B. A sensitive gelatin agglutination test for detecting antisperm agglutinins. Fertil Steril. 1952;3:430–8.
21. Friberg J. A simple and sensitive micro-method for demonstration of sperm-agglutinating activity in serum from infertile men and women. Acta Obstet Gynecol Scand Suppl. 1974;36: 21–9.
22. Ackermann SB, Graff D, van Uem JFHM, Swanson RJ, Veeck LL, Acosta AA, Garcia JE. Immunologic infertility and in vitro fertilization. Fertil Steril. 1984;42:474–7.
23. Hellema HW, Samuel T, Rumke P. Sperm autoantibodies as a consequence of vasectomy. II. Long-term follow-up studies. Clin Exp Immunol. 1979;38:31–6.
24. Clarke GN, McBain JC, Lopata A, Johnston WIH. In vitro fertilization results for women with antisperm antibodies in plasma and follicular fluid. Am J Reprod Immunol. 1985;8:130–1.
25. Sugimoto Y, Hasegawa A, Yokoyama K, Ikeda Y, Bessho T, Shigeta M, Ikuma K, Taira S, Koyama K, Isojima S. Successful application of in vitro fertilization and embryo replacement in the treatment of infertile women with sperm immobilizing antibody. Acta Obstet Gynaecol Jpn. 1987;38:1135–6.
26. Isojima S, Koyama K. Quantitative estimation of sperm immobilizing antibody in the sera of women with sterility of unknown etiology; the 50% sperm immobilization unit (SI_{50}). Excerpta Medica Int Congr Ser. 1974;370:10–5.
27. Koyama K, Kubota K, Ikuma K, Shigeta M, Isojima S. Application of the quantitative sperm immobilization test for follow-up study of sperm-immobilizing antibody in the sera of sterile women. Int J Fertil. 1988;33:201–6.
28. Daitoh T, Kamada M, Yamano S, Murayama S, Kobayashi T, Maegawa M, Aono T. High implantation rate and consequently high pregnancy rate by in vitro fertilization-embryo transfer treatment in infertile women with antisperm antibody. Fertil Steril. 1995;63:87–91.
29. Taya T, Shibahara H, Komori S, Shigeta M, Koyama K, Isojima S. Changes of sperm-immobilizing antibody activities during pregnancy and its passage through placenta. Adv Obstet Gynecol. 1992;44:129–31. (in Japanese).
30. Tsuji Y, Clausen H, Nudelman E, Kaizu T, Hakomori S, Isojima S. Human sperm carbohydrate antigens defined by an antisperm human monoclonal antibody derived from an infertile woman bearing antisperm antibodies in her serum. J Exp Med. 1988;168:343–56.
31. Kameda K, Tsuji Y, Koyama K, Isojima S. Comparative studies of the antigens recognized by sperm-immobilizing monoclonal antibodies. Biol Reprod. 1992;46:349–57.
32. Koyama K, Ikuma K, Kubota K, Isojima S. Effects of antisperm antibodies on sperm migration through cervical mucus. Excerpta Med Int Congr Ser. 1980;512:705–8.
33. Shibahara H, Shiraishi Y, Hirano Y, Kasumi H, Koyama K, Suzuki M. Relationship between level of serum sperm immobilizing antibody and its inhibitory effect on sperm migration through cervical mucus in immunologically infertile women. Am J Reprod Immunol. 2007;57:142–6.
34. Shibahara H, Shigeta M, Toji H, Koyama K. Sperm immobilizing antibodies interfere with sperm migration from the uterine cavity through the fallopian tubes. Am J Reprod Immunol. 1995;34:120–4.
35. Shibahara H, Koriyama J, Shiraishi Y, Hirano Y, Suzuki M, Koyama K. Diagnosis and treatment of immunologically infertile women with sperm-immobilizing antibodies in their sera. J Reprod Immunol. 2009;83:139–44.

Part II
Anti-Sperm Antibody (ASA): Immune Infertility Associated with ASA in Men

Chapter 7
Historical Background of Immune Infertility Associated with Anti-sperm Antibody (ASA) in Men

Hiroaki Shibahara

Abstract Anti-sperm antibodies (ASAs) can be located on the surface of sperm and are also detected in seminal plasma and serum in infertile men. Several authors reported the conflicting relationship between ASA and male infertility from the middle of the 1950s to the beginning of the twenty-first century.

We investigated the reasons for such a confusion. Finally, we found that there was a diversity of ASA, which caused the misunderstanding of the clinical significance of the antibodies in men. For example, some of the sperm-bound antibodies are associated with complement-dependent sperm-immobilizing antibodies, indicating that there exists a heterogeneity of sperm-bound antibodies. Therefore, we offered that the ASA testing should be carried out as a routine test for the appropriate diagnosis of immunological infertility in men. There is no doubt that the most useful diagnostic method for detecting ASA is one that can assess the antibodies on the surface of ejaculated sperm.

For the appropriate diagnosis of immunological infertility in males, the direct immunobead test (IBT) or the mixed antiglobulin reaction (MAR) test should be carried out as a routine test in the infertility clinic. Then, post-coital test (PCT) and the hemizona assay (HZA) should be carried out as an initial screening in infertile men with ASA and used as the basis for decision-making.

7.1 Introduction

As similar as in women, immunity of sperm has been considered as one of the possible causes of infertility in men. In infertile men, anti-sperm antibodies (ASAs) can be located on the surface of sperm and are also detected in seminal plasma and serum. The condition of having ASA has also been found in men with congenital

H. Shibahara (✉)
Department of Obstetrics and Gynecology, School of Medicine, Hyogo Medical University, Nishinomiya, Hyogo, Japan
e-mail: sibahara@hyo-med.ac.jp

© The Editor(s) (if applicable) and The Author(s), under exclusive license to Springer Nature Singapore Pte Ltd. 2022
H. Shibahara, A. Hasegawa (eds.), *Gamete Immunology*,
https://doi.org/10.1007/978-981-16-9625-1_7

103

absence of the vas [1], vasectomy [2], mumps orchitis [3], varicocele [4], testicular trauma [5], homosexuality [6], and spinal cord injury [7].

ASA in men were firstly found by Rumke [8] and Wilson [9]. They reported that the presence of serum sperm-agglutinating activity in the sera and seminal plasma from certain men suffered from unexplained infertility in 1954. Since then, several authors reported the conflicting relationship between ASA and male infertility until the beginning of the twenty-first century. Therefore, a standardized and universally accepted assay for the detection of ASA, a consensus about the clinical consequences of ASA, and a mechanistic explanation of how ASA impairs conception in immunologically infertile men remain unestablished, leading to an argument that the routine testing for ASA in men may not be always necessary [10].

About half a century has passed, our group examined the reasons for such a confusion. Finally, we found that there was a diversity of ASA, which caused the misunderstanding of the clinical significance of the antibodies in men [11, 12]. In brief, we have offered that the ASA testing should be carried out as a routine test for the appropriate diagnosis of immunological infertility in men. There is no doubt that the most useful diagnostic method for detecting ASA is one that can assess the antibodies on the surface of ejaculated sperm. For this purpose, the direct immunobead test (IBT) [13] or the mixed antiglobulin reaction (MAR) test [14] should be carried out as a routine test in the infertility clinic. Then, post-coital test (PCT) and the hemizona assay (HZA) [15] should be carried out as an initial screening in infertile men with ASA. These results can be used as the basis for the decision-making [10].

7.2 History of the Anti-sperm Antibody (ASA) Testing in Infertile Men

In 1959, Rumke et al. [16] investigated several thousand individuals and showed that 3% of them had spermagglutinins in titers of 1:32 or more, while a sufficiently large group of fertile men had zero incidence of such activity. Fjallbrant et al. [17] also found a correlation between sperm-agglutination and infertility in men. However, it is important to understand the difficulty of discriminating true antibody-specific sperm-agglutination from nonspecific agglutination.

Later, Menge et al. [18] tested 698 infertile couples for circulating ASA and analyzed subsequent pregnancy rates. They found that the pregnancy rate (7.1%) was significantly lower when men had agglutinin titers above 1:16 compared with that (42.7%) in the group with negative or low antibody levels. They also found a relatively lower frequency (10.7%) of good cervical mucus penetration among the 28 men with agglutinin titers above 1:128 than that (64.6%) amongst 65 men with the titers below 1:128. Ayvaliotis et al. [19] reported that in the group with more than 50% of antibody-coated sperm, only 4 (15%) of 26 couples conceived during the follow-up period of up to 3 years vs 6 (67%) of 9 couples with less than 50% of coated sperm. Francavilla et al. [20] treated 19 couples with severe sperm autoimmunity using intrauterine insemination (IUI) for a total of 110 cycles, and no pregnancies resulted. In a control of 86 patients treated by IUI for a total of 411 cycles, they obtained 23 pregnancies. These earlier studies indicated that ASA testing in men seemed to be useful to predict conception by timed intercourse and IUI.

As for the effects of ASA on fertilization, Bronson et al. [21] reported that the binding of head-directed immunoglobulin (Ig)G or IgA ASA to sperm reduced sperm binding to human zona pellucida. Junk et al. [22] reported that 72 couples, including 15 with ASA in the semen, were studied in a program of in vitro fertilization and embryo transfer (IVF-ET). Cases were further subclassified as normospermic or oligospermic, and ASAs were assessed with categorization into the respective human Ig classes as determined using the indirect IBT. They found that fertilization was significantly reduced ($P < 0.001$) only if both IgA and IgG antibodies were present in semen but there was no reduction if either class was present alone. The fertilization rate of oocytes was significantly reduced ($P < 0.001$) by sperm from oligospermic samples, and there was a further reduction in those cases with combined IgA/IgG ASA. Liu et al. [23] observed that antibody-coated sperm from seven men with sperm autoimmunity had an impaired ability to bind to the human zona pellucida but that oolemma binding was unaffected. Yeh et al. [24] investigated the impact of Ig isotypes and their location on the human sperm surface on IVF in 48 couples (80 IVF cycles) with men showing positive ASA on the sperm surface by IVF at the Norfolk Program. They found that IgG and IgA antibody levels had no significant correlation with total fertilization rate. IgM, present in 44% of the couples, had a strong correlation with fertilization. When IgA showed very high levels of binding (>68%) and IgM binding was >40%, the fertilization rate dropped significantly. A strong correlation between the presence of ASA and fertilization rate was seen when IgM was directed to the head or tail tip of the sperm. IgA induced a statistically significant reduction of fertilization only when it was present on the head.

On the contrary, there have been some papers demonstrating the negative impact of ASA in infertile men on reproduction. Culligan PJ et al. [25] found that the fertilization rates were similar in 119 couples whose husbands were and were not tested for ASA (64% versus 68%). Four (25%) of the 16 couples whose husbands had ASA fertilized ≦50% of oocytes, compared with 31 (30%) of the 103 couples whose husbands did not have these antibodies. Overall, 21 couples (8.4%) experienced complete fertilization failure. In a program that included ASA testing for selected couples and intracytoplasmic sperm injection (ICSI) for those who tested positive, it would cost $11,735 to prevent a fertilization failure, whereas it would cost $9250 to perform ICSI in a second IVF cycle for those who initially failed. They concluded that ASA testing had low sensitivity in predicting low or no fertilization and did not appear to be cost-effective when selectively ordered as part of an IVF workup. Zini et al. [26] performed a meta-analysis to further evaluate the relationship between ASA and pregnancy after IVF or ICSI. The serious problem of this study that they suggested was that the study characteristics (including the ASA cutoff values) were heterogeneous. They found that the combined OR for failure to achieve a pregnancy using IVF or ICSI in the presence of positive semen ASA was 1.22 (95% CI, 0.84, 1.77) and 1.00 (95% CI, 0.72, 1.38), respectively. The overall (IVF + ICSI) combined OR was 1.08 (95% CI, 0.85, 1.38). They concluded that additional, well-designed prospective studies using appropriate ASA cutoff levels are needed to further address this issue.

7.3 Heterogeneity of Anti-sperm Antibodies (ASAs) in Infertile Men

The presence of ASA in males can reduce fecundity; however, the relationship between the two had been disputed as mentioned above. We investigated if there is diversity of ASA bound to sperm surface using the direct IBT combined with complement-dependent sperm-immobilization test (SIT) (Table 7.1) [11]. The ASAs bound to sperm surface were detected using the direct IBT in 275 semen samples. In some cases with ASA detected by the direct IBT, sperm-immobilizing antibodies bound to sperm surface were also evaluated using the direct SIT. The incidence of the IgG, IgA, and IgM classes of ASA detected by direct IBT were 2.5%, 1.8%, and 0.4%, respectively. Totally, 9 (3.3%) of 275 infertile men had ASA on the sperm surface. Direct SIT was tested positive in four (66.7%) of six cases with ASA assessed by direct IBT. In conclusion, some of the sperm-bound antibodies are associated with complement-dependent sperm-immobilizing antibodies, indicating that there exists a heterogeneity of sperm-bound antibodies. This result might be one of the reasons for the controversy about the relationship between ASA and immunological infertility in men.

7.4 Strategy for the Diagnosis and Treatment of Infertile Men with Anti-sperm Antibodies (ASAs)

Based on our studies [11, 12], a strategy for the diagnosis and treatment of infertile men with ASA was established (Fig. 7.1) [10]. For the appropriate diagnosis of immunological infertility in males, direct IBT should be carried out as a routine test in the infertility clinic. Then, PCT and the hemizona assay (HZA) [15] should be

Table 7.1 Diversity of anti-sperm antibodies (ASA) bound to sperm surface in infertile men

Heading	Diversity
Ig class	IgG, IgA, IgM
Localization	Head, midpiece, tail
Titers	No. of IB bound
Biological activity	Sperm immobilization
Production	Systemic, local

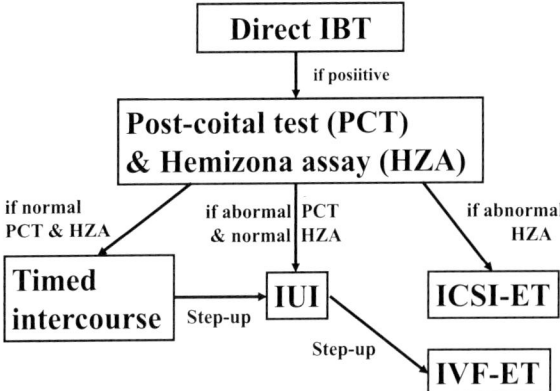

Fig. 7.1 A strategy for the diagnosis and the treatment of infertile males with anti-sperm antibodies (ASAs) [10]. In order to diagnose immunological infertility in infertile men, direct immunobead test (IBT) should be carried out as a routine test in the infertility clinic. Then, the post-coital test (PCT) and the hemizona assay (HZA) should be performed as an initial screening for infertile males with ASA diagnosed by direct IBT, as a basis for decision-making. If a patient with ASA has an abnormal hemizona index (HZI), it seems reasonable to advise selecting intracytoplasmic sperm injection-embryo transfer (ICSI-ET) as the primary treatment. For patients with ASA and normal HZI, selection of timed intercourse (TI) or intrauterine insemination (IUI) should be advised depending on the PCT result

carried out as an initial screening in infertile males with ASA diagnosed by direct IBT and used as the basis for decision-making.

If the patient with ASA has an abnormal hemizona index (HZI), it seems reasonable to advise selecting intracytoplasmic sperm injection (ICSI)-embryo transfer (ET) as a primary treatment. However, it was also shown that some immunologically infertile males with normal fertilizing ability established pregnancy by using timed intercourse (TI) or IUI. In such patients with ASA and a normal HZI, whether TI or IUI should be selected can be judged by the results of PCT. Therefore, the treatment strategy for males with ASA is similar to that for infertile males with oligozoospermia or asthenozoospermia.

In conclusion, it should be emphasized that a diversity of ASA exists and has effects on fertility in infertile males. Although there is an argument that the routine testing for ASA in males is not always necessary, one should be aware that in some cases of failed IUI or IVF, afterward ICSI is selected because of the diagnosis of ASA. Clinicians should judge the emotional cost to couples that have gone through the procedures and failed to conceive.

References

1. Patrizio P, Silber S, Ord T, Moretti-Rojas I, Asch RH. Relationship of epididymal sperm antibodies in their in vitro fertilization capacity in men with congenital absence of the vas deference. Fertil Steril. 1992;58:1006–10.
2. Shulman S, Zappi E, Ahmed U, David JE. Immunologic consequences of vasectomy. Contraception. 1972;5:269–78.
3. Shulman A, Shohat B, Gillis D, Yavetz H, Homounnai ZT, Paz G. Mumps orchitis among soldiers: frequency, effect on sperm quality, and sperm antibodies. Fertil Steril. 1992;57:1344–6.
4. Golomb J, Vardinon N, Homonnai ZT, Braf Z, Yust I. Demonstration of antispermatozoal antibodies in varicocele-related infertility with an enzyme-linked immunosorbent assay (ELISA). Fertil Steril. 1986;45:397–402.
5. Hjort T, Husted S, Linnet-Jepsen P. The effect of testis biopsy on autosensitization against spermatozoa antigens. Clin Exp Immunol. 1974;18:201–5.
6. Witkin SS, Sonnabend J. Immune responses to spermatozoa in homosexual men. Fertil Steril. 1983;39:397–402.
7. Siosteen A, Steen Y, Forssman L, Sullivan L. Auto-immunity to spermatozoa and quality of semen in men with spinal cord injury. Int J Fertil. 1993;38:117–22.
8. Rumke PH. The presence of sperm antibodies in the serum of two patients with oligospermia. Vox Sang. 1954;4:135–40.
9. Wilson L. Sperm agglutinins in human semen and blood. Proc Soc Exp Biol Med. 1954;85:652–5.
10. Shibahara H, Shiraishi Y, Suzuki M. Diagnosis and treatment of immunologically infertile males with antisperm antibodies. Reprod Med Biol. 2005;4:133–41.
11. Shibahara H, Tsunoda T, Taneichi A, Hirano Y, Ohno A, Takamizawa S, Yamaguchi C, Tsunoda H, Sato I. Diversity of antisperm antibodies bound to sperm surface in male immunological infertility. Am J Reprod Immunol. 2002;47:146–50.
12. Shibahara H, Shiraishi Y, Hirano Y, Suzuki T, Takamizawa S, Suzuki M. Diversity of the inhibitory effects on fertilization by anti-sperm antibodies bound to the surface of ejaculated human sperm. Hum Reprod. 2003;18:1469–73.
13. Bronson RA, Cooper GW, Rosenfeld D. Detection of sperm specific antibodies on the spermatozoa surface by immunobead binding. Arch Androl. 1982;9:61.
14. Jager S, Kremer J, van Slochteren-Draaisma T. A simple method of screening for antisperm antibodies in the human male. Detection of spermatozoal surface IgG with the direct mixed antiglobulin reaction carried out on untreated fresh human semen. Int J Fertil. 1978;23:12–21.
15. Burkman LJ, Coddington CC, Franken DR, Kruger TF, Rosenwaks Z, Hodgen GD. The hemizona assay (HZA)- development of a diagnostic test for the binding of human spermatozoa to the human hemizona pellucida to predict fertilization potential. Fertil Steril. 1988;49:688–97.
16. Rumke P, Hellinga G. Autoantibodies against spermatozoa in sterile men. Am J Clin Pathol. 1959;32:357.
17. Fjallbrant B, Obrant O. Clinical and seminal findings in men with sperm antibodies. Acta Obstet Gynecol Scand. 1968;47:451.
18. Menge AC, Medley NE, Mangione CM, Dietrich JW. The incidence and influence of antisperm antibodies in infertile human couples on sperm-cervical mucus interactions and subsequent fertility. Fertil Steril. 1982;38:439–46.
19. Ayvaliotis B, Bronson R, Rosenfeld D, Cooper G. Conception rates in couples where autoimmunity to sperm is detected. Fertil Steril. 1985;43:739–42.
20. Francavilla F, Romano R, Santucci R, Marrone V, Vorraa G. Failure of intrauterine insemination in male immunological infertility in cases in which all spermatozoa are antibody-coated. Fertil Steril. 1992;58:587–92.
21. Bronson RA, Cooper GW, Rosenfeld DL. Sperm-specific isoantibodies and autoantibodies inhibit the binding of human sperm to the human zona pellucida. Fertil Steril. 1982;38:724–9.

22. Junk SM, Matson PL, Yovich JM, Bootsma B, Yovich JL. The fertilization of human oocytes by spermatozoa from men with antispermatozoal antibodies in semen. J In Vitro Fert Embryo Transf. 1986;3:350–2.
23. Liu DY, Clarke GN, Baker HW. Inhibition of human sperm-zona pellucida and sperm-oolemma binding by antisperm antibodies. Fertil Steril. 1991;55:440–2.
24. Yeh WR, Acosta AA, Seltman HJ, Doncel G. Impact of immunoglobulin isotype and sperm surface location of antisperm antibodies on fertilization in vitro in the human. Fertil Steril. 1995;63:1287–92.
25. Culligan PJ, Crane MM, Boone WR, Allen TC, Price TM, Blauer KL. Validity and cost-effectiveness of antisperm antibody testing before in vitro fertilization. Fertil Steril. 1998;69: 894–8.
26. Zini A, Fahmy N, Belzile E, Ciampi A, Al-Hathal N, Kotb A. Antisperm antibodies are not associated with pregnancy rates after IVF and ICSI: systematic review and meta-analysis. Hum Reprod. 2011;26:1288–95.

Chapter 8
Detection of Anti-sperm Antibody (ASA) in Men

Hiroaki Shibahara

Abstract The presence of anti-sperm antibodies (ASAs) in men can reduce fecundity; however, the relationship between the two had been disputed. It should be emphasized that a diversity of ASA exists and has effects on fertility in infertile men. Although there is an argument that the routine testing for ASA in men is not always necessary, one should be aware that in some cases of failed intrauterine insemination (IUI) or in vitro fertilization (IVF), afterward intracytoplasmic sperm injection (ICSI) is selected because of the diagnosis of ASA.

For the appropriate diagnosis of immunological infertility in men, ASA testing should be carried out as a routine test in the infertility clinic. Although several assay methods that detect ASA in infertile men have been described, there is no doubt that the most useful diagnostic method for detecting ASA is one that can assess the antibodies on the surface of ejaculated sperm. The important factors to consider in selecting the appropriate ASA assay are the availability of material being tested (motile sperm or immotile sperm), the sensitivity and specificity of the test, the ability to define immunoglobulin isotypes and location of attachment to the sperm, the quantitation of the level of ASA, and the ease of performing the test. For this purpose, the World Health Organization (WHO) has recommended the two tests, such as the mixed antiglobulin reaction (MAR) test and direct immunobead test (direct IBT).

From our investigations, for example, some of the sperm-bound antibodies are associated with complement-dependent sperm-immobilizing antibodies, indicating that there exists a heterogeneity of sperm-bound antibodies. This result might be one of the reasons for the controversy about the relationship between ASA and immunological infertility in men.

H. Shibahara (✉)
Department of Obstetrics and Gynecology, School of Medicine, Hyogo Medical University, Nishinomiya, Hyogo, Japan
e-mail: sibahara@hyo-med.ac.jp

8.1 Methods for Anti-sperm Antibody Testing in Infertile Men

Although several assay methods that detect anti-sperm antibodies (ASAs) in infertile men have been described, there is no doubt that the most useful diagnostic method for detecting ASA is one that can assess the antibodies on the surface of ejaculated sperm. For this purpose, assays which directly detect ASA on sperm membrane such as the mixed antiglobulin reaction (MAR) test and direct immunobead test (direct IBT) are recommended by the World Health Organization (WHO) [1].

Direct IBT is widely used as a screening test for ASA [2–4], and the inhibitory effects by ASA on sperm motion and fertilizing ability are evaluated to make a decision for the strategy of infertility treatments [5]. Recently, the IBT was discontinued of its production. However, the ASA test results obtained by the immunospheres (IS) assay were shown to be in agreement with the results obtained with the IBT [6]. (See Chap. 2.)

The important issue of the ASA testing is that not all of the ASAs detected by the MAR test or direct IBT are associated with the cause of infertility; they can only detect ASAs that react with the surface antigens against motile sperm. Thereafter the physicians should assess the biological effects of ASA detected whether they affect sperm function such as sperm penetrating ability through cervical mucus and fertilizing ability.

In this chapter, several direct tests for the detection of ASA for infertile men are described.

8.2 Available Tests for the Direct Detection of Anti-sperm Antibodies (ASAs)

8.2.1 Immunobead Test (IBT)

One of the methods that can detect antibodies bound to motile sperm is the direct IBT [2–4]. Immunobeads (IBs; IgG, IgA, and IgM; Irvine Scientific, Santa Ana, CA, USA) are polyacrylamide spheres with covalently bound rabbit anti-human immunoglobulin (Ig). The method for the indirect IBT was shown in Chap. 2.

For the direct IBT, suspensions of IB and a drop of washed sperm are mixed on a glass slide, and the mixture is allowed to react for 10 min. The preparation is then examined at $400\times$ magnification with a phase-contrast microscope. The IBs adhere to the motile sperm that have surface-bound antibodies (Fig. 8.1). The percentage of motile sperm with surface antibodies is determined, the pattern of binding is noted, and the Ig class of these antibodies can be identified using different sets of IB (Fig. 8.2). A test mixture in this direct test is scored positive if at least 20% of the sperm cells show IB binding [4].

However, based on reports that sperm penetration into the cervical mucus and in vivo fertilization (IVF) tend not to be significantly impaired unless 50% or more

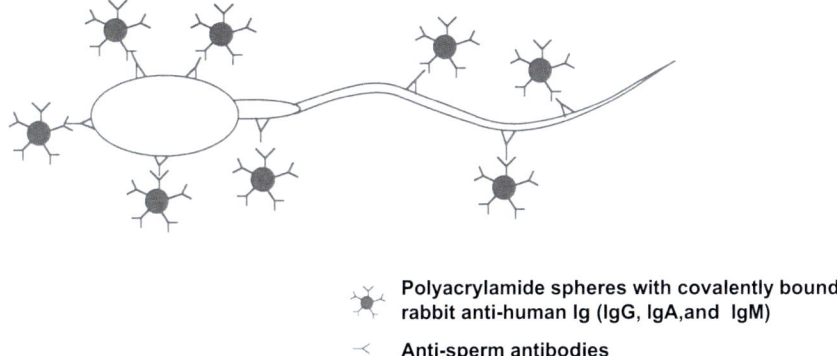

>   **Polyacrylamide spheres with covalently bound**
>   **rabbit anti-human Ig (IgG, IgA,and IgM)**
>
> ≺   **Anti-sperm antibodies**

Fig. 8.1 Immunobead test (IBT). Immunobeads (IBs) are polyacrylamide spheres with covalently bound rabbit anti-human Ig. The IBs adhere to the motile sperm that have surface-bound antibodies. A test mixture in this direct test is scored positive if at least 20% of the sperm cells show IB binding [4] that is recently supported by Koriyama et al. [7]

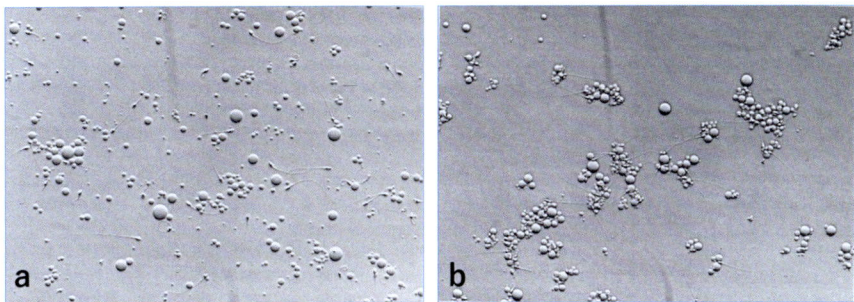

Fig. 8.2 Judgment of the immunobead test (IBT). The percentage of motile sperm with surface antibodies is determined, the pattern of binding is noted, and the Ig class of these antibodies can be identified using different sets of immunobeads (IBs). A test mixture in this direct test is scored positive if at least 20% of the sperm cells show IB binding. (**a**) A negative control. (**b**) An infertile man with ASA in sperm. The IBs are attached to the head and/or tail of the motile sperm

of the motile sperm have antibodies bound to sperm, a cutoff level of 50% was adopted in the laboratory manual published by the WHO [1]. On the contrary, Koriyama et al. [7] strongly suggested that the direct IBT is a screening test, and the value of 20% initially suggested by Bronson et al. [4] seems to be more appropriate than that of 50% in the criteria defined by the WHO [1]. The subjects they selected were 26 direct IBT-positive and 140 direct IBT-negative men. The results of post-coital tests (PCTs) for each subject were examined. A significant difference was observed in abnormal PCTs between values <20% and those ≥20% ($P = 0.02$). However, there was no significant difference in abnormal PCTs between values <50% and those ≥50% ($P = 0.084$). They also found that a cutoff value of

20% was correlated with the possibility of conception by the treatment with intra-uterine insemination (IUI).

As for the incidence of ASA detected by the direct IBT, we have reported 18 (3.54%) out of 509 infertile men had at least 1 Ig class of ASA detected by IBT with the cutoff value of 20% [8]. Clarke et al. reported that the incidence of ASA detected by IgG and/or IgA class of IBT was 7.8% in 813 infertile men [9].

8.2.2   Mixed Antiglobulin Reaction (MAR) Test

In 1978, Jager et al. reported a simple and rapid test, named the mixed antiglobulin reaction (MAR) test, for the detection of ASA of the IgG class on freely swimming sperm in fresh human semen [10]. The test is based on a modification of the famed Coombs test [11]. Motile mixed agglutinates between erythrocytes sensitized with incomplete anti-Rh antibodies and freely swimming sperm with surface ASA are formed after mixing both cell types together with anti-IgG antiserum (Fig. 8.3). Agglutination of the red blood cells serves as an internal control.

Said et al. [12] described the simple initial version of the assay entailed mixing of three ingredients as single drop and covering them with a cover slip. The semen sample is mixed with a suspension of Group O, Rh-positive, human red cells of R_1 R_2 type, sensitized with human IgG in addition to rabbit or goat, undiluted, mono-specific anti-IgG antiserum. The reaction is then observed after 10 min of incubation. Since the red cells are coated with IgG as well as the sperm cells if they have antibodies on them, the added anti-IgG antiserum will then link together the two kinds of cells. Agglutination can be seen under a light microscope as mixed clumps

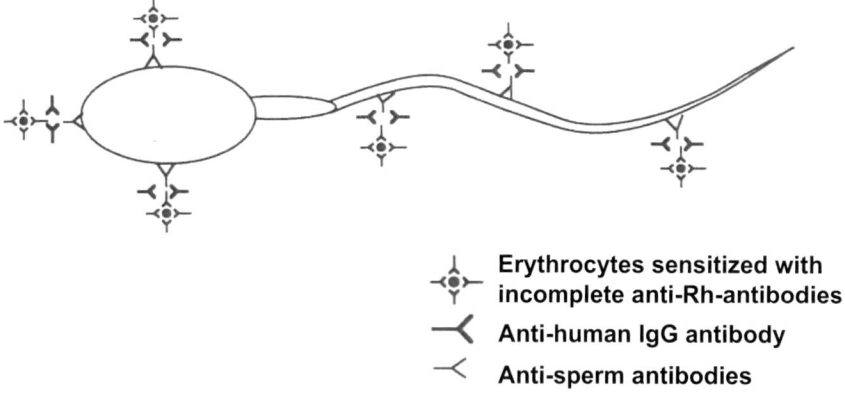

Erythrocytes sensitized with
incomplete anti-Rh-antibodies
Anti-human IgG antibody
Anti-sperm antibodies

Fig. 8.3 Mixed antiglobulin reaction (MAR) test. The formation of motile mixed agglutinates between erythrocytes sensitized with incomplete anti-Rh antibodies and freely swimming sperm with surface ASA, after mixing both cell types together with anti-IgG antiserum. Agglutination of the red blood cells serves as an internal control

of sperm and red blood cells with a slow "shaky" movement. Results of the MAR test are indicated as percentages of motile sperm incorporated into the mixed agglutinates. The site of attachment could be also noted. No interpretation of the test was given unless agglutination of red blood cells and the presence of sufficient motile spermatozoa was observed. A MAR test was considered positive and of clinical significance when >50% agglutination is seen [13].

The test can be applied on ejaculates with sperm concentrations down to one million per ml, provided the motility is sufficient. The percentage of motile sperm found to be coated with ASA of the IgG class, and the extent of the coating, proved to be correlated with the agglutination titer of circulating ASA and with the inhibition of sperm penetration into cervical mucus. The test can be used as a screening for the presence of anti-sperm autoantibodies in serum and semen.

The MAR test correlates with most other ASA tests, such as SIT and IBT [14]. Although MAR test is considered an ideal method for screening of ASA, it is not without limitations [15]. The assay cannot be used in patients with oligozoospermia, asthenozoospermia, and azoospermia. Also it must be performed on a fresh sample and can be difficult to quantitate due to the presence of debris, semen viscosity, mucus, and microbial factors. Commercially available SpermMAR kit uses an antiserum against human IgG to induce mixed agglutination between antibody-coated latex beads conjugated with human IgG [16]. The SpermMAR kit can be considered a superior alternative to erythrocyte MAR since it is time- and cost-effective. One formulation of the kit contemplates the assessment of IgA as well as IgG classes. The assay can be successfully used for the evaluation of male partners of infertile couples if included routinely in semen analysis [17]. An indirect MAR using serum or seminal plasma samples and donor sperm can be considered in cases with azoospermia. However, it has been reported as difficult to interpret. Therefore, ASA should be rather detected using another approach in these cases.

8.2.3 Panning Test

In 1984, Hancock RJ et al. [18] described the use of "panning" procedures as shown in Fig. 8.4 to detect immunoglobulin on the sperm surface. Wells in leukocyte migration trays were coated with anti-immunoglobulin by leaving the appropriate antibody diluted in phosphate-buffered saline (PBS) in the wells at room temperature for 1–2 h and then washing the trays five times in PBS. Sperm were centrifuged at $550 \times g$ for 5 min, resuspended in ASA diluted in PBS plus 5% fetal calf serum (FCS) or control media, and incubated for 1 h at 37 °C. Sperm were then diluted in PBS and centrifuged, and the cells were resuspended in PBS plus 5% FCS to a concentration of approximately five million/mL. Following this, 0.25 mL volumes were incubated in the antibody-coated wells for 1 h at room temperature. The contents of the wells were then shaken out and the wells gently washed by rocking trays in a bath of PBS. The degree to which sperm remained attached to the wells was determined by microscopy using an inverted microscope.

Fig. 8.4 Panning test.
Diagram summarizes
rationale of the panning test.
Sperm with ASA at surface
will bind to anti-
immunoglobulin-coated
wells and remain attached
when the wells are washed,
while sperm without ASA
will not

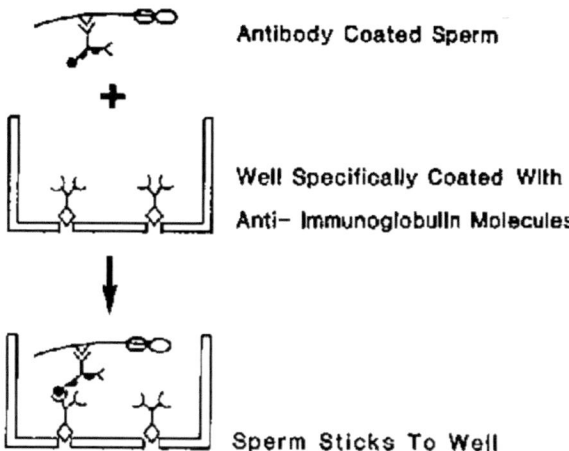

Antibody Coated Sperm

Well Specifically Coated With
Anti- Immunoglobulin Molecules

Sperm Sticks To Well

Later, Hancock RJ [19] developed monospecific and cross-reactive human monoclonal antibodies to HLA antigens. He used the panning assays, and the results
correlated with those of agglutination assays, but not immunofluorescence or
enzyme-linked immunosorbent assays (ELISA). He concluded that panning assays
can be used for the detection of antibodies to sperm and can be useful where target
sperm are of poor motility. The assays may be more sensitive for antibodies to
superficial antigens.

8.2.4 Direct Sperm-Immobilization Test (Direct SIT)

As there is a diversity of ASA themselves, their biological activities are varied. We
developed a direct sperm-immobilization test (SIT) to detect sperm-immobilizing
antibodies on the sperm surface [20] and demonstrated that sperm-immobilizing
antibody bound to the sperm surface is one of the causes of asthenozoospermia [21].

For the direct SIT, 12 μL of a suspension of the patient's sperm (40×10^6 sperm/
mL) and 1 μL of active or heat-inactivated guinea pig serum (Sigma, Tokyo, Japan)
as complement source are mixed in a Terasaki plate (Greiner, Frickenhausen,
Germany) and incubated at 32 °C for 60 min. Heat-inactivated guinea pig serum is
used as a control. The sperm-immobilization value (SIV) is calculated by dividing
the sperm motility in the control containing heat-inactivated complement by that
containing active complement. The SIV of two or more is considered positive. In our
study, 9 out of 13 males with ASA tested positive in D-SIT, giving a positive rate of
69% [8]. All males with sperm-immobilizing antibodies were diagnosed as having
asthenozoospermia by computer-aided sperm analysis (CASA) (Table 8.1).

Table 8.1 Direct tests for the detection of anti-sperm antibodies (ASA) bound to sperm surface in infertile men

1. Immunobead test (IBT)
2. Mixed antiglobulin reaction (MAR) test
3. Panning test
4. Direct sperm-immobilization test (direct SIT)

References

1. World Health Organization, editor. WHO laboratory manual for the examination and processing of human semen. 5th ed. Geneva: WHO; 2010.
2. Clarke GN, Baker HWG. Treatment of sperm antibodies in infertile men: sperm antibody coating and mucus penetration. In: Bratanov K, editor. Proc Int Symp immunology of reproduction. Sofia: Bulgarian Academy of Sciences Press; 1982. p. 337.
3. Clarke GN, Stojanoff A, Cauchi MN. Immunoglobulin class of sperm-bound antibodies in seman. In: Bratanov K, editor. Proc Int Symp immunology of reproduction. Sofia: Bulgarian Academy of Sciences Press; 1982. p. 482.
4. Bronson RA, Cooper GW, Rosenfeld DL. Correlation between regional specificity of antisperm antibodies to the spermatozoan surface and complement-mediated sperm immobilization. Am J Reprod Immunol. 1982;2:222–4.
5. Shibahara H, Shiraishi Y, Suzuki M. Diagnosis and treatment of immunologically infertile males with antisperm antibodies. Reprod Med Biol. 2005;4:133–41.
6. Centola GM, Andolina E, Deutsch A. Comparison of the immunobead binding test (IBT) and immunospheres (IS) assay for detecting serum antisperm antibodies. Am J Reprod Immunol. 1997;37:300–3.
7. Koriyama J, Shibahara H, Ikeda T, Hirano Y, Suzuki T, Suzuki M. Toward standardization of the cut-off value for the direct immunobead test using the postcoital test in immunologically infertile males. Reprod Med Biol. 2012;12:21–5.
8. Shibahara H, Shiraishi Y, Hirano Y, Suzuki T, Takamizawa S, Suzuki M. Diversity of the inhibitory effects on fertilization by anti-sperm antibodies bound to the surface of ejaculated human sperm. Hum Reprod. 2003;18:1469–73.
9. Clarke GN, Elliot PJ, Smaila C. Detection of sperm antibodies in semen using the immunobead test: a survey of 813 consecutive patients. Am J Reprod Immunol. 1985;7:118–23.
10. Jager S, Kremer J, van Slochteren-Draaisma T. A simple method of screening for antisperm antibodies in the human male. Detection of spermatozoal surface IgG with the direct mixed antiglobulin reaction carried out on untreated fresh human semen. Int J Fertil. 1978;23:12–21.
11. Coombs RR, Marks J, Bedford D. Specific mixed agglutination: mixed erythrocyte platelet antiglobulin reaction for the detection of platelet antibodies. Br J Haematol. 1956;2:84–94.
12. Salid TM, Agarwal A. Tests for sperm antibodies. In: Krause WKH, Naz RK, editors. Immune infertility. Cham: Springer; 2017. p. 197–207.
13. Ackerman S, McGuire G, Fulgham DL, Alexander NJ. An evaluation of a commercially available assay for the detection of antisperm antibodies. Fertil Steril. 1988;49:732–4.
14. Kallen CB, Arici A. Immune testing in fertility practice: truth or deception? Curr Opin Obstet Gynecol. 2003;15:225–31.
15. Mortimer D. Clinical relevance of diagnostic procedures. In: Mortimer D, editor. Practical laboratory andrology. New York: Oxford University Press; 1994. p. 241–67.
16. Ke RW, Dockter ME, Majumdar G, Buster JE, Carson SA. Flow cytometry provides rapid and highly accurate detection of antisperm antibodies. Fertil Steril. 1995;63:902–6.

17. Witkin SS, Zelikovsky G, Good RA, Day NK. Demonstration of 11S IgA antibody to spermatozoa in human seminal fluid. Clin Exp Immunol. 1981;44:368–74.
18. Hancock RJ, Faruki S. Detection of antibody-coated sperm by 'panning' procedures. J Immunol Methods. 1984;66:149–59.
19. Hancock RJ. From antibodies to sperm to antibodies in transplantation. Urology. 1998;51 (5A Suppl):146–9.
20. Shibahara H, Tsunoda T, Taneichi A, Hirano Y, Ohno A, Takamizawa S, Yamaguchi C, Tsunoda H, Sato I. Diversity of antisperm antibodies bound to sperm surface in male immunological infertility. Am J Reprod Immunol. 2002;47:146–50.

Chapter 9
Causes of Immune Infertility in Men with Anti-sperm Antibody (ASA)

Hiroaki Shibahara

Abstract The presence of anti-sperm antibodies (ASAs) in men can reduce fertility as similar as that in women. However, the association between the presence of ASA in men and infertility continues to be disputed, in part because a standardized and universally accepted assay for the detection of ASA, a consensus about the clinical consequences of ASA, and a mechanistic explanation of how ASA impairs conception in infertile men remain to be unestablished.

As for the effect on sperm motility by ASA bound to sperm surface, there exists significant inhibitory effect of sperm-immobilizing antibodies in ejaculated human sperm on sperm motility in immunologically infertile men.

ASAs bound to sperm surface in infertile men also show significant inhibitory effect on sperm passage through cervical mucus (CM). However, it was also proved by our study that the ability of sperm passage through CM is not inhibited in two-fifths of men with ASA. It indicates there is a diversity of ASA bound to sperm surface in infertile men.

It has been shown that ASAs in men have an inhibitory effect on fertilization. However, others have arrived at the opposite conclusion. These results also indicate that there is a diversity of the blocking effects on fertilization by ASA bound to sperm surface. It has been also shown that ICSI demonstrates comparable fertilization, pregnancy, implantation, and miscarriage rates in female partners of men with and without ASA.

Based on our study, a sperm-zona pellucida binding assay such as the hemizona assay (HZA) should be performed for appropriate decision-making in infertile men with ASA.

H. Shibahara (✉)
Department of Obstetrics and Gynecology, School of Medicine, Hyogo Medical University, Nishinomiya, Hyogo, Japan
e-mail: sibahara@hyo-med.ac.jp

9.1 Introduction

The presence of anti-sperm antibodies (ASAs) in men can reduce fertility as similar as that in women. In infertile women with ASA, it has been demonstrated that ASA, especially sperm-immobilizing antibodies [1, 2], can impair sperm migration in the female genital tract, in particular in the cervical mucus [3, 4] and the uterine cavity through the fallopian tubes [5]. We also reported the inhibitory effects of sperm-immobilizing antibodies on fertilization [6–10] and post-fertilization events [9–12].

However, the association between the presence of ASA in men and infertility continues to be disputed [13, 14], in part because a standardized and universally accepted assay for the detection of ASA, a consensus about the clinical consequences of ASA, and a mechanistic explanation of how ASA impairs conception in infertile men remain to be unestablished.

Our group has demonstrated that there is a diversity of ASA bound to the sperm surface, including different immunoglobulin (Ig) classes of ASA, differing localization of the corresponding antigens for ASA, and different biological activities of ASA, in infertile men [15, 16]. These findings strongly suggested that the heterogeneity of ASA could be one of the reasons for the confusion of the diagnosis and treatment for immunologically infertile men.

Here, the causes of immune infertility in men with ASA are described.

9.2 Inhibitory Effect on Sperm Motility by Anti-sperm Antibodies (ASAs) Bound to Sperm Surface

Diverse ASAs are bound to sperm surfaces in men [15, 16]. Moreover, immunologically infertile men had a relatively high incidence of asthenozoospermia. Previously, we reported the effect of sperm-immobilizing antibodies bound to the surface of ejaculated human sperm on sperm motility [17].

In brief, semen samples were collected from 449 infertile men. Sperm concentration and sperm motility were assessed by using computer-aided sperm analysis (CASA) [18]. To detect ASA bound to the sperm surface, the direct immunobead test (IBT) [19] was performed. Sixteen of the 449 men (3.56%) had at least one of the Ig classes (IgG, IgA, IgM) of ASA. On the basis of the descriptive criteria by the World Health Organization [20], 4 (25.0%) of 16 immunologically infertile men were diagnosed with normozoospermia, 10 (62.5%) were diagnosed with asthenozoospermia, and 2 (12.5%) were diagnosed with teratoasthenozoospermia. In some cases, direct sperm-immobilization test (SIT) [15] was carried out as described in Chap. 9. Twelve infertile men who had ASA bound to sperm surface diagnosed according to direct IBT were also tested for sperm-immobilizing antibodies bound to sperm surface. Eight (66.7%) of these 12 infertile men were also positive on the direct SIT. The incidence of asthenozoospermia according to CASA was compared between eight men with sperm-immobilizing antibodies and four men

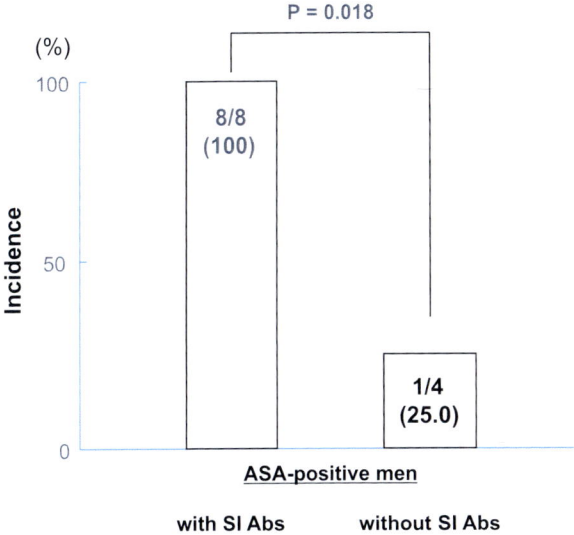

Fig. 9.1 Comparison of the incidence of asthenozoospermia in immunologically infertile men with and without sperm-immobilizing antibodies (SI Abs) bound to sperm surface. Eight of 12 infertile men who had ASA bound to sperm surface diagnosed by direct immunobead test (IBT) were also positive on the direct sperm-immobilization test (SIT). The incidence of asthenozoospermia according to CASA was compared between eight men with sperm-immobilizing antibodies and four men without the antibodies. All men who were positive on the direct SIT and one (25%) of the four men who were negative on this test were diagnosed with asthenozoospermia. The incidence between the groups differed significantly ($P = 0.018$)

without the antibodies bound to sperm surface. All men who were positive on the direct SIT and one (25%) of the four men who were negative on this test were diagnosed with asthenozoospermia (Fig. 9.1). The incidence between the groups differed significantly ($P = 0.018$). No correlation was found between asthenozoospermia and the Ig class of ASA or the sites on the sperm bound by ASA. Moreover, no specific clinical features were identified in men with a positive result on the direct SIT.

These findings indicated the significant effect of sperm-immobilizing antibodies bound to the surface of ejaculated human sperm on sperm motility in immunologically infertile men. Sperm-immobilizing antibodies display activity only with complement [1]. Scanning electron microscopy has revealed that the sperm plasma membrane peeled out after sperm was treated with sperm-immobilizing antibodies and complement as shown in Fig. 8.5, Chap. 8 [21]. Moreover, the necessary amounts of two key complement components, C3 and C4, exist in seminal plasma [22]. These findings strongly support the above conclusion.

9.3   Inhibitory Effect on Sperm Passage Through Cervical Mucus (CM) by ASA Bound to Sperm Surface

It has been shown that sperm that are antibody-bound over most of their surfaces are unable to enter the cervical mucus (CM), antibody binding to the sperm tail tip being an exception [23]. Ayvaliotis et al. [24] also reported that when all sperm were coated with Ig, it was rare to find sperm within the CM, despite the presence of hundreds of millions of motile sperm in the ejaculate. The number of motile sperm observed in the CM increased as the proportion of antibody-coated sperm declined below 50%, as judged by immunobead binding. This observed impairment of the ability of sperm to penetrate through the CM appears to be mediated by the effector region (Fc) portion of the Ig molecules [25, 26]. The Ig possess both an antigen recognition region (Fab) and an Fc that binds to various leukocytes through specific surface receptors (FcR) [27]. Sperm exposed to the Fab portion of IgG of ASA can swim through the CM, whereas those exposed to intact IgG of ASA cannot.

We also investigated whether ASAs bound to sperm surface in infertile men show inhibitory effect on sperm passage through CM. Sixty-five infertile men were assessed on their ability of sperm passage through CM using the post-coital test (PCT). All of them had normal semen characteristics, and the direct IBT was also performed. The results of PCT were compared between 10 infertile men with ASA and 55 infertile men without ASA. Six (60.0%) of 10 infertile men with ASA and 13 (23.6%) of 55 infertile men had abnormal PCT results. There was a significant difference between the two groups ($P = 0.02$) (Fig. 9.2).

These findings suggested that infertile men with ASA bound to sperm surface show inhibitory effect on sperm passage through CM. However, it was also proved that the ability of sperm passage through CM is not inhibited in two-fifths of men even with ASA in this study. It indicates there is a diversity of ASA bound to sperm surface in infertile men.

9.4   Inhibitory Effect on Fertilization by ASA Bound to Sperm Surface

It has been shown that ASAs in men have an inhibitory effect on fertilization [28–32]. However, others have arrived at the opposite conclusion [33]. These results might indicate that there is a diversity of the blocking effects on fertilization by ASA bound to sperm surface. Previously, we reported the diverse effects of ASA bound to the surface of ejaculated human sperm on fertilization [16].

In brief, ASAs were detected using the direct IBT in 509 semen samples. In some cases, the direct SIT was carried out. The fertilizing ability of infertile men with ASA was determined in the following three categories:

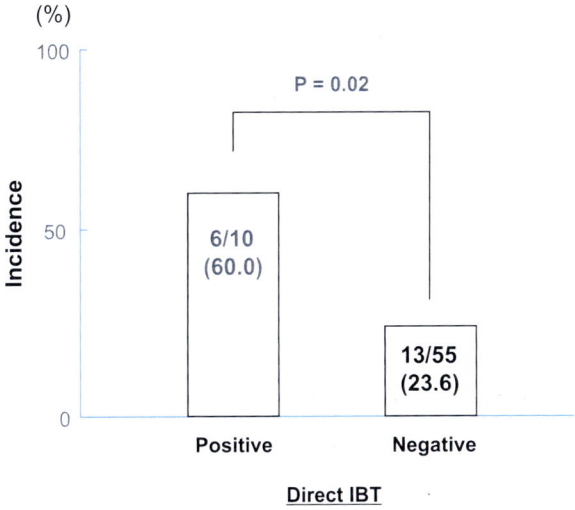

Fig. 9.2 Comparison of the incidence of abnormal PCT results in infertile men with ASA assessed by the direct immunobead test (IBT). Sixty-five infertile men were assessed on their ability of sperm passage through CM using the post-coital test (PCT). All of them had normal semen characteristics, and the direct IBT was also performed. The results of PCT were compared between 10 infertile men with ASA and 55 infertile men without ASA. Six (60.0%) of 10 infertile men with ASA and 13 (23.6%) of 55 infertile men had abnormal PCT results. There was a significant difference between the two groups ($P = 0.02$). These findings suggested that infertile men with ASA bound to sperm surface show inhibitory effect on sperm passage through CM. However, it was also proved that the ability of sperm passage through CM is not inhibited in two-fifths of men even with ASA

An IVF fertilization rate of $\geqq 50\%$
A hemizona index (HZI) [6–8] of $\geqq 50\%$
Pregnancy established without the use of ART

In total, 18 (3.54%) infertile men had ASA on the sperm surface. Except for 1 man with an absolute indication for ICSI because of severe asthenozoospermia and 2 men who dropped out of this study, fertilizing ability in 15 men could be determined. Four (26.7%) men did not satisfy the above criteria. The existence of sperm-immobilizing antibodies on the surface of ejaculated sperm had no impact on fertilization. As shown in Fig. 9.3, in four (57.1%) of seven patients who had IB-bound sperm of $\geqq 80\%$, fertilizing ability was inhibited, while none of the eight patients who had $<80\%$ IB-bound sperm had an inhibitory effect on fertilization. There was a significant difference between the two groups ($P = 0.01$). Some sperm-bound antibodies are related to the inhibitory effects on fertilization, indicating that a diversity of sperm-bound antibodies exists in men.

It was also demonstrated by a few investigators that the IVF rate significantly decreased when the inseminated sperm were coated with higher numbers of ASA [29, 31]. These findings also suggest that in infertile men having positive but lower levels of ASA on the ejaculated sperm, ASA-free sperm might contribute to higher

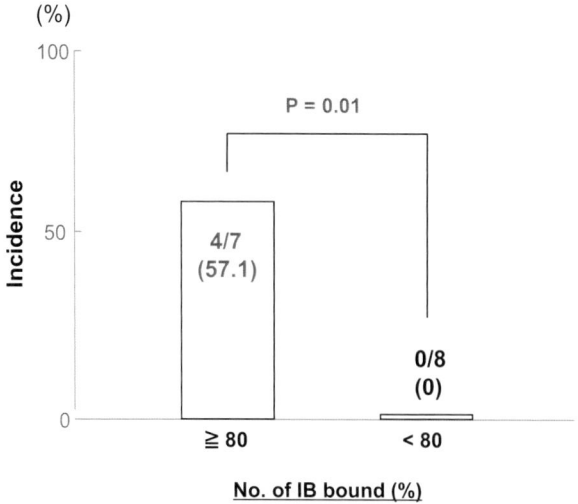

Fig. 9.3 Association of the incidence of the inhibitory effects on fertilization with the number of immunobeads (IBs) bound. Except for 1 man with an absolute indication for ICSI because of severe asthenozoospermia and 2 men who dropped out of this study, fertilizing ability in 15 men could be determined. In four (57.1%) of seven patients who had IB-bound sperm of ≧80%, fertilizing ability was inhibited, while none of the eight patients who had <80% IB-bound sperm had an inhibitory effect on fertilization. There was a significant difference between the two groups (*P* = 0.01)

fertilization rates in IVF. In contrast, those with higher levels of ASA have fewer ASA-free sperm, which leads to a significant reduction in fertilization. If so, a higher sperm concentration for insemination might contribute to acceptable fertilization rates in IVF for such immunologically infertile men with ASA.

These findings might be one of the reasons for the controversy of the relationship between ASA and male immunological infertility. Based on our study, a sperm-zona pellucida binding assay such as the hemizona assay (HZA) [34] should be performed for appropriate decision-making in infertile men with ASA [16, 35].

9.5   Usefulness of ICSI for Infertile Men with ASA Bound to Sperm Surface

ICSI can easily bypass pre-fertilization biological hindrances caused by ASA. In fact, ICSI has shown to be an effective technique that yields comparable clinical outcomes as those cases with negative ASA in infertile men, and the presence of ASA bound to sperm does not show any negative impact to ICSI outcomes [36, 37].

It has been also shown that ICSI demonstrates comparable fertilization, pregnancy, implantation, and miscarriage rates in female partners of men with and without ASA [36, 37].

References

1. Isojima S, Li TS, Ashitaka Y. Immunologic analysis of sperm immobilizing factor found in sera of women with unexplained sterility. Am J Obstet Gynecol. 1968;101:677–83.
2. Isojima S, Tsuchiya K, Koyama K, Tanaka C, Naka O, Adachi H. Further studies on sperm immobilizing antibody found in sera of unexplained cases of sterility in women. Am J Obstet Gynecol. 1972;112:199–207.
3. Koyama K, Ikuma K, Kubota K, Isojima S. Effects of antisperm antibodies on sperm migration through cervical mucus. Excerpta Med Int Congr Series. 1979;512:705–8.
4. Shibahara H, Shiraishi Y, Hirano Y, Kasumi H, Koyama K, Suzuki M. Relationship between level of serum sperm immobilizing antibody and its inhibitory effect on sperm migration through cervical mucus in immunologically infertile women. Am J Reprod Immunol. 2007;57:142–6.
5. Shibahara H, Shigeta M, Toji H, Koyama K. Sperm immobilizing antibodies interfere with sperm migration from the uterine cavity through the fallopian tubes. Am J Reprod Immunol. 1995;34:120–4.
6. Shibahara H, Shigeta M, Koyama K, Burkman LJ, Alexander NJ, Isojima S. Inhibition of sperm-zona pellucida tight binding by sperm immobilizing antibodies as assessed by the hemizona assay (HZA). Acta Obstet Gynaecol Jpn. 1991;43:237–8.
7. Shibahara H, Burkman LJ, Isojima S, Alexander NJ. Effects of sperm-immobilizing antibodies on sperm-zona pellucida tight binding. Fertil Steril. 1993;60:533–9.
8. Shibahara H, Shigeta M, Inoue M, Hasegawa A, Koyama K, Alexander NJ, Isojima S. Diversity of the blocking effects of antisperm antibodies on fertilization in human and mouse. Hum Reprod. 1996;11:2595–9.
9. Taneichi A, Shibahara H, Hirano Y, Suzuki T, Obara H, Fujiwara H, Takamizawa S, Sato. Sperm immobilizing antibodies in the sera of infertile women cause low fertilization rates and poor embryo quality in vitro. Am J Reprod Immunol. 2002;47:46–51.
10. Taneichi A, Shibahara H, Takahashi K, Sasaki S, Kikuchi K, Sato I, Yoshizawa M. Effects of sera from infertile women with sperm immobilizing antibodies on fertilization and embryo development in vitro in mice. Am J Reprod Immunol. 2003;49:146–51.
11. Koyama K, Hasegawa A, Isojima S. Effects of antisperm antibody on the in vitro development of rat embryos. Gamete Res. 1984;10:143–52.
12. Shibahara H, Mitsuo M, Ikeda Y, Shigeta M, Koyama K. Effects of sperm immobilizing antibodies on pregnancy outcome in infertile women treated with IVF-ET. Am J Reprod Immunol. 1996;36:96–100.
13. Eggert-Kruse W, Christman M, Gerhard I, Pohl S, Klingla K, Runnebaum B. Circulating antisperm antibodies and fertility prognosis: a prospective study. Hum Reprod. 1989;4:513–20.
14. Collins JA, Burrows EA, Yeo J, YoungLai EV. Frequency and predictive value of antisperm antibodies among infertile couples. Hum Reprod. 1993;8:592–8.
15. Shibahara H, Tsunoda T, Taneichi A, Hirano Y, Ohno A, Takamizawa S, Yamaguchi C, Tsunoda H, Sato I. Diversity of antisperm antibodies bound to sperm surface in male immunological infertility. Am J Reprod Immunol. 2002;47:146–50.
16. Shibahara H, Shiraishi Y, Hirano Y, Suzuki T, Takamizawa S. Diversity of the inhibitory effects on fertilization by anti-sperm antibodies bound to the surface of ejaculated human sperm. Hum Reprod. 2003;18:1469–73.
17. Shibahara H, Hirano Y, Takamizawa S, Sato I. Effect of sperm-immobilizing antibodies bound to the surface of ejaculated human spermatozoa on sperm motility in immunologically infertile men. Fertil Steril. 2003;79:641–2.
18. Hirano Y, Shibahara H, Obara H, Suzuki T, Takamizawa S, Yamaguchi C, Tsunoda H, Sato I. Relationships between sperm motility characteristics assessed by the computer-aided sperm analysis (CASA) and fertilization rates in vitro. J Assist Reprod Genet. 2001;18:213–8.
19. Bronson RA, Cooper GW, Rosenfeld D. Detection of sperm specific antibodies on the spermatozoa surface by immunobead binding. Arch Androl. 1982;9:61.

20. World Health Organization, editor. WHO laboratory manual for the examination and processing of human semen. 5th ed. Geneva: WHO; 2010.
21. Isojima S, Koyama K, Ikuma K, Kubota K, Shigeta M. Female immunological infertility: a new immunological approach to contraception. In: Cortes-Prieto J, editor. Research on fertility and sterility. MTP Press Limited; International Medical Publishers: Lancaster; 1980. p. 339–54.
22. Rahimi A, Sepehri H, Pakravesh J, Bahar K. Quantification of C3 and C4 in infertile men with antisperm antibody in their seminal plasma. Am J Reprod Immunol. 1999;41:330–6.
23. Wang C, Baker HW, Jennings MG, Burger HG, Lutjen P. Interactions between cervical mucus and sperm surface antibodies. Fertil Steril. 1985;44:484–8.
24. Ayvaliotis B, Bronson R, Rosenfeld D, Cooper G. Conception rates in couples where auto-immunity to sperm in detected. Fertil Steril. 1985;43:739–42.
25. Jager S, Kremer J, Kuiken J, van Slochteren-Draaisma T, Mulder I, de Wilde-Janssen IW. Induction of the shaking phenomenon by pretreatment of spermatozoa with sera containing antispermatozoal antibodies. Fertil Steril. 1981;36:784–91.
26. Jager S, Kremer J, Kuiken J, Mulder I. The significance of the Fc part of antispermatozoal antibodies for the shaking phenomenon in the sperm-cervical mucus contact test. Fertil Steril. 1981;36:792–7.
27. Goodman JW. Immunoglobulin structure and function. In: Stites DP, Terr Jr AI, editors. Basic human immunology. Norwalk: Appleton & Lang; 1991. p. 109–21.
28. Mandelbaum SL, Diamond SP, DeCherney AH. Relationship of antisperm antibodies to oocyte fertilization in in-vitro fertilization-embryo transfer. Fertil Steril. 1987;47:644–51.
29. de Almeida M, Gazagne I, Jeulin C, Herry M, Belaisch-Allart J, Frydman R, Jouannet P, Testart J. In-vitro processing of sperm with autoantibodies and in-vitro fertilization results. Hum Reprod. 1989;4:49–53.
30. Rajah SV, Parslow JM, Howell RJ, Hendry WF. The effects on in-vitro fertilization of autoantibodies to spermatozoa in subfertile men. Hum Reprod. 1993;8:1079–82.
31. Lahteenmaki A. In-vitro fertilization in the presence of antisperm antibodies detected by the mixed antiglobulin reaction (MAR) and the tray agglutination test (TAT). Hum Reprod. 1993;8: 84–8.
32. Yeh WR, Acosta AA, Seltman HJ, Doncel G. Impact of immunoglobulin isotype and sperm surface location of antisperm antibodies. Fertil Steril. 1995;63:1287–92.
33. Sukcharoen N, Keith J. The effect of the antisperm auto-antibody bound sperm on in vitro fertilization outcome. Andrologia. 1995;27:281–9.
34. Burkman LJ, Coddington CC, Franken DR, Krugen TF, Rosenwaks Z, Hogen GD. The hemizona assay (HZA): development of a diagnostic test for the binding of human spermatozoa to the human hemizona pellucida to predict fertilization potential. Fertil Steril. 1988;49:688–97.
35. Shibahara H, Shiraishi Y, Suzuki M. Diagnosis and treatment of immunologically infertile males with antisperm antibodies. Reprod Med Biol. 2005;4:133–41.
36. Check ML, Check JH, Katsoff D, Summers-Chase D. ICSI as an effective therapy for male factor with antisperm antibodies. Arch Androl. 2000;45:125–30.
37. Esteves SC, Schneider DT, Verza S Jr. Influence of antisperm antibodies in the semen on intracytoplasmic sperm injection outcome. Int Braz J Urol. 2007;33:795–802.

Chapter 10
Production of Anti-sperm Antibody (ASA) in Men

Hiroaki Shibahara

Abstract The blood-testis barrier (BTB) creates favorable conditions for the spermatogenesis and sperm survival in the testicular fluid. It prevents the occurrence of autoimmunity after puberty. The alteration of the BTB allows the production of anti-sperm antibodies (ASAs) and may lead to infertility. This leads to autoimmune-mediated testicular damage, germ cell loss, and impaired sperm function.

The possible risk factors of ASA production in men have been shown such as congenital or acquired chronic obstruction of the male reproductive tract (MRT), infection in the MRT, varicocele, cryptorchidism, testicular injuries, testicular tumors, homosexual men, spinal cord injury, and autoimmune diseases. However, the relationship between these risk factors and the production of ASA is still controversial. Recently, it was suggested that vasectomy followed by vasovasostomy is the only clinical condition that shows almost permanently high titers of ASA in numerous clinical series.

In this chapter, the possible risk factors for the production of ASA in men are discussed.

10.1 Introduction

Testicular tissue and sperm health are important factors in clinical determination of male fertility [1, 2]. As the site of sperm production (spermatogenesis), a healthy testis is vital to male reproduction. There are several key cell types involved in spermatogenesis in the testes. The first is the Sertoli cells, which are immunosuppressive cells that form the blood-testis barrier (BTB) and provide nutritional support for spermatogenesis. The second is the Leydig cells, which are interstitial androgen-producing cells. The third is the spermatogenic stem cells, which are progenitor cells

H. Shibahara (✉)
Department of Obstetrics and Gynecology, School of Medicine, Hyogo Medical University, Nishinomiya, Hyogo, Japan
e-mail: sibahara@hyo-med.ac.jp

from where all sperm arise. The fourth is the peritubular myoid cells, which are basement membrane-forming cells. And the last is the testicular macrophages, which are potently immunosuppressive cells [3].

The testis is an immune-privileged site at which immunogenic germ cells are protected from the detrimental effects of immune responses [4]. The most commonly recognized mechanism for the immunological privilege is the BTB, limiting the access of germ cell antigens to interstitial immune cells and the passage of antibodies from the interstitium to the tubular lumen. Immunological privilege outside the BTB involves secretion of immunosuppressive factors mainly by macrophages, Sertoli cells, peritubular cells, and Leydig cells [5–9].

The BTB and immune privilege are critical for fertility, as breakdown can result in the generation of autoimmune T and B lymphocytes and production of anti-sperm antibodies (ASAs) [10]. This leads to autoimmune-mediated testicular damage, germ cell loss, and impaired sperm function [10, 11]. ASA can affect male fertility by various mechanisms [12]. Some of them relate with sperm-agglutination and/or complement-dependent sperm-immobilization [13]. ASAs also impair sperm penetration through cervical mucus and interfere with sperm-egg interaction. (See Chap. 9).

In this chapter, the existing data were evaluated to identify the possible risk factors for the production of ASA in men.

10.2   Risk Factors for the Production of Anti-sperm Antibodies (ASAs)

The possible risk factors of ASA production in men have been shown such as congenital or acquired chronic obstruction of the male reproductive tract (MRT), infection in the MRT, varicocele, cryptorchidism, testicular injuries, testicular tumors, homosexual men, spinal cord injury, and autoimmune diseases (Table 10.1). However, the relationship between these risk factors and the production of ASA is still controversial. Recently, Marconi et al. [14] suggested that vasectomy followed by vasovasostomy (VV) is the only clinical condition that shows almost permanently high titers of ASA in numerous clinical series.

10.2.1   Chronic Obstruction in the Male Reproductive Tract (MRT)

10.2.1.1   Congenital Obstruction

To test the hypothesis that puberty is a necessary factor in the pathogenesis of autoimmunity to sperm in men with cystic fibrosis (CF), Bronson et al. [15] studied prepubertal and postpubertal men with CF versus an age-matched group of men with type 1 diabetes as controls. Sera from CF and diabetic men were tested by indirect

Table 10.1 Risk factors for the production of anti-sperm antibodies in men

Chronic obstruction in the male reproductive tract
Congenital obstruction
Cystic fibrosis
Congenital bilateral absence of the vas deferens (CBAVD)
Acquired obstruction
Vasectomy and vasovasostomy
Infection in the male reproductive tract
Varicocele
Cryptorchidism
Testicular injuries
Testicular trauma
Testicular surgery including testicular sperm extraction (TESE)
Testicular torsion
Testicular tumors
Testicular microlithiasis
Testicular carcinoma
Homosexual men
Spinal cord injury
Autoimmune diseases

immunobead test (IBT). Autoimmunity to sperm, as detected by humoral ASA, was documented solely in postpubertal males, as judged by hormonal and clinical criteria. Eighty-three percent of sexually mature CF males and 6.3% (1 of 16) of diabetic males exhibited autoantibodies to sperm. They demonstrated that ASAs were detected after the onset of puberty in a cohort of men with CF and congenital bilateral absence of the vas deferens (CBAVD). D'Cruz OJ et al. [16] also reported that ASAs were also detected in men with CF, who also have infertility from CBAVD. It was suggested that puberty, and presumably active spermatogenesis, is a requirement for the development of autoimmunity to sperm in men with CF.

Matsuda et al. [17] found a high rate of ASA in men with vasal obstruction from hernia repair as an infant. In men with CBAVD, ASAs were found in 29% of serum specimens, 16% of epididymal fluid samples, and 35% of the patients' sperm surface [18]. There was no difference in pregnancy rates with IVF of aspirated specimens between those with and those without ASA [18].

10.2.1.2 Acquired Obstruction

By using a rat model, Frickinger et al. [19] have studied temporal patterns of IgM and IgG autoantibodies to sperm proteins by Western blot analysis at intervals after bilateral vasectomy and vasectomy followed 1 month later by vasovasostomy. They showed the high incidence of IgM antibodies in the earliest sample, taken 2 weeks after vasectomy, suggests that the initial immunizing event takes place within about a week after the operation. Vasovasostomy did not bring about a decrease in ASA. Instead, some animals demonstrated an increased reaction to certain antigens after reversal of vasectomy, even though the vasovasostomies were anatomically successful.

In human, vasectomy has been also shown as the most common identifiable cause of ASA [20–24]. ASAs are present in the patients' sera following vasectomy in 34–74% [25–27], and those bound to sperm are found following successful vasovasostomy in 38–60% [26, 28, 29]. The preoperative status of the serum correctly identified men who would have positive direct IBT following vasovasostomy in 69% of cases. Although prereversal status cannot accurately predict postoperative semen quality, a subgroup of men with very high serum antibody titers had a very poor prognosis for future pregnancy [26].

Gubin et al. [30] found that prior vas reversal was significantly associated with the presence of ASA ($P = 0.0002$) by mixed agglutination reaction (MAR) or IBT with a fivefold increased relative risk (95% confidence interval, 1.97–12.38). They concluded that vas reversal is one of the significant risk factors for the development of ASA.

Pregnancy rates following vasovasostomy are altered by the presence of ASA bound to sperm. Matson et al. [29] reported that the pregnancy rates were 67% if no antibodies were detected, 67% if only IgA was present, 20% with both IgA and IgG, and only 11% with pure IgG. Conversely, Meinertz et al. [31] found that IgA was a more potent inhibitor of conception. In men with pure IgG antibodies, the pregnancy rate was 86% versus 43% in men with IgA.

10.2.2 Infection in the Male Reproductive Tract (MRT)

Ohl et al. [32] previously suggested that genitourinary infection has been associated with ASA. Several studies have shown a higher than expected incidence of ASA in men being infected with nonspecific urethritis [33, 34], chlamydia [35–38], mycoplasma [36], and gonorrhea [34]. Acute epididymitis led to autoimmunity in 27% of a series of 27 men [39]. Gubin et al. [30] also reported that prior infections, independent of vas reversal, were significantly associated with ASA. They concluded that prior infection is one of the significant risk factors for the development of ASA. Experimental orchitis in the rabbit led to development of ASA and contralateral histologic change [40]. ASAs have been found in 25–56% of men with chronic prostatitis [41, 42].

On the contrary, Marconi et al. [24, 43] reported controversial data exist on the association between inflammation/infection of the MRT and ASA. In a series of 79 infertile patients with inflammatory/infectious diseases of the MRT, the comparative results for ASA detection in seminal plasma demonstrated no clear role of this association for male infertility [24]. In a second series of 365 patients with documented inflammation/infection of the MRT, such as chronic bacterial prostatitis

(CBP), inflammatory chronic prostatitis/chronic pelvic pain syndrome (CP/CPPS), non-inflammatory CPPS, chronic urethritis, and chronic epididymitis, they found no association between ASA formation and these diseases [43].

Marconi et al. [14] recently showed four mechanisms by infection/inflammation of the MRT that related with formation of ASA in men: first, obstruction of the MRT because of inflammatory and post-inflammatory changes; second, tearing of the BTB because of local inflammation; third, decrease of the immunomodulatory factors (cellular and humoral) present in seminal plasma that normally prevent sperm autoimmunization; and fourth, cross-reactivity between antigens of the microorganisms responsible for MRT infections (i.e., *Chlamydia trachomatis*) and sperm antigens. In the hypothetical scenario that inflammatory/infectious diseases of the MRT are associated with ASA formation, for anatomical reasons previously discussed, the epididymis would be the most probable site of ASA formation. However, the prostate gland is another site where a localized immune response can be induced, since prostatic fluids have been identified to contain specific IgA antibody against *Escherichia coli* and sperm [44].

10.2.3   Varicocele

The pathophysiological mechanisms of male infertility associated with varicocele remain unclear. One of the potential pathogenic factors that adversely affect infertility in varicocele patients is anti-sperm immune response [45]. Whether or not varicocele leads to ASA has been shown to be controversial [32]. Some studies showed a significant number of varicocele patients with autoimmunity [46–50], and sperm antibodies can be induced in an animal model by inducing varicocele [51]. However, other studies show no increased incidence of antibodies with varicocele [30, 52–54]. One of these latter studies showed no sperm antibodies prior to repairing the varicocele but a 16% incidence following surgery [53]. When antibodies are present, deficiencies in the semen analysis have been noted [47, 48], but the autoimmune status has not been shown to alter the degree of improvement from surgery [47].

Bozhedomov et al. [45] recently performed a multivariate analysis of the clinical and laboratory data of 1729 men of reproductive age. The OR of immune infertility after a testicular trauma in varicocele patients increases twofold. In varicocele patients, the autoimmune anti-sperm reaction is accompanied by a more significant decrease in the semen quality, acrosome reaction disorders, and an increase in the proportion of sperm with DNA fragmentation. These disorders correlate with the level of sperm oxidative stress; reactive oxygen species (ROS) production in ASA-positive varicocele patients is 2.8 and 3.5 times higher than in ASA-negative varicocele patients and fertile men, respectively. They found that the varicocele was not an immediate cause of anti-sperm immune response, but is a cofactor increasing ASA risk.

Marconi et al. [14] recently found that the basic research evidence is also controversial. Shook et al. [51] demonstrated in an animal model that a surgically induced varicocele triggers ASA formation. However, Turner et al. [55], working also with surgically induced varicocele model in rats, demonstrated that the BTB was not damaged in these animals, suggesting that the impairment of spermatogenesis in this disease is not immunologically mediated. Interestingly, in patients with varicocele and ASA in the ejaculate, these immunoglobulins are also present in testicular biopsies, more specifically inside the seminiferous tubule, suggesting that if in fact there is an association between these two conditions, the most probable site of formation would be the testis [56].

10.2.4 Cryptorchidism

Cryptorchidism is defined as a condition in which one or both testes fail to descend to the scrotal position [14]. The incidence of this condition varies from 1.4 to 2.7% in male births and is increased in premature birth [57]. Of the 3% of full-term infants affected, most will have testes that will descend normally within a few months. The remaining 1%, who have a cryptorchid condition that persists, should consider medical or surgical intervention [57] because it is a major risk for male infertility and for testicular malignancy in adulthood [58].

Several studies have reported an increased incidence of ASA (up to 28%) in patients with history of cryptorchidism either treated or untreated by orchidopexy [58–60]. However, others arrived at the opposite conclusion [61, 62]. Most of these studies included prepubertal population where ASA has been only tested in serum, not addressing the important issue of the presence of ASA in the ejaculate. Moreover, there is a high probability that an undefined percentage of the patients who have undergone orchidopexy develop, as a complication of surgery, some degree of obstruction at the epididymal or ductal level. Those patients have a high probability of ASA formation, but the etiology would be falsely classified to cryptorchidism and not to chronic obstruction [14]. Sinisi et al. [60] retrospectively evaluated serum ASA in cryptorchidism before and after orchidopexy. Their results indicated that cryptorchidism may elicit an autoimmune response against sperm antigen in childhood independent of testis location and orchidopexy. Moreover, patients of pubertal age appear to be at higher risk for ASA development. In other study, they suggested that in these patients, the sperm surface antigens are already present before meiosis and the BTB is either immature or impaired by heat due to the abnormal position [63]. In this point, it was demonstrated by the experimental rat models of cryptorchidism that the BTB remains competent under this situation [64, 65].

In conclusion, the association between cryptorchidism and ASA remains controversial. Marconi et al. [14] suggested if the association is real and the bias from surgical treatment complications, namely, iatrogenic obstruction of the vas deferens, is excluded, the most probable site of ASA production in these patients would be the testis.

10.2.5 Testicular Injuries

Testicular injuries include testicular trauma, testicular surgery, and testicular torsion. It seems logical that every condition where the BTB is breached should constitute a clear risk factor for ASA production, since the immune system establishes a direct contact with the antigens present in the sperm surface.

10.2.5.1 Testicular Trauma

Kukadia et al. [66] evaluated the presence of ASA in the ejaculate, using the direct IBT, in eight patients with a history of severe testicular trauma who underwent surgical exploration. Only one (13%) patient had ASA, the levels of which were probably low enough to be clinically insignificant. There was definite evidence of subfertility as assessed by abnormal semen analyses and atrophic testes following testicular trauma. However, the subfertility did not appear to be immune mediated nor did the patients present with infertility.

Pogorelic et al. [67] reported seven patients who were operated on because of testicular rupture. Mean age at the time of the accident was 15 years. Semen analysis showed normospermia 6 months after surgery, and ASA count was within normal limits in all patients.

These findings might suggest that there is no association between testicular trauma and production of ASA in men.

10.2.5.2 Testicular Surgery Including Testicular Sperm Extraction (TESE)

Harrington et al. [68] reported that needle or open testicular biopsy did not result in the formation of new or increased levels of ASA in the seminal fluid or serum when patients were tested 6 months postoperatively.

Komori et al. [69] assayed for the presence of serum ASA for 1 year after conventional multiple TESE in 13 men or microdissection TESE in 12 men and compared postoperative testicular damage between procedures. No incidence of new ASA formation was identified in the present study. Ozturk et al. [70] also reported that none of the 37 patients developed significant levels of ASA in their sera as a result of the TESE procedure.

Leonhartsberger et al. [71] found that organ-sparing surgery does not lead to greater ASA levels in sera than standard orchidectomy, and patients are therefore at no greater risk of developing an autoimmune infertility.

Moreover, in surgeries that do not directly compromise the testis but the nearby structures, such as inguinal hernia repair with and without mesh, Stula et al. [72] have also reported no association to ASA formation.

In conclusion, surgical procedures to the testis could disrupt the BTB; however, a risk for ASA formation may not increase.

10.2.5.3 Testicular Torsion

Testicular torsion theoretically may cause a breakdown of the BTB due to ischemic injury. It is unclear whether these abnormalities are due to an autoimmune process that occurs after the rupture of the hematotesticular barrier leading to the formation of ASA or as a result of reperfusion-induced injury to the testis [73, 74].

Studies in animal models generally confirm the presence of ASA [73]. Such animal models usually show histologic lesions occurring in the contralateral testis and abnormal levels of ASA after experimental testicular torsion [75–78]. However, animal models for testicular torsion do not reproduce the conditions found in human, such as the anatomy of the testis, known to be abnormally loose inside the vaginal layer [79]; therefore, it was suggested that care should be taken in extrapolating these data [80].

In human, testicular torsion is a surgical emergency, which requires prompt diagnosis and immediate treatment. Testicular torsion occurring before puberty has a very low risk for future development of immunological infertility [81]. Testicular torsion has been shown to cause an immune response in postpubertal animals [78], but firm documentation in human is lacking, and this phenomenon remains theoretical.

Anderson et al. [82] studied ASA formation in 16 postpubertal patients following testicular torsion. Nine were treated with detorsion and bilateral orchidopexy (detorsion group), and seven were treated with ipsilateral orchiectomy and contralateral orchidopexy (orchiectomy group). It was unable to find an increased rate of semen ASA detection in these patients.

Arap et al. [80] evaluated ASA formation in the ejaculate of patients with history of testicular torsion. Of 24 patients, 15 were treated with orchiectomy, and 9 were treated with orchidopexy. There were no significant differences in the seminal ASA levels between patients and proven fertile controls, regardless of the treatment applied. Identifying the risk factors for ASA production in a population of male patients, Heidenreich et al. [23] also concluded that testicular torsion is not associated with this condition.

Taken together, testicular injuries including trauma, surgery, TESE, and torsion of the testis do not seem to constitute a risk factor for production of ASA. Nevertheless, the BTB is clearly disturbed in all these cases. ASA formation is not regularly triggered; this fact demonstrates that the pathophysiology of ASA formation is still unclear [14].

10.2.6 Testicular Tumors

10.2.6.1 Testicular Microlithiasis

Testicular microlithiasis is a pathological event characterized by the presence of microliths within the testicular entities, and such calcium deposition is thought to have deleterious impacts on the structure of BTB.

Patel et al. [83] examined testicular microlithiasis and ASA production following testicular biopsy in 112 boys with cryptorchidism. By using direct IBT, no patient exhibited evidence of ASA. Jiang et al. [84] also found no presence of ASA using direct IBT in semen samples from 22 patients with testicular microlithiasis. They suggested that the presence of testicular microlithiasis may cause disruption in the BTB, it seems that testicular microlithiasis has no capacity to increase the risk for autoimmune disorders, and the discovery of more potential determinants involved in testicular microlithiasis-associated infertility will further aid in understanding the pathogenesis of testicular microlithiasis and even improving its treatment.

10.2.6.2 Testicular Carcinoma

Testicular cancer constitutes approximately 1% of all cancers in men but is the most frequent cancer in men aged 15–39 years. Because testicular cancer directly affects the gamete production site, it has been postulated as a trigger for an autoimmune response. The presence of ASA may in fact be explained by local effects caused by the cancer, such as raised scrotal temperature connected with blood flow alterations or disruption of the BBT, with a massive release of sperm antigens that stimulate anti-sperm immunization [85].

In the earlier studies, some investigators have reported that testicular tumors might be a risk factor for ASA formation [86–88]. The incidence of ASA in these patients ranges from 18 to 73%. However, most studies were biased in that only serum ASAs have been evaluated and most importantly did not include a control group to evaluate if the detection rates were significantly higher. In this point, Paoli et al. [85] also suggested that those studies used immunofluorescence and immunoenzymatic assays, which can give numerous false-positive and false-negative results and are therefore not considered as reference methods.

In the report by Paoli et al. [85], they evaluated 190 patients who underwent orchiectomy for testicular cancer. ASAs were examined by direct IBT for the sperm surface and also indirect IBT and the gelatin agglutination test for serum. One month after surgery, they found only a 5.8% prevalence of ASA in serum. They concluded that testicular cancer might not be a possible cause of ASA formation.

With the available evidence, the link between testicular tumors and ASA remains questionable, and the larger studies including healthy fertile controls are required.

10.2.7 Homosexual Men

It has been suggested that the introduction of sperm into the rectum may lead to the development of a humoral immune response in experimental animal models [89, 90]. Therefore, it seems logical that unprotected anal intercourse in homosexual men could constitute a risk factor for ASA formation. ASAs have also been detected in the sera of healthy homosexual men or those who presented with human immunodeficiency virus (HIV) infection [91–93]. However, many of these studies either lacked information on the subjects' HIV status or used methods for measuring ASA that were only poorly reproducible [94].

Naz et al. [95] found that 70% of HIV-positive men produced sperm-specific antibodies. Adams et al. [93] also found a high incidence in HIV-positive individuals. The high incidence of antibodies has commonly been thought due to receptive anal intercourse [92]. Mulhall et al. [94] also found a correlation between antibodies and a history of unprotected receptive anal intercourse in the previous 6 months.

However, in a mouse model, anal immunization with human sperm was far less efficient than intraperitoneal injection [96]. Contradicting the previous results, Sands et al. [97] found no significant difference in the serum ASA titers between sexually active heterosexual men and homosexual men with or without HIV infection, concluding that ASA levels are not higher in homosexual men.

In conclusion, there is not enough evidence to support homosexuality as a risk factor for ASA formation. However, taking into account clinical evidence and basic research studies, it seems highly probable that if this association exists, the primary site of ASA production would be the distal gastrointestinal mucosa [14].

10.2.8 Spinal Cord Injury

Spinal cord injury has been shown in humans and rats to induce leukocytospermia, with the presence of inflammatory cytokines, ASA, and reactive oxygen species found within the ejaculate. Because spinal cord injury induces ejaculatory dysfunction in nearly all patients, the question arises whether this "physiologic obstruction" leads to antibody development.

Beretta et al. [98] tested seminal parameters in 142 paraplegic/quadriplegic men. They found that the occurrence of ASA in men with spinal cord injury was frequent enough to be considered a causal factor in the reduction of fertility. Hirsch et al. [99] found ASA response in the seminal plasma of five of seven patients with spinal cord injury by using enzyme-linked immunosorbent assay (ELISA). They described that sperm autoimmunity should be considered among the important causes underlying their seminal dysfunction.

On the contrary, Dahlberg et al. [100] reported none of 16 men with spinal cord injury had ASA examined by the MAR test and the tray agglutination test. They concluded that anejaculation and sperm retention in men with spinal cord injury, even of 30 years' duration, does not result in ASA formation. Similarly, Menge et al. [101] also found none of the serum samples from 73 men unable to ejaculate naturally were found to be positive by sperm-agglutination or sperm-immobilization methods. ASAs were detected by an IBT on the sperm cells of only 1 of the 13 men examined. Siosteen et al. [102] tested the presence of ASA in serum and on sperm in 30 men with spinal cord injury. Eight of 30 exhibited sperm-agglutinating antibodies, but with low titers. Low titers of serum ASA and absence of attached IgA antibodies indicate autoimmunity is not the cause of low sperm motility and penetration capacity in these patients.

10.2.9   Autoimmune Diseases

We investigated if systemic autoimmune diseases could be one of the risk factors for developing ASA in men [103]. ASAs in the sera of 70 men with systemic autoimmune diseases and 80 healthy controls were examined by using the indirect IBT. Their autoimmune diseases were as follows: 32 patients with rheumatoid arthritis, 14 patients with Behcet's disease, 8 patients with systemic lupus erythematosus, 5 patients with polymyalgia rheumatica, 4 patients with vasculitis syndrome, 3 patients with progressive systemic sclerosis, 2 patients with psoriatic arthritis, 1 patient with mixed connective tissue disease, and 1 patient with adult-onset Still's disease. Their average age was 56.2 years old (from 27 to 77 years old). The sperm-immobilization test (SIT) was also performed to detect sperm-immobilizing antibodies to the patients who were positive in the indirect IBT. Among 70 males with systemic autoimmune diseases, 5 (7.1%) were positive in the indirect IBT. However, no positives existed in 80 healthy males. There was a significant difference between the two groups ($P = 0.02$). None of these five ASA-positive patients had sperm-immobilizing antibodies. Therefore, systemic autoimmune diseases may be one of the risk factors for developing ASA in men. Based on this study, we concluded that further studies will be needed to delineate the mechanism of the relations between autoimmune disorders and the development of ASA in men.

References

1. Moazenchi M, Totonchi M, Salman Yazdi R, Hratian K, Mohseni Meybodi MA, Ahmadi Panah M, Chehrazi M, Mohseni MA. The impact of chlamydia trachomatis infection on sperm parameters and male fertility: a comprehensive study. Int J STD AIDS. 2017;29(5):466–73.
2. Dohle GR, Elzanaty S, van Casteren NJ. Testicular biopsy: clinical practice and interpretation. Asian J Androl. 2012;14:88–93.
3. Bryan ER, Redgrove KA, Mooney AR, Mihalas BP, Sutherland JM, Carey AJ, Armitage CW, Trim LK, Kollipara A, Mulvey PBM, Palframan E, Trollope G, Bogoevski K, McLachlan R, McLaughlin EA, Beagley KW. Chronic testicular *Chlamydia muridarum* infection impairs mouse fertility and offspring development. Biol Reprod. 2020;102:888–901.

4. Qu N, Itoh M, Sakabe K. Effects of chemotherapy and radiotherapy on spermatogenesis: the role of testicular immunology. Int J Mol Sci. 2019;20:957. https://doi.org/10.3390/ijms20040957.
5. Head JR, Billingham RE. Immune privilege in the testis. II. Evaluation of potential local factors. Transplantation. 1985;40:269–75.
6. Pollanen P, Maddocks S. Macrophages, lymphocytes and MHC II antigen in the ram and rat testis. J Reprod Fertil. 1988;82:437–45.
7. Wyatt CR, Law L, Magnuson JA, Griswold MD, Magnuson NS. Suppression of lymphocyte proliferation by cultured Sertoli cells. J Reprod Immunol. 1988;14:27–40.
8. Itoh M, Terayama H, Naito M, Ogawa Y, Tainosho S. Tissue microcircumstances for leukocytic infiltration into the testis and epididymis in mice. J Reprod Immunol. 2005;67: 57–67.
9. Bhushan S, Meinhardt A. The macrophages in testis function. J Reprod Immunol. 2017;119: 107–12.
10. Fijak M, Iosub R, Schneider E, Linder M, Respondek K, Klug J, Meinhardt A. Identification of immunodominant autoantigens in rat autoimmune orchitis. J Pathol. 2005;207:127–38.
11. Redgrove KA, McLaughlin EA. The role of the immune response in chlamydia trachomatis infection of the male genital tract: a double edged sword. Front Immunol. 2014;5:534.
12. WHO. World Health Organization manual for the standardized investigation and diagnosis of the infertile couple. 2nd ed. Cambridge: Cambridge University Press; 2000.
13. Shibahara H, Hirano Y, Takamizawa S, Sato I. Effect of sperm-immobilizing antibodies bound to the surface of ejaculated human spermatozoa on sperm motility in immunologically infertile men. Fertil Steril. 2003;79:641–2.
14. Marconi M, Weidner W. Site and risk factors of antisperm antibodies production in the male population. In: Krause WKH, Naz RK, editors. Immune infertility. Cham: Springer; 2017. p. 133–47.
15. Bronson RA, O'Connor WJ, Wilson TA, Bronson SK, Chasalow FI, Droesch K. Correlation between puberty and the development of autoimmunity to spermatozoa in men with cystic fibrosis. Fertil Steril. 1992;58:1199–204.
16. D'Cruz OJ, Haas GG Jr, de La Rocha R, Lambert H. Occurrence of serum antisperm antibodies in patients with cystic fibrosis. Fertil Steril. 1991;56:519–27.
17. Matsuda T, Muguruma K, Horii Y, Ogura K, Yoshida O. Serum antisperm antibodies in men with vas deferens obstruction caused by childhood inguinal herniorrhaphy. Fertil Steril. 1993;59:1095–7.
18. Patrizio P, Silber SJ, Ord T, Moretti-Rojas I, Asch RH. Relationship of epididymal sperm antibodies to their in vitro fertilization capacity in men with congenital absence of the vas deferens. Fertil Steril. 1992;58:1006–10.
19. Flickinger CJ, Howards SS, Bush LA, Baker LA, Herr JC. Temporal recognition of sperm autoantigens by IgM and IgG autoantibodies after vasectomy and vasovasostomy. J Reprod Immunol. 1994;27:135–50.
20. Jarow JP, Sanzone JJ. Risk factors for male partner antisperm antibodies. J Urol. 1992;148: 1805–7.
21. Sinisi AA, Di Finizio B, Pasquali D, Scurini C, D'Apuzzo A, Bellastella A. Prevalence of antisperm antibodies by sperm MAR test in subjects undergoing a routine sperm analysis for infertility. Int J Androl. 1993;16:311–4.
22. Mohan H, Yadav S, Singh U, Kadian A, Mohan P. Circulating iso- and autoantibodies to human spermatozoa in infertility. Indian J Pathol Microbiol. 1990;33:161–5.
23. Heidenreich A, Bonfig R, Wilbert DM, Strohmajer WL, Engelmann UH. Risk factors for antisperm antibodies in infertile men. Am J Reprod Immunol. 1994;31:69–76.
24. Marconi M, Nowotny A, Pantke P, Diemer T, Weidner W. Antisperm antibodies detected by MAR and immunobead test are not associated with inflammation and infection of the seminal tract. Andrologia. 2008;40:227–34.

25. Fisch H, Laor E, BarChama N, Witkin SS, Tolia BM, Reid RE. Detection of testicular endocrine abnormalities and their correlation with serum antisperm antibodies in men following vasectomy. J Urol. 1989;141:1129–32.
26. Broderick GA, Tom R, McClure RD. Immunological status of patients before and after vasovasostomy as determined by the immunobead antisperm antibody test. J Urol. 1989;142:752–5.
27. Jarow JP, Goluboff ET, Chang TS. Relationship between antisperm antibodies and testicular histologic changes in humans after vasectomy. Urology. 1994;43:521–4.
28. Newton RA. IgG antisperm antibodies attached to sperm do not correlate with infertility following vasovasostomy. Microsurgery. 1988;9:278–80.
29. Matson PL, Junk SM, Masters JR, Pryor JP, Yovich JL. The incidence and influence upon fertility of antisperm antibodies in seminal fluid following vasectomy reversal. Int J Androl. 1989;12:98–103.
30. Gubin DA, Dmochowski R, Kutteh WH. Multivariant analysis of men from infertile couples with and without antisperm antibodies. Am J Reprod Immunol. 1998;39:157–60.
31. Meinertz H, Linnet L, Fogh-Anderson P, Hjort T. Antisperm antibodies and fertility after vasovasostomy: a follow-up study of 216 men. Fertil Steril. 1990;54:315–21.
32. Ohl DA, Naz RK. Infertility due to antisperm antibodies. Urology. 1995;46:591–602.
33. Clarke GN. Sperm antibodies in normal men: association with a history of nongonococcal urethritis (NGU). Am J Reprod Immunol Microbiol. 1986;12:31–2.
34. Shahmanesh M, Stedronska J, Hendry WF. Antispermatozoal antibodies in men with urethritis. Fertil Steril. 1986;46:308–11.
35. Close CE, Wang SP, Roberts PL, Berger RE. The relationship of infection with Chlamydia trachomatis to the parameters of male fertility and sperm autoimmunity. Fertil Steril. 1987;48:880–3.
36. Soffer Y, Ron-El R, Golan A, Herman A, Caspi E, Samra Z. Male genital mycoplasmas and Chlamydia trachomatis culture: its relationship with accessory gland function, sperm quality, and autoimmunity. Fertil Steril. 1990;53:331–6.
37. Micic S, Petrovic S, Dotlic R. Seminal antisperm antibodies and genitourinary infection. Urology. 1990;35:54–6.
38. Witkin SS, Jeremias J, Grifo JA, Ledger WJ. Detection of Chlamydia trachomatis in semen by the polymerase chain reaction in male members of infertile couples. Am J Obstet Gynecol. 1993;168:1457–62.
39. Ingerslev HJ, Walter S, Andersen JT, Brandenhoff P, Eldrup J, Geerdsen JP, Scheibel J, Tromholt N, Jensen HM, Hjort T. A prospective study of antisperm antibody development in acute epididymitis. J Urol. 1986;136:162–4.
40. Lekili M, Tekgul S, Ergen A, Tasar C, Hascelik G. Acute experimental unilateral orchitis in the rabbit and its effect on fertility. Int Urol Nephrol. 1992;24:291–7.
41. Witkin SS, Zelikovsky G. Immunosuppression and sperm antibody formation in men with prostatitis. J Clin Lab Immunol. 1986;21:7–10.
42. Jarow JP, Kirkland JA Jr, Assimos DG. Association of antisperm antibodies with chronic nonbacterial prostatitis. Urology. 1990;36:154–6.
43. Marconi M, Pilatz A, Wagenlehner F, Diemer T, Weidner W. Are really antisperm antibodies associated with inflammatory/infectious diseases of the male reproductive tract. Eur Urol. 2009;56:708–15.
44. Witkin SS, Toth A. Relationship between genital tract infections, sperm antibodies in seminal fluid and infertility. Fertil Steril. 1983;40:805–8.
45. Bozhedomov VA, Lipatova NA, Rokhlikov IM, Alexeev RA, Ushakova IV, Sukhikh GT. Male fertility and varicocoele: role of immune factors. Andrology. 2014;2:51–8.
46. Ozen H, Asar G, Gungor S, et al. Varicocele and antisperm antibodies. Int Urol Nephrol. 1985;17:97–101.

47. Knudson G, Ross L, Stuhldreher D, Houlihan D, Bruns E, Prins G. Prevalence of sperm bound antibodies in infertile men with varicocele: the effect of varicocele ligation on antibody levels and semen response. J Urol. 1994;151:1260–2.
48. Gilbert BR, Witkin SS, Goldstein M. Correlation of sperm-bound immunoglobulins with impaired semen analysis in infertile men with varicoceles. Fertil Steril. 1989;52:469–73.
49. Marmar JL. The pathophysiology of varicoceles in the light of current molecular and genetic information. Hum Reprod Update. 2001;7:461–72.
50. Will MA, Swain J, Fode M, Sonksen J, Christman GM, Ohl D. The great debate: varicocele treatment and impact on fertility. Fertil Steril. 2011;95:841–52.
51. Shook TE, Nyberg LM, Collins BS, Mathur S. Pathological and immunological effects of surgically induced varicocele in juvenile and adult rats. Am J Reprod Immunol Microbiol. 1988;17:141–4.
52. Oshinsky GS, Rodriguez MV, Mellinger BC. Varicocele-related infertility is not associated with increased sperm-bound antibody. J Urol. 1993;150:871–3.
53. Cetinkaya M, Memis A, Adsan O, Beyribey S, Ozturk B. Antispermatozoal antibody values after varicocelectomy. Int Urol Nephrol. 1994;26:89–92.
54. Verajankorva E, Laato M, Pollanen P. Analysis of 508 infertile male patients in south-western Finland in 1980–2000: hormonal status and factors predisposing to immunological infertility. Eur J Obstet Gynecol Reprod Biol. 2003;111:173–8.
55. Turner IT, Jones CE, Roddy MS. Experimental varicocele does not affect the blood-testis barrier, epididymal electrolyte concentrations, or testicular blood gas concentrations. Biol Reprod. 1987;36:926–31.
56. Isitmangil G, Yildirim S, Orhan I, Kadioglu A, Akinci M. A comparison of the sperm mixed-agglutination reaction test with the peroxidase-labelled protein A test for detecting antisperm antibodies in infertile men with varicocele. BJU Int. 1999;84:835–8.
57. Trussell JC, Lee PA. The relationship of cryptorchidism to fertility. Curr Urol Rep. 2004;5: 142–8.
58. Kurpisz M, Nakonechnyy A, Niepieklo-Miniewska W, Havrylyuk A, Kamieniczna M, Nowakowska B, Chopyak V, Kusnierczyk P. Weak association of anti-sperm antibodies and strong association of familial cryptorchidism/infertility with HLA-DRB1 polymorphisms in prepubertal Ukrainian boys. Reprod Biol Endocrinol. 2011;9:129.
59. Urry RL, Carrell DT, Starr NT, Snow BW, Middleton RG. The incidence of antisperm antibodies in infertility patients with a history of cryptorchidism. J Urol. 1994;151:381–3.
60. Sinisi AA, Pasquali D, Papparella A, Valente A, Orio F, Esposito D, Cobellis G, Cuomo A, Angelone G, Martone A, Fioretti GP, Bellastella A. Antisperm antibodies in cryptorchidism before and after surgery. J Urol. 1998;160:1834–7.
61. Mirilas P, De Almeida M. Absence of antisperm surface antibodies in prepubertal boys with cryptorchidism and other anomalies of the inguinoscrotal region before and after surgery. J Urol. 1999;162:177–81.
62. Mirilas P, Mamoulakis C, De Almeida M. Puberty does not induce serum antisperm surface antibodies in patients with previously operated cryptorchidism. J Urol. 2003;170:2432–5.
63. Sinisi AA, D'Apuzzo A, Pasquali D, Venditto T, Esposito D, Pisano G, De Bellis A, Ventre I, Papparella A, Perrone L, Bellastella A. Antisperm antibodies in prepubertal boys treated with chemotherapy for malignant or non-malignant diseases and in boys with genital tract abnormalities. Int J Androl. 1997;20:23–8.
64. Hagenas L, Ploen L, Ritzen EM, Ekwall H. Blood-testis barrier: maintained function of inter-sertoli cell junctions in experimental cryptorchidism in the rat, as judged by a simple lanthanum-immersion technique. Andrologia. 1977;9:250–4.
65. Stewart RJ, Boyd S, Brown S, Toner PG. The blood-testis barrier in experimental unilateral cryptorchidism. J Pathol. 1990;160:51–5.
66. Kukadia AN, Ercole CJ, Gleich P, Hensleigh H, Pryor JL. Testicular trauma: potential impact on reproductive function. J Urol. 1996;156:1643–6.

67. Pogorelic Z, Juric I, Biocic M, Furlan D, Budimir D, Todoric J, Milunovic KP. Management of testicular rupture after blunt trauma in children. Pediatr Surg Int. 2011;27:885–9.
68. Harrington TG, Schauer D, Gilbert BR. Percutaneous testis biopsy: an alternative to open testicular biopsy in the evaluation of the subfertile man. J Urol. 1996;156:1647–51.
69. Komori K, Tsujimura A, Miura H, Shin M, Takada T, Honda M, Matsumiya K, Fujioka H. Serial follow-up study of serum testosterone and antisperm antibodies in patients with non-obstructive azoospermia after conventional or microdissection testicular sperm extraction. Int J Androl. 2004;27:32–7.
70. Ozturk U, Ozdemir E, Dede O, Sagnak L, Goktug HN. Assessment of anti-sperm antibodies in couples after testicular sperm extraction. Clin Invest Med. 2011;34:E179–83.
71. Leonhartsberger N, Gozzi C, Akkad T, Springer-Stoehr B, Bartsch G, Steiner H. Organ-sparing surgery does not lead to greater antisperm antibody levels than orchidectomy. BJU Int. 2007;100:371–4.
72. Stula I, Druzijanic N, Sapunar A, Perko Z, Bosnjak N, Kraljevic D. Antisperm antibodies and testicular blood flow after inguinal hernia mesh repair. Surg Endosc. 2014;28:3413–20.
73. Becker EJ, Prillaman HM, Turner TT. Microvascular blood flow is altered after repair of testicular torsion in the rat. J Urol. 1997;157:1493–8.
74. Lievano G, Nguyen L, Radhakrishnan J, Fornell L, John E. New animal model to evaluate testicular blood flow during testicular torsion. J Pediatr Surg. 1999;34:1004–6.
75. Harrison RG, Lewis-Jones DI, Moreno de Marval MJ, Connolly RC. Mechanism of damage to the contralateral testis in rats with an ischaemic testis. Lancet. 1981;2:723–5.
76. Cosentino MJ, Nishida M, Rabinowitz R, Cockett AT. Histological changes occurring in the contralateral testes of prepubertal rats subjected to various durations of unilateral spermatic cord torsion. J Urol. 1985;133:906–11.
77. Henderson JA 4th, Smey P, Cohen MS, Davis CP, Payer AF, Parkening TA, Warren MM. The effect of unilateral testicular torsion on the contralateral testicle in prepubertal Chinese hamsters. J Pediatr Surg. 1985;20:592–7.
78. Kurpisz M, Mazurkiewicz I, Fernandez N. Morphological and immunological observations in experimentally induced torsion of testis in rats. Am J Reprod Immunol Microbiol. 1985;9:129–35.
79. Anderson JB, Williamson RC. Fertility after torsion of the spermatic cord. Br J Urol. 1990;65:225–30.
80. Arap MA, Vicentini FC, Cocuzza M, Hallak J, Athayde K, Lucon AM, Arap S, Srougi M. Late hormonal levels, semen parameters, and presence of antisperm antibodies in patients treated for testicular torsion. J Androl. 2007;28:528–32.
81. Puri P, Barton D, O'Donnell B. Prepubertal testicular torsion: subsequent fertility. J Pediatr Surg. 1985;20:598–601.
82. Anderson MJ, Dunn JK, Lipshultz LI, Coburn M. Semen quality and endocrine parameters after acute testicular torsion. J Urol. 1992;147:1545–50.
83. Patel RP, Kolon TF, Huff DS, Carr MC, Zderic SA, Canning DA, Snyder HM 3rd. Testicular microlithiasis and antisperm antibodies following testicular biopsy in boys with cryptorchidism. J Urol. 2005;174:2008–10.
84. Jiang H, Zhu W-J. Testicular microlithiasis is not a risk factor for the production of antisperm antibody in infertile males. Andrologia. 2013;45:305–9.
85. Paoli D, Gilio B, Piroli E, Gallo M, Lombardo F, Dondero F, Lenzi A, Gandini L. Testicular tumors as a possible cause of antisperm autoimmune response. Fertil Steril. 2009;91:414–9.
86. Guazzieri S, Lembo A, Ferro G, Artibani W, Merlo F, Zanchetta R, Pagano F. Sperm antibodies and infertility in patients with testicular cancer. Urology. 1985;26:139–42.
87. Foster RS, Rubin LR, McNulty A, Bihrle R, Dfonohue JP. Detection of antisperm-antibodies in patients with primary testicular cancer. Int J Androl. 1991;14:179–85.
88. Hobarth K, Klingler HC, Maier U, Kollaritsch H. Incidence of antisperm antibodies in patients with carcinoma of the testis and in subfertile men with normogonadotropic oligoasthenoteratozoospermia. Urol Int. 1994;52:162–5.

89. Witkin SS, Sonnabend J, Richards J, et al. Induction of antibody to asialo-GMI by spermatozoa and its occurrence in the sera of homosexual men with the Acquired Immune-deficiency Syndrome (AIDS). Clin Exp Immunol. 1983;4:346–50.
90. Richards J, Bedford T, Witkin S. Rectal insemination modifies immune responses in rabbits. Science. 1984;224:390–2.
91. Witkin S, Sonnabend J. Immune responses to spermatozoa in homosexual men. Fertil Steril. 1983;39:337–42.
92. Wolff H, Schill WB. Antisperm antibodies in infertile and homosexual men: relationship to serological and clinical findings. Fertil Steril. 1985;44:673–7.
93. Adams LE, Donovan-Brand R, Fieldman-Kein A, el Ramahi K, Hess EV. Sperm and seminal plasma antibodies in Acquired Immune Deficiency Syndrome (AIDS) and other associated syndromes. Clin Immunol Immunopathol. 1988;46:442–9.
94. Mulhall BP, Fieldhouse S, Clark S, Carter L, Harrison L, Donovan B, Short RV. Anti-sperm antibodies in homosexual men: prevalence and correlation with sexual behaviour. Genitourin Med. 1990;66:5–7.
95. Naz RK, Ellaurie M, Phillips TM, Hall J. Antisperm antibodies in human immunodeficiency virus infection: effects on fertilization and embryonic development. Biol Reprod. 1990;42:859–68.
96. Stern JE, Nelson TS, Gibson SH, Colby E. Antisperm antibodies in female mice: responses following intrauterine immunization. Am J Reprod Immunol. 1994;31:211–8.
97. Sands M, Phair JP, Hyprikar J, Hanses C, Brown RB. A study on antisperm antibody in homosexual men. J Med. 1985;16:483–91.
98. Beretta G, Zanollo A, Chelo E, Livi C, Scarselli G. Seminal parameters and auto-immunity in paraplegic/quadraplegic men. Acta Eur Fertil. 1987;18:203–5.
99. Hirsch IH, Sedor J, Callahan HJ, Staas WE. Antisperm antibodies in seminal plasma of spinal cord-injured men. Urology. 1992;39:243–7.
100. Dahlberg A, Hovatta O. Anejaculation following spinal cord injury does not induce sperm-agglutinating antibodies. Int J Androl. 1989;12:17–21.
101. Menge AC, Ohl DA, Denil J, Korte MK, Keller L, McCabe M. Absence of antisperm antibodies in anejaculatory men. J Androl. 1990;11:396–8.
102. Siosteen A, Steen Y, Forssman L, Sullivan L. Auto-immunity to spermatozoa and quality of semen in men with spinal cord injury. Int J Fertil. 1993;38:117–22.
103. Shiraishi Y, Shibahara H, Koriyama J, Hirano Y, Okazaki H, Minota S, Suzuki M. Incidence of antisperm antibodies in males with systemic autoimmune diseases. Am J Reprod Immunol. 2009;61:183–9.

Chapter 11
Classical Treatments for Infertile Men with Anti-sperm Antibody (ASA)

Hiroaki Shibahara

Abstract In the era before the introduction of assisted reproductive technology (ART) for clinical use in infertile men with ASA, several treatments have been used to overcome the potentially deleterious effects of ASA-mediated infertility, including methods to decrease ASA production by immunosuppressive therapies using corticosteroids or cyclosporine A, or to remove ASA bound to sperm.

From the results of the non-randomized studies, the various effects of corticosteroid treatment on immunosuppression of ASA titers and pregnancies were observed. To establish the validity of this therapy, appropriate placebo control studies were required. However, the RCTs have not found a therapeutic significance for the corticosteroid therapy for infertile men with ASA. Moreover, as corticosteroids are well known to have potential side effects, important clinical considerations in the use of corticosteroids to treat infertile patients with ASA are mandatory. As for the effect of cyclosporine A treatment, no definite conclusions can be drawn so far.

The purpose of the technique for removing ASA bound to sperm for infertile men with ASA is the effort to remove sperm-bound antibody from sperm. However, the dissociation of tightly bound antibodies has been found to be difficult. There have been several reports, including simple sperm washing, split ejaculation into medium containing serum, immune absorption of ASA, and enzymatic treatment. The simple sperm washing did not remove ASA from the sperm surface. The addition of serum to sample collection pots may improve the oocyte fertilization rate. The latter two methods have not been clinically applied yet.

After the introduction of intracytoplasmic sperm injection (ICSI) for the treatment of infertility, the clinical importance of these classical treatments for infertile men with ASA has been positioned more less than before.

H. Shibahara (✉)
Department of Obstetrics and Gynecology, School of Medicine, Hyogo Medical University, Nishinomiya, Hyogo, Japan
e-mail: sibahara@hyo-med.ac.jp

© The Editor(s) (if applicable) and The Author(s), under exclusive license to
Springer Nature Singapore Pte Ltd. 2022
H. Shibahara, A. Hasegawa (eds.), *Gamete Immunology*,
https://doi.org/10.1007/978-981-16-9625-1_11

143

11.1 Introduction

Anti-sperm antibody (ASA) is associated with decreased fertility in both men and women. Treatment for immunological infertility has always caused concern, especially in the era before the introduction of assisted reproductive technology (ART) for clinical use.

In infertile men with ASA, several treatments have been used to overcome the potentially deleterious effects of ASA-mediated infertility, including methods to decrease ASA production by immunosuppressive therapies using corticosteroids or cyclosporine A, or to remove ASA bound to sperm, as well as ART. Various treatment options for infertile men with ASA are summarized in Table 11.1 [1].

Similar to other autoimmune diseases, suppression of the immune system could be a key for the treatment of autoimmune infertility. The most common medicine for immune suppression is corticosteroids, which prevent the chemotaxis of inflammatory cells, impede cytokine release, decreased antibody production, and even weaken antigen-antibody association [2]. However, there is evidence suggesting that the use of systemic corticosteroid treatment for immunosuppression to decrease ASA production is relatively ineffectual, benefiting only approximately 20% of treated men [3].

Methods that attempt to remove ASA bound to sperm have been investigated. Simple sperm washing does not remove ASA from the sperm surface [4]. However, ejaculation directly into a washing buffer appears to be beneficial for maximizing sperm recovery as well as for minimizing the amount of ASA coating sperm [5, 6]. Techniques that lead to dissociation of ASA-Ig complexes are associated with irreversible loss of sperm motility. In this chapter, these classical treatments for infertile men with ASA are introduced.

Table 11.1 Treatment options reported for anti-sperm antibodies in infertile men

Treatment options	Effectiveness
1. Decrease ASA production	
Systemic corticosteroid	Relatively insufficient
Cyclosporine A	No definite conclusion
2. Remove ASA bound to sperm	
Simple sperm washing	Relatively insufficient
Split ejaculation into medium containing serum	Beneficial in some cases
Immune absorption	No clinical application
Enzymatic treatment	No clinical application
3 ART	
IUI	Beneficial in some cases
IVF	Beneficial in some cases
ICSI	Excellent chances

ASA anti-sperm antibodies, *ART* assisted reproductive technology, *ICSI* intracytoplasmic sperm injection, *IUI* intrauterine insemination, *IVF* in vitro fertilization

11.2 Treatment for Immunosuppression to Decrease ASA Production

In 1978, Shulman et al. first introduced a new immunosuppressive method for the treatment of men with ASA [7, 8], followed by several reports suggesting that corticosteroids suppress antibody levels and thus enhance fertility in some patients [9, 10]. Later, Bouloux et al. [11] reported the effect of cyclosporine A in male autoimmune infertility. However, well-designed studies are limited or lacking for the effects of these treatments.

11.2.1 Corticosteroid Therapy

11.2.1.1 Non-randomized Study

Shulman et al. [7, 8] first reported the usefulness of immunosuppressive therapy for infertile men with ASA. In brief, they administered 96 mg of methylprednisolone per day for 7–14 days to 15 men, which resulted in five men causing pregnancies. The pregnancy rate was 33%.

Hendry et al. [9] found 15 mg/day of prednisone administered for 3–12 months improved sperm counts in some infertile men with ASA. However, they found that the treatment only slightly affected ASA levels. For reduction of circulating ASA titers, they suggested that intermittent high-dose therapy might be necessary.

Alexander et al. [10] carried out a cohort study comparing 24 patients with ASA treated by prednisone and 26 patients with similar level of ASA without any treatment served as controls. The treated group comprised 19 men and 5 women, while the control group comprised 25 men and 1 woman. ASAs were evaluated with the sperm-agglutinating assay by Kibrick as described by Rumke [12] and the sperm-immobilization test by Isojima [13]. Some were given 60 mg of prednisone daily for 7 days. Others were given two 7-day treatments. All treatment and nontreatment couples were actively attempting to achieve a pregnancy. As a result, prednisone did not cause change in sperm count, motility, or normal forms in men. However, it did cause significant reductions in levels of circulating sperm-immobilizing antibody ($P < 0.05$) and sperm-agglutinating antibody ($P < 0.001$) in both men and women. As for the pregnancy rates, 14 pregnancies occurred during the 4-month posttreatment study period. Significantly more pregnancies occurred in the treated group (45%) than in the control group (12%) ($P < 0.05$).

Our group has also investigated the efficacy of corticosteroid treatments for infertile men with sperm-immobilizing antibodies in their sera [14]. Sperm-immobilization test (SIT) was carried out for 1126 infertile men who visited our hospital between January 1985 and December 1989. Twenty-six (2.3%) of them were tested positive in SIT. In them, five infertile men were treated using cortico-steroids with the informed consent. They were given 40–60 mg prednisolone for

Fig. 11.1 Effects of corticosteroid therapy on quantitative sperm-immobilizing antibody (SI_{50}) titers in the sera of infertile men. Five infertile men with sperm-immobilizing antibodies in their sera were treated using corticosteroids with the informed consent. They were given 40–60 mg prednisolone for 3 days followed by a 3-day tapering of the medicine on the basis of their wives' menstrual cycle. After the corticosteroid therapy, SI_{50} titers decreased in all cases

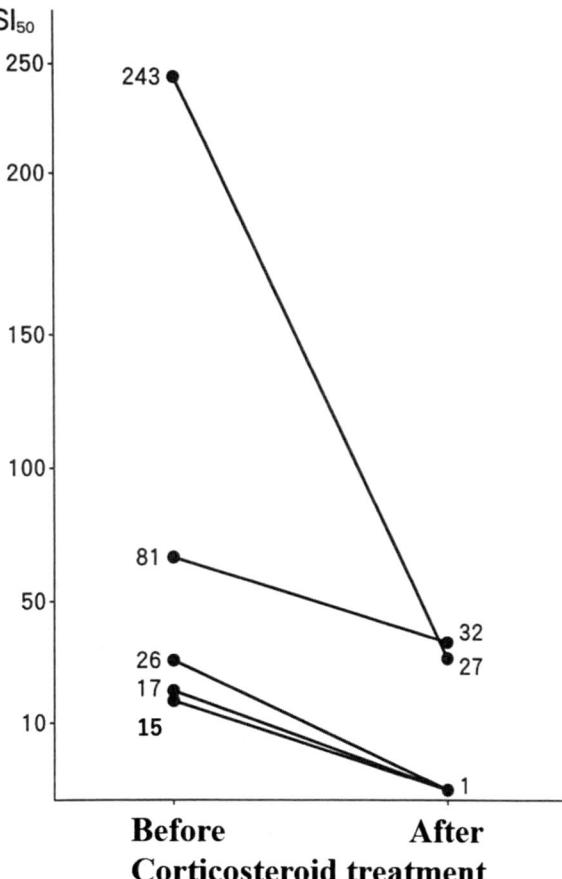

3 days followed by a 3-day tapering of the medicine on the basis of their wives' menstrual cycle. As shown in Fig. 11.1, the quantitative sperm-immobilizing antibody (SI_{50}) titers decreased in all cases. An infertile couple successfully conceived following the corticosteroid therapy combined with intrauterine insemination (IUI) at the time when SI_{50} titer decreased (Fig. 11.2).

The suppressive effect for the production of sperm-immobilizing antibodies by corticosteroid in them lasted as long as 3 months (Fig. 11.3). By evaluating sperm-immobilizing antibody titers, it should be considered for the repeated prescription of the corticosteroid definitely with careful consideration of the side effects.

From the results of these studies, the various effects of corticosteroid treatment on immunosuppression of ASA titers and pregnancies were observed. To establish the validity of this therapy, appropriate placebo control studies were required.

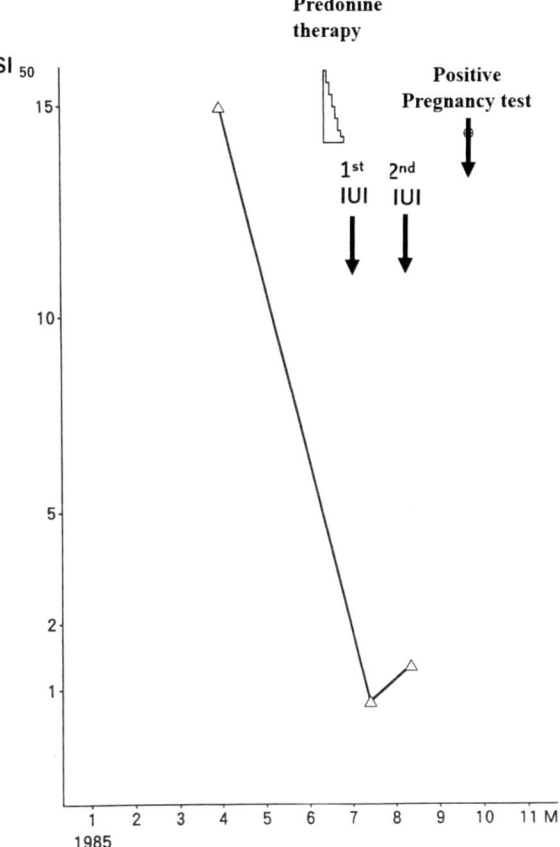

Fig. 11.2 Successful pregnancy in an infertile couple with sperm-immobilizing antibody in the husband's serum. An infertile couple successfully conceived following the corticosteroid therapy combined with intrauterine insemination (IUI) at the time when SI_{50} titer decreased. Before the treatment, the husband's SI_{50} titer was 15.0. After the treatment, SI_{50} titer decreased to only 1.0; however, the first attempt of IUI failed. One month later, they conceived in the second attempt of IUI when SI_{50} titer was 1.3. Both the sperm concentration and sperm motility of the patient were within normal limit

11.2.1.2   Randomized-Controlled Trial (RCT)

The first randomized-controlled trial (RCT) of the corticosteroid therapy for male infertility associated with ASA was reported by De Almeida et al. [15]. In brief, ten infertile men with significant sperm-agglutinating titers were entered into a small RCT to test the efficacy of corticosteroid therapy. Prednisolone was given orally at a dose of 1 mg/kg/day for 9 days repeated over three cycles (their wives' menstrual cycles). They concluded that this treatment had no significant effect, when compared to placebo, on fertility, serum antibody levels, or semen characteristics.

Haas GG Jr. et al. [16] carried out a RCT of the use of methylprednisolone in infertile men with sperm-associated immunoglobulins. Forty-three men with antibody-mediated infertility identified by measuring sperm-associated immuno-globulin with a direct radiolabeled antiglobulin assay were enrolled to judge the efficacy of corticosteroids. In the active drug group, men were given 96 mg meth-ylprednisolone in three divided doses for 7 days followed by a 2-day tapering of the drug. The control group was given placebo in the same manner. Both regimens were

Fig. 11.3 Undulation of SI_{50} titers in the patient's serum during the corticosteroid therapy. In an infertile man with the highest SI_{50} titer in his serum, evaluation of SI_{50} titer was performed through the corticosteroid therapy. After the first treatment, the SI_{50} titers decreased to 100.0 from 243.0. The suppressive effect for the production of sperm-immobilizing antibodies by corticosteroid lasted as long as 3 months. Then the second treatment was carried out by a higher dose of corticosteroid, and the SI_{50} titer decreased to 24.0. In this case, no pregnancy was established by timed intercourse. By evaluating the SI_{50} titers, it should be considered for the repeated prescription of the corticosteroid definitely with careful consideration of the side effects

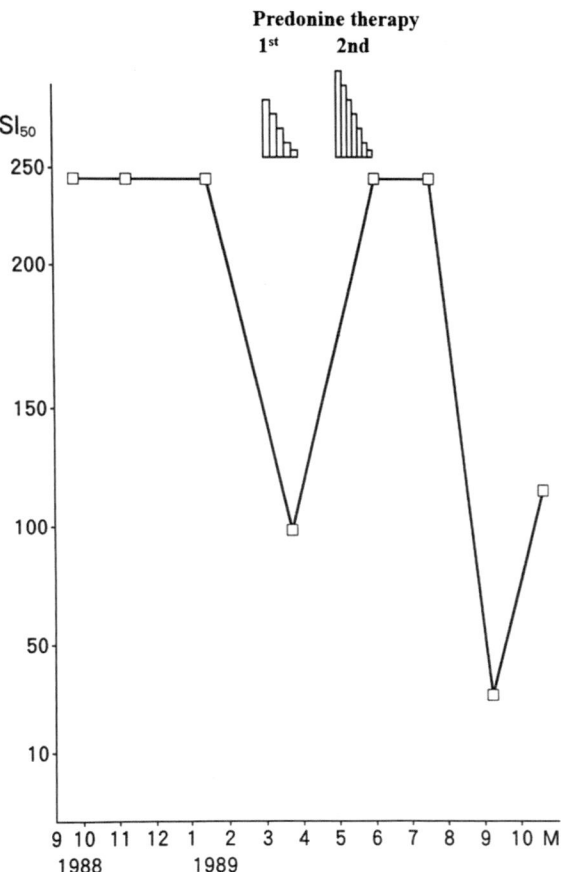

repeated three times approximately 1 month apart on the basis of their female partner's menstrual cycle. There was not a favorable effect of corticosteroids on the semen parameters, the results of assays for sperm-associated IgA and plasma IgG ASA, or the men's subsequent fertility.

Bals-Pratsch et al. [17] performed a RCT for 20 infertile men with ASA detected by the mixed antiglobulin reaction (MAR) test in the ejaculate and by the tray agglutination test (TAT) in serum. No pregnancy occurred during prednisolone or placebo treatment. In eight of 12 patients, postcoital testing showed little improvement and antibody titers decreased in seven of 16 patients. Five patients' partners conceived at a later stage by IUI ($n = 3$) or by in vitro fertilization (IVF; $n = 2$). They concluded that high-dose corticosteroid therapy was ineffective in achieving pregnancies induced by infertile men positive for ASA.

Rasanen et al. [18] investigated the effect of short-term, low-dose prednisolone treatment on sperm-bound antibody levels using an objective flow cytometric (FCM) measurement. A RCT was performed for 20 infertile men with ASA examined by the mixed antiglobulin reaction (MAR) test. They were treated with either 20 mg/day prednisolone ($n = 10$) or placebo ($n = 10$) for 3 weeks. The study showed that, overall, the antibody levels measured before and after treatment with low-dose prednisolone or with placebo were not statistically different. They suggested the prednisolone treatment might be effective for infertile men with low to moderate amounts of sperm-bound antibodies, while sperm with high levels of ASA had remained resistant to this treatment.

Taken together, all these RCTs have not found a therapeutic significance for the corticosteroid therapy for infertile men with ASA.

11.2.1.3 The Potential Side Effects of Using Corticosteroids

Corticosteroids are well known to have potential side effects. The main short-term side effects are acne, stomach irritation, and an increased risk of infection. Side effects can be more severe with long-term administration. The most clinical consideration in the use of corticosteroids is increased risk of infection. Symptoms of an infection may not be as obvious or typical while taking corticosteroids.

As for long-term side effects of corticosteroids such as taken daily for long periods of time, they can cause adrenal gland suppression. Prolonged use has also been associated with cataracts and glaucoma, immunosuppression, muscle wasting, bone changes, fluid shifts, and personality changes. High blood pressure, an increased appetite leading to weight gain, menstrual irregularities, and an increased growth of body hair have also been reported. Skin discoloration, thinning, and easy bruising can occur after topical corticosteroids are applied repeatedly to the skin. They may also precipitate sudden mood swings, cause fluid retention, worsen diabetes, and lead to a condition known as Cushing syndrome with a moon face and a buffalo hump. Corticosteroids are medicines that reduce inflammation by mimicking the hormone cortisol that is produced by the adrenal gland.

Therefore, important clinical considerations in the use of corticosteroids to treat infertile patients with ASA are mandatory. Indeed, Hendry et al. (1990) reported that there occurred mild side effects including acne, dyspepsia, skin rashes, fluid retention, and mood changes in approximately 60% of patients [19]. Unfortunately, there have been at least two reported cases of aseptic necrosis of the hip [20, 21]. Both occurred in patients on high-dose methylprednisolone. However, aseptic necrosis is generally considered as an idiosyncratic reaction that is not dose-related.

11.2.2 Cyclosporine A Therapy

Cyclosporine A is a cyclic undecapeptide with strong immunosuppressive potency. Firstly marketed in the mid-1980s, cyclosporine A was widely used in transplantation and greatly improved the survival rates of patients and grafts after solid organ transplantation. Unfortunately, cyclosporine A administration can be associated with a number of side effects due to its high toxicity. Its common side effects include tremor, kidney damage, hypertension, infection, headache, nausea, male pattern hair growth in women, excessive hair growth, female reproductive disorder, swollen/red/painful gums, increased triglycerides, abdominal discomfort, upper respiratory infection, diarrhea, indigestion, leg cramps, numbness and tingling, vomiting, stomach upset, dizziness, and flushing. Less common side effects of cyclosporine include acne, convulsions, itching, hyperkalemia, hypomagnesemia, pancreatitis, liver damage, and flu-like syndrome. Other side effects include leukopenia, thrombocytopenia, anaphylaxis, glomerular capillary thrombosis, migraine, and hyponatremia. Serious side effects of cyclosporine include fever, sweating, chills, sores in the mouth, weight loss, problems speaking or walking, and decreased vision. Therefore, important clinical considerations in the use of cyclosporine A to treat infertile patients with ASA are also essential.

Bouloux et al. [11] reported the case series investigating the effect of cyclosporine A in male autoimmune infertility. In brief, they treated nine infertile men with ASA using cyclosporine A for 6 months with a dose of 5–10 mg/kg/day. Seminal plasma and serum ASA fell in three (33%) subjects on treatment, and sperm count and motility increased substantially in one (11%). Three (33%) successful pregnancies occurred in the study group: one on treatment, one in the first cycle of IUI after treatment (twin infants), and one 3 months after cessation of treatment (twin infants). Successful conceptions with cyclosporine A were unrelated to falls in ASA titer.

As this study did not have placebo controls, no definite conclusions can be drawn.

11.3   Technique for Removing Anti-sperm Antibody (ASA) Bound to Sperm

The purpose of the technique for removing ASA bound to sperm for infertile men with ASA is the effort to remove sperm-bound antibody from sperm. However, the dissociation of tightly bound antibodies has been found to be difficult. There have been several reports, including simple sperm washing, split ejaculation into medium containing serum, immune absorption of ASA, and enzymatic treatment.

11.3.1   Simple Sperm Washing

Methods that attempt to remove ASA bound to sperm have been investigated. Haas GG [4] reported the effect of sperm washing on the stability of sperm-bound immunoglobulin G (IgG) antibody derived from plasma from four patients, and also IgG bound in vivo on the sperm of four other men was quantitatively evaluated. In the first series of experiments, human sperm were incubated with an IgG antibody-containing plasma and subjected to 18 cycles of sperm washing. In the second set of experiments, sperm from men positive for sperm-bound IgG were subjected to four cycles of sperm washing. The amount of residual antibody bound to a constant number of sperm was quantitated by a radiolabeled antiglobulin assay during and following the washing procedures. There was no significant loss of sperm-bound antibody due to the washing procedures. They concluded that the simple sperm washing does not remove ASA from the sperm surface [4].

11.3.2   Split Ejaculation into Medium Containing Serum

Immunoglobulins may bind to sperm after ejaculation, with a time-dependent passive transfer of antibodies to the sperm surface during liquefaction of semen. A longer time interval between ejaculates can be associated with higher levels of antibody binding [5].

Elder et al. [6] carried out a retrospective analysis which studied IVF results for couples in whom the male partner had ASA judged by a positive MAR test in seminal plasma. Of the 59 couples studied over a total of 113 cycles, 30 had IVF semen samples collected into sterile, dry pots (group A). Thirty-eight couples had samples collected into Earle's medium containing 50% maternal serum (group B).

Whereas 21.6% of oocytes fertilized in patients whose samples were collected into dry pots (group A), the addition of serum doubled this rate to 43.7% (group B). Only 5% of group B cycles (samples collected in serum) had complete failure of fertilization, compared with 22% in the absence of serum (group A, $P < 0.01$). In group A, there were only three pregnancies (5.1% per cycle), of which one aborted before the sixth week of gestation. Group B achieved a pregnancy rate of 24.1% per cycle ($P < 0.05$).

They concluded that the results demonstrate that the addition of serum to sample collection pots significantly improves the oocyte fertilization rate and is followed by a greater chance of conception for these couples.

11.3.3   Immune Absorption of ASA

Kiser GC, et al. [22] reported in vitro immune absorption of ASA with immunobead-rise, immunomagnetic, and immunocolumn separation techniques. Sperm from 14 men with ASA were less able to penetrate an overlaying buffer layer than sperm from a fertile control. Addition of immunobeads to the specimen was of little use, because few motile sperm could swim into the overlaying buffer; retained immunobeads were noted in the buffer layer of 18-h capacitated specimens. Magnetic isolation of antibody-coated sperm from antibody-free sperm avoids potential damage to fragile sperm through centrifugation. Viable sperm were isolated from magnetite-complexed sperm, but the motility of the isolated sperm deteriorated rapidly during the subsequent capacitation period. Passage of diluted ejaculate through a column of dextran beads for ASA processing was associated with superior sperm quality and fertilizing potential. The use of ASA processing resulted in good sperm velocity and linearity and improved sperm function, as measured with the hamster egg penetration test. They concluded that sperm from men with immunologically mediated infertility can be processed through the ASA processing and used for IUI of their partners or in an IVF program. Subsequently, Foresta [23] and Viganò [24] agreed the usefulness for insemination in infertile men with ASA. So far, however, there have not been shown clinical application of this procedure.

11.3.4 Enzymatic Treatment

Pattinson et al. [25] investigated the possibility of enzymatic disagglutination by using four proteases. Subtilisin had prohibitively detrimental effects even at 10 U/ mL. However, chymotrypsin (less than or equal to 500 U/mL), trypsin (less than or equal to 500 U/mL), and papain (less than or equal to 50 U/mL) had no adverse effects. One or more of these latter three enzymes was found to disagglutinate sperm which had previously been incubated with sperm-agglutinating antibody-positive sera in 87% of cases. They concluded that the enzymatic disagglutination may be beneficial for the treatment of immunologically mediated spermagglutination.

Then they assessed the functional potential of disagglutinated sperm [26]. Chymotrypsin (500 U/mL) and papain (50 U/mL) resulted in impairment of oocyte penetration in the zona-free hamster egg penetration test. Trypsin (500 U/mL), while having no effect on egg penetration of normal sperm, significantly improved oocyte penetration of sperm which had been previously incubated with spermagglutinating antibody-positive sera. Sperm-mucus interaction was not improved, however, by trypsin treatment of agglutinated sperm. This technique may be of value in conjunction with IVF in situations where spermagglutination exists, and also possibly with IUI if improved fertilizing ability can be confirmed in vitro. So far, trypsin treatment has not been applied for clinical use.

Bronson et al. [27] documented IgA1 and IgA2 ASA within seminal fluid. ASA of the IgA class, when bound to sperm, impair the sperm's ability to penetrate cervical mucus. IgA proteases offer a potential treatment of autoimmunity to sperm in infertile men by enzymatically degrading immunoglobulins on the sperm surface. As IgA1 but not IgA2 is cleaved by IgA proteases, they determined through use of IgA subclass-specific monoclonal antibodies the presence and relative proportions of anti-sperm IgA1 and IgA2 in seminal fluid. This ratio varies substantially between men, perhaps reflecting differences in the etiology of autoimmunity to sperm.

Later, Kutteh et al. [28] investigated to determine the effect of an IgA I protease on the binding of IgA ASA. In serum, the IgA1 subclass ASA was predominate (91%) when compared with IgA2 (9%) subclass. Direct sperm-bound antibodies displayed a distribution more characteristic of the secretory immune system with IgA1 accounting for 63% and IgA2 accounting for 37% of the total IgA ASA.

Although IgA I direct sperm-bound antibodies were dominant, when compared with serum, there was a higher proportion of IgA2 subclass, which suggests a local production of IgA.

Enzyme treatment dramatically reduced the amount of serum IgA antibodies bound to sperm ($P < 0.05$). Similarly, a significant reduction in direct sperm-bound antibodies was observed after enzymatic treatment with no loss in sperm motility. Specific IgA I protease treatment is capable of reducing the amount of immunobead-detectable IgA on sperm. They considered the hamster oocyte sperm penetration assays to determine if this treatment might improve sperm penetration rates; however, there have been no reports about this.

References

1. Shibahara H, et al. Diagnosis and treatment of immunologically infertile males with antisperm antibodies. Reprod Med Biol. 2005;4:133–41.
2. Rosse WF. Quantitative immunology of immune hemolytic anemia; II. The relationship of cell-bound antibody to hemolysis and the effect of treatment. J Clin Invest. 1971;50:734–43.
3. Hendry WF, Hughes L, Scammell G, et al. Comparison of prednisolone and placebo in subfertile men with antibodies to spermatozoa. Lancet. 1990;I:85–8.
4. Haas GG Jr, D'Cruz OJ. Effect of repeated washing on sperm-bound immunoglobulin G. J Androl. 1988;9:190–6.
5. Bronson RA. Immunity in sperm and in vitro fertilization. J In Vitro Fertil Embryo Transf. 1987;4:195–7.
6. Elder KT, Wick KL, Edwards RG. Seminal plasma anti-sperm antibodies and IVF; the effect of semen sample collection into 50% serum. Hum Reprod. 1990;5:179–84.
7. Shulman S, Mininberg DT, Davis LE. Significant immunologic factors in male infertility. J Urol. 1978;119:231–4.
8. Shulman S, Harlin B, Davis P. New method of treatment of immune infertility. Urology. 1978;12:582–6.
9. Hendry WF, Stedronska J, Hughes L, et al. Steroid treatment of male subfertility caused by antisperm antibodies. Lancet. 1979;2:498–501.
10. Alexander NJ, Sampson JH, Fulgham DL. Pregnancy rates in patients treated for antisperm antibodies with prednisone. Int J Fertil. 1983;28:63–7.
11. Bouloux PM, Wass JA, Parslow JM, et al. Effect of cyclosporine A in male autoimmune infertility. Fertil Steril. 1986;46:81–5.
12. Rumke P. Sperm antibodies and their action upon human spermatozoa. Ann Inst Pasteur. 1970;118:525–8.
13. Isojima S, Li YS, Ashitaka Y. Immunologic analysis of sperm-immobilizing factor found in sera of women with unexplained sterility. Am J Obstet Gynecol. 1968;101:677–83.
14. Henmi T, Shibahara H, Kobayashi S, et al. Efficacy of corticosteroid therapy for infertile men with sperm-immobilizing antibodies in their sera. Adv Obstet Gynecol. 1990;42:461. (in Japanese).
15. De Almeida M, Feneux D, Rigaud C, et al. Steroid therapy for male infertility associated with antisperm antibodies. Results of a small randomized clinical trial. Int J Androl. 1985;8:111–7.
16. Haas GG Jr, Manganiello P. A double-blind, placebo-controlled study of the use of methyl-prednisolone in infertile men with sperm-associated immunoglobulins. Fertil Steril. 1987;47:295–301.

17. Bals-Pratsch M, Doren M, Karbowski B, et al. Cyclic corticosteroid immunosuppression is unsuccessful in the treatment of sperm antibody-related male infertility: a controlled study. Hum Reprod. 1992;7:99–104.
18. Rasanen M, Lahteenmaki A, Agrawal YP, et al. A placebo-controlled flow cytometric study of the effect of low-dose prednisolone treatment on sperm-bound antibody levels. Int J Androl. 1996;19:150–4.
19. Hendry WF, Hughes L, Scammell G, et al. Comparison of prednisolone and placebo in subfertile men with antibodies to spermatozoa. Lancet. 1990;335:85–8.
20. Hendry WF. Bilateral aseptic necrosis of femoral heads following intermittent high-dose steroid therapy. Fertil Steril. 1982;38:120.
21. Shulman JF, Shulman S. Methylprednisolone treatment of immunologic infertility. Fertil Steril. 1982;38:591–9.
22. Kiser GC, Alexander NJ, Fuchs EF, Fulgham DL. In vitro immune absorption of antisperm antibodies with immunobead-rise, immunomagnetic, and immunocolumn separation techniques. Fertil Steril. 1987;47:466–74.
23. Foresta C, Varotto A, Caretto A. Immunomagnetic method to select human sperm without sperm surface-bound autoantibodies in male autoimmune infertility. Arch Androl. 1990;24:221–5.
24. Viganò P, Fusi FM, Brigante C, et al. Immunomagnetic separation of antibody-labelled from antibody-free human spermatozoa as a treatment for immunologic infertility. A preliminary report. Andrologia. 1991;23:367–71.
25. Pattinson HA, Mortimer D, Curtis EF, et al. Treatment of spermagglutination with proteolytic enzymes. I. Sperm motility, vitality, longevity and successful disagglutination. Hum Reprod. 1990;5:167–73.
26. Pattinson HA, Mortimer D, Taylor PJ. Treatment of spermagglutination with proteolytic enzymes, II. Sperm function after enzymatic disagglutination. Hum Reprod. 1990;5:174–8.
27. Bronson RA, Cooper GW. Documentation of IgA1 and IgA2 antisperm antibodies within seminal fluid. Am J Reprod Immunol Microbiol. 1988;18:7–10.
28. Kutteh WH, Kilian M, Ermel LD, et al. Antisperm antibodies (ASAs) in infertile males: subclass distribution of IgA antibodies and the effect of an IgA1 protease on sperm-bound antibodies. Am J Reprod Immunol. 1994;31:77–83.

Chapter 12
Assisted Reproductive Technology (ART) as A Treatment for Infertile Men with ASA

Hiroaki Shibahara

Abstract There have been reported other options for the treatments of infertile men with ASA by using assisted reproductive technology (ART). So far, the usefulness of intrauterine insemination (IUI), in vitro fertilization-embryo transfer (IVF-ET), and intracytoplasmic sperm injection (ICSI) for the treatments of infertile men with ASA has been reported. However, it should be emphasized that a diversity of ASA exists and has various effects on fertility in infertile men.

For the appropriate strategy to infertile men with ASA, it is necessary to evaluate fertilizing ability of the patient's ejaculated sperm before starting the IUI treatment. If there is no inhibitory effect on fertilization by ASA, the treatment using IUI might be useful for some men with ASA who have the negative postcoital test results. For the patient with ASA who is diagnosed to have abnormal fertilizing ability based on the hemizona assay, it is necessary to advise the infertile couple for selecting ICSI-ET as a primary treatment.

Although there is an argument that the routine testing for ASA in infertile men is not always necessary, one should be aware that in some cases of failed IUI or IVF, afterward ICSI is selected because of the diagnosis of ASA. Clinicians should judge the emotional cost to couples that have gone through the procedures and failed to conceive.

12.1 Introduction

In the former chapter, it was demonstrated that some infertile men with anti-sperm antibody (ASA) might be treated with the classical treatments such as immunosuppressive therapies using corticosteroids or cyclosporine to reduce production of

H. Shibahara (✉)
Department of Obstetrics and Gynecology, School of Medicine, Hyogo Medical University, Nishinomiya, Hyogo, Japan
e-mail: sibahara@hyo-med.ac.jp

© The Editor(s) (if applicable) and The Author(s), under exclusive license to
Springer Nature Singapore Pte Ltd. 2022
H. Shibahara, A. Hasegawa (eds.), *Gamete Immunology*,
https://doi.org/10.1007/978-981-16-9625-1_12

ASA. Methods that attempt to remove ASA bound to sperm have also been described.

There have been reported other options for the treatments of infertile men with ASA by using assisted reproductive technology (ART). So far, the usefulness of intrauterine insemination (IUI), in vitro fertilization-embryo transfer (IVF-ET), and intracytoplasmic sperm injection (ICSI) for the treatments of infertile men with ASA has been reported. However, it should be emphasized that a diversity of ASA exists and has various effects on fertility in infertile men.

In this chapter, a strategy, established by investigations of the diverse inhibitory effects of ASA on fertility, for the ideal diagnosis and treatment of infertile men with ASA is described. For infertile men with ASA, further diagnosis using the postcoital test (PCT) and the hemizona assay (HZA) should be carried out as the basis for decision-making [1, 2]. If the patient with ASA is diagnosed to have abnormal fertilizing ability based on the hemizona index (HZI) below 50 [3–5], it is necessary to advise the infertile couple for selecting ICSI-ET as a primary treatment. However, it has been shown that some immunologically infertile men with normal fertilizing ability were able to establish pregnancy by timed intercourse (TI) or IUI. In such patients with ASA having normal HZI, TI or IUI can be selected based on the PCT result.

12.2 Intrauterine Insemination (IUI)

The treatment using IUI appears to be useful for some men with ASA. It was reported that the conception rate per cycle was 3–10% by IUI using untreated sperm [6, 7]. Ombelet et al. [8] used ejaculate collected in sterile medium from men who had high levels of ASA; a higher pregnancy rate of 64% was achieved following three cycles of IUI, with a 47% conception rate following the first cycle.

However, some ASA bound to ejaculated sperm show inhibitory effects on fertilization, such as sperm-zona pellucida binding, penetration through zona-pellucida, and oolemma. For the appropriate strategy to infertile men with ASA, it is necessary to evaluate fertilizing ability of the patient's ejaculated sperm before starting the IUI treatment.

12.3 In Vitro Fertilization (IVF)

Several studies have examined the use of IVF and ICSI when IUI fails, and these methods currently appear to offer excellent chances of pregnancy for infertile men with ASA.

First, IVF circumvents the problem that the diminished number of ASA-coated sperm entering the fallopian tubes after IUI markedly lowers the chance that the inseminated sperm and the ovulated egg will meet. In some reports, fertilization rates achieved by men with ASA have been high, indicating that impaired sperm transport in the female genital tract was the primary basis of their infertility [8–11] In contrast, other studies led to the opposite conclusion. Rajah et al. [12] reported that the fertilization rates of IVF for men with ASA were significantly lower than those for men without ASA. Zouari et al. [13] concluded based on a retrospective analysis of IVF treatments for infertile men with ASA that the patients with successful fertilization had significantly more ASA binding to the tail than the head. This result might indicate that ASA binding to the acrosome or sperm head inhibits fertilization more potently than ASA bound to the tail or midpiece. The results will also depend on whether the sperm antigens to which ASA are directed have important roles in fertilization. Immunodominant sperm surface antigens that bind to sera containing ASA from infertile men and women have been identified by 2-D gel electrophoresis by our group [14, 15]. Some of the sperm surface antigens identified might be related to sperm-egg interaction.

In our previous experience [16], four (57.1%) of seven patients who had immunobead (IB)-bound sperm of $\geq$80%, the fertilizing ability was inhibited, while none of the eight patients who had <80% IB-bound sperm had an inhibitory effect on fertilization. There was a significant difference between the two groups ($P = 0.01$).

Therefore, it might be recommended that infertile men who have IB-bound sperm of $\geq$80% had better be advised for selecting ICSI-ET as a primary treatment if the HZA is not available. Of course, it is strongly suggested that the HZA should be mastered in each of the infertility facility as a routine examination for male factor infertile patients including ASA.

12.4   Intracytoplasmic Sperm Injection (ICSI)

ICSI has been shown to be useful for men with ASA to avoid fertilization failure secondary to an autoimmune mechanism. For 22 infertile men with ASA who demonstrated a poor fertilization rate (6%) in previous IVF treatments, the fertilization rate (79%) and cleavage rate (89%) obtained after ICSI were similar to those for ASA-negative men [17]. However, the embryo quality was lower in men with ASA compared with that in those without ASA. They concluded that ICSI offers a good chance of fertilization for couples with male immunological infertility. However, postfertilization events may compromise these results because of factors not yet clearly understood.

Nagy et al. [18] also showed by a retrospective analysis of 55 ICSI cycles with high levels of ASA-bound sperm. The mean normal fertilization rate was 75.7% in these 55 cycles, which was significantly higher than the fertilization rate in another 1767 ICSI cycles (69.2%) performed over the same period and where ASA-negative semen was used for microinjection. Embryonic development was comparable, but a higher proportion of poor-quality embryos was obtained with ASA-positive than with ASA-negative semen samples. Out of the 55 patients, 53 had embryos replaced

(96.4%), and a fetal sac was detected by ultrasonography in 14 patients (26.4%). The data indicate that fertilization, embryo development, and pregnancy rates after ICSI are not influenced significantly by the proportion of ASA-bound sperm, nor by the dominant type of antibodies present, nor by the location of the ASA on the sperm. They concluded that ICSI should be the primary choice for patients who have high numbers of ASA present in their semen.

12.5 A Strategy for the Ideal Diagnosis and the Treatment of Infertile Men with ASA

There have never been reported useful strategy for the management of infertile men with ASA. The reason was that there exist problems that include understanding the diversity of ASAs themselves and also requiring to perform further clinical tests that can predict sperm function such as sperm penetration through the cervical mucus and sperm-fertilizing ability.

We previously suggested a strategy for the diagnosis and the treatment of infertile men with ASA (Fig. 12.1) [1]. For the diagnosis, both of the PCT and the HZA are

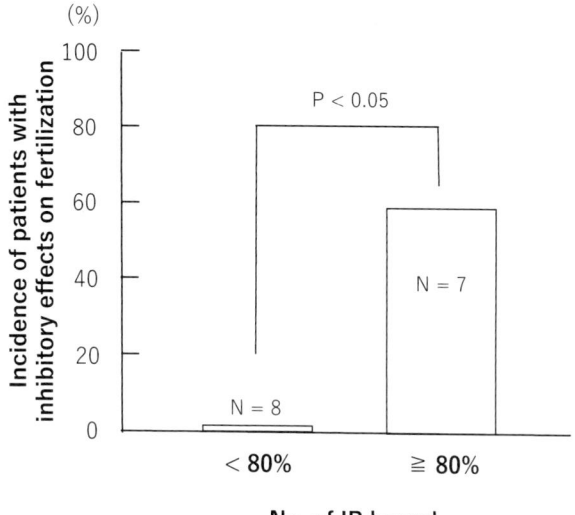

Fig. 12.1 Association of the incidence of the inhibitory effects on fertilization with the number of immunobeads (IB) bound. In four (57.1%) of seven patients who had immunobead (IB)-bound sperm of $\geqq 80\%$, the fertilizing ability was inhibited, while none of the eight patients who had $<80\%$ IB-bound sperm had an inhibitory effect on fertilization. There was a significant difference between the two groups ($P = 0.01$). Therefore, it might be recommended that infertile men who have IB-bound sperm of $\geqq 80\%$ had better be advised for selecting ICSI-ET as a primary treatment if the HZA is not available

recommended to be carried out as the basis for decision-making. The PCT is able to assess the inhibitory effects of ASA on sperm motility and sperm penetration through the cervical mucus. The HZA is able to evaluate the blocking effects of ASA on fertilization. The method for the HZA is described in detail in Chap. 2. If the patient with ASA is diagnosed to have abnormal fertilizing ability based on the HZI, it is necessary to advise the infertile couple for selecting ICSI-ET as a primary treatment. However, in such patients with ASA having normal HZI, TI or IUI can be selected based on the PCT result.

12.6   Our Clinical Experiences Based on the Established Strategy

Based on the established strategy, we have reported the clinical experiences for 509 infertile men with ASA treated with TI, IUI, and IVF/ICSI-ET [16]. In total, 18 (3.54%) infertile men had ASA, detected using D-IBT, on the sperm surface. According to the results of PCT, six (60.0%) out of ten infertile men with ASA had impaired sperm penetration in the cervical mucus. Moreover, four (26.7%) out of 15 men with ASA were judged to have impaired fertilizing ability assessed by the HZA or IVF. Excluding two cases that dropped out, 12 pregnancies in ten cases were established using TI in four cases, IUI in two cases, IVF-ET in one case, and ICSI-ET in five cases (Table 12.1). In total, conception was achieved for ten out of 16 infertile males with ASA, giving a pregnancy rate of 62.5%. Therefore, the treatment strategy for men with ASA is similar to that for infertile men with oligozoospermia or asthenozoospermia.

In conclusion, it should be emphasized that a diversity of ASA exists and has effects on fertility in infertile men. Although there is an argument that the routine testing for ASA in infertile men is not always necessary, one should be aware that in some cases of failed IUI or IVF, afterward ICSI is selected because of the diagnosis of ASA. Clinicians should judge the emotional cost to couples that have gone through the procedures and failed to conceive (Fig. 12.2).

Table 12.1 Pregnancies established in 12 treatment cycles from ten infertile men with ASA

Treatment	No. of patients conceived
TI	4
IUI	2
IVF-ET	1
ICSI-ET	5

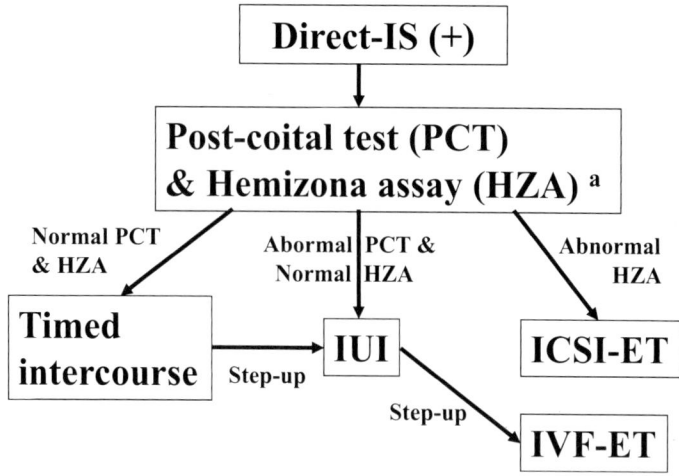

Fig. 12.2 A strategy for the diagnosis and the treatment of infertile men with ASA. To diagnose immunological infertility in men, the direct-ImmunoSpheres (D-IS) test should be carried out as a routine test in the infertility clinic. Then, the postcoital test (PCT) and the hemizona assay (HZA) should be performed as initial screening for infertile men with ASA diagnosed by D-IS, as a basis for decision-making. If a patient with ASA has an abnormal hemizona index (HZI), it seems reasonable to advise selecting intracytoplasmic sperm injection-embryo transfer (ICSI-ET) as the primary treatment. For the patients with ASA and normal HZI, selection of timed intercourse (TI) or intrauterine insemination (IUI) should be advised depending on the PCT result. ᵃ: If the HZA is not available, infertile men who have IB-bound sperm of ≧80% had better be advised for selecting ICSI-ET as a primary treatment

References

1. Shibahara H, Shiraishi Y, Suzuki M. Diagnosis and treatment of immunologically infertile males with antisperm antibodies. Reprod Med Biol. 2005;4:133–41.
2. Burkman LJ, Coddington CC, Franken DR, et al. Hemizona assay (HZA): development of a diagnostic test for the binding of human spermatozoa to the human hemizona pellucida to predict fertilization potential. Fertil Steril. 1988;49:688–97.
3. Shibahara H, Shigeta M, Koyama K, et al. Inhibition of sperm-zona pellucida tight binding by sperm immobilizing antibodies as assessed by the hemizona assay (HZA). Acta Obstet Gynaecol Jpn. 1991;43:237–8.
4. Shibahara H, Burkman LJ, Isojima S, et al. Effects of sperm-immobilizing antibodies on sperm-zona pellucida tight binding. Fertil Steril. 1993;60:533–9.
5. Shibahara H, Shigeta M, Inoue M, et al. Diversity of the blocking effects of antisperm antibodies on fertilization in human and mouse. Hum Reprod. 1996;11:2595–9.
6. Haas GG Jr. Male infertility and immunity. In: Lipshutz LI, Howards Jr SS, editors. Infertility in the male, vol. 2. St. Louis: Mosby Yearbook; 1991. p. 277–96.
7. Francavilla F, Romano R, Santucci R, et al. Failure of intrauterine insemination in male immunological infertility in cases in which all spermatozoa are antibody-coated. Fertil Steril. 1992;58:587–92.
8. Ombelet W, Vandeput H, Janssen M, et al. Treatment of male infertility due to sperm surface antibodies: IUI or IVF? Hum Reprod. 1997;12:1165–70.

9. Lahteenmaki A. In-vitro fertilization in the presence of antisperm antibodies detected by the mixed antiglobulin reaction (MAR) and the tray agglutination test (TAT). Hum Reprod. 1993;8: 84–8.

10. Sukcharoen N, Keith J. The effect of the antisperm autoantibody bound sperm on in vitro fertilization outcome. Andrologia. 1995;199527:281–9.

11. Pagidas K, Hemmings R, Falcone T, et al. The effect of antisperm autoantibodies in male or female partners undergoing in vitro fertilization-embryo transfer. Fertil Steril. 1994;62:363–9.

12. Rajah SV, Parslow JM, Howell RJ, et al. The effects on in-vitro fertilization of autoantibodies to spermatozoa in subfertile men. Hum Reprod. 1993;8:1079–82.

13. Zouari R, DeAlmeida M, Rodrigues D, et al. Localization of antibodies on spermatozoa and sperm movement characteristics are good predictors of in vitro fertilization success in cases of male autoimmune infertility. Fertil Steril. 1993;59:606–12.

14. Shetty J, Naaby-Hansen S, Shibahara H, et al. Human sperm proteome: immunodominant sperm surface antigens identified with sera from infertile men and women. Biol Reprod. 1999;61:61–9.

15. Shibahara H, Sato I, Shetty J, et al. Two-dimensional electrophoretic analysis of sperm antigens recognized by sperm immobilizing antibodies detected in infertile women. J Reprod Immunol. 2002;53:1–12.

16. Shibahara H, Shiraishi Y, Hirano Y, et al. Diversity of the inhibitory effects on fertilization by anti-sperm antibodies bound to the surface of ejaculated human sperm. Hum Reprod. 2003;18: 1469–73.

17. Lahteenmaki A, Reima I, Hovatta O, et al. Treatment of severe male immunological infertility by intracytoplasmic sperm injection. Hum Reprod. 1995;10:2824–8.

18. Nagy ZP, Verheyen G, Liu J, et al. Results of 55 intracytoplasmic sperm injection cycles in the treatment of male immunological infertility. Hum Reprod. 1995;10:1775–80.

Part III
Anti-Sperm Antibody (ASA): Development of Immunocontraceptives by Using Target Antigens Against ASA

Chapter 13
Development of Immunocontraceptives in Female

Hiroaki Shibahara

Abstract Contraception is one of the ways of controlling population growth; however, contraceptive methods used today are not available to many individuals due to sociological, financial, or educational limitations. Immunocontraception may provide alternative to the current contraceptive methods due to its reversibility and the fact that most of the developing countries where population growth is highest have service infrastructure for delivery of disease vaccines into which contraceptive vaccines could be incorporated.

Historically, immunization of female mice or humans with either sperm or their extracts led to the production of anti-sperm antibodies (ASA) and infertility. However, these studies had a drawback of the presence of several proteins that were shared by other somatic cells. It is critical that antibodies developed by contraceptive vaccine based on sperm-specific proteins should not react with any other somatic cells. Candidate immunogens should either have a fertility-related function that can be blocked by antibody or be located on the sperm surface for ASA to affect sperm motility in the oviduct, uterus, cervix, or vagina.

Many sperm antigens identified have potential; however, no single protein fulfills all the criteria necessary for a human vaccine. Therefore, some researchers have opted for construction of multi-antigen vaccines as a mechanism to overcome the limitations of a single molecule. There is also a new idea to engineer monoclonal antibody-based multipurpose prevention technology (MPT) products to address serious problems in female reproductive health, such as highly prevalent sexually transmitted infections and unplanned pregnancies.

H. Shibahara (✉)
Department of Obstetrics and Gynecology, School of Medicine, Hyogo Medical University, Nishinomiya, Hyogo, Japan
e-mail: sibahara@hyo-med.ac.jp

© The Editor(s) (if applicable) and The Author(s), under exclusive license to
Springer Nature Singapore Pte Ltd. 2022
H. Shibahara, A. Hasegawa (eds.), *Gamete Immunology*,
https://doi.org/10.1007/978-981-16-9625-1_13

13.1   Introduction

So far, the onset, diagnosis, and treatments of infertility in men and women by the anti-sperm antibodies (ASA) were described. Research has been conducted to develop a contraceptive vaccine by taking the ASA production that causes this refractory infertility in the opposite hand. Contraceptive vaccines have been proposed for controlling the growing human population. This idea is also expected to be applied to population regulation in order to prevent forest damage caused by wild birds and animals.

As shown in Table 13.1, the conditions mostly required for contraceptive vaccines include safety, reversible contraceptive effects, and mass production. Moreover, it is better to have an advantage over the present contraceptive methods. Of course, ethical or religious problems also need to be overcome. Sperm-specific proteins that are involved in their development and functions leading to successful fertilization also provide an exciting option to develop contraceptive vaccines. Many sperm antigens identified have potential; however, no single protein fulfills all the criteria necessary for a human vaccine. Therefore, some researchers have opted for construction of multi-antigen vaccines as a mechanism to overcome the limitations of a single molecule. There is also a new idea to engineer monoclonal antibody (mAb)-based multipurpose prevention technology (MPT) products to address serious problems in female reproductive health, such as highly prevalent sexually transmitted infections and unplanned pregnancies.

Here, the history and present state of contraceptive vaccine development using sperm antigen in female are presented.

13.1.1   World Overpopulation in the Near Future

The global population growth represents one of the greatest threats to human race today. In 1800, the world population was one billion, and it doubled in 1900. As shown in Fig. 13.1, it reached six billion in 1999 and grew steadily to around 7.7 billion in 2020. The prospected number is 9.1 billion in 2050 by United Nations. The countries in which more than half of the increased population by 2050 are predicted to focus on the next nine: India, Nigeria, Pakistan, Democratic Republic of the Congo, Ethiopia, Tanzania, Indonesia, Egypt, and the USA. Ninety-five percent of this global population growth is concentrated in the developing countries [1]. As a

Table 13.1 Conditions required for contraceptive vaccines	
1.	Safety
2.	Reversible contraceptive effects
3.	Mass production, leading to be established as a business
4.	Advantage over the present contraceptive methods
5.	Ethical or religious problems also need to be overcome

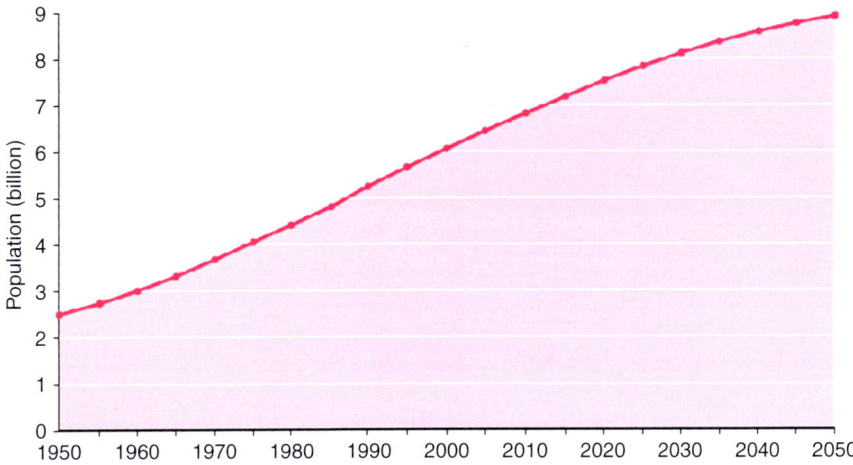

Fig. 13.1 Prospected world population (1950 to 2050) by United Nations. The global population growth represents one of the greatest threats to human race today. In 1800, the world population was 1 billion, and it doubled in 1900. It reached 6 billion in 1999 and grew steadily to around 7.7 billion in 2020. The prospected number is 9.1 billion in 2050 by United Nations

result, 80% of the total world population will be concentrated in the developing countries.

Contraception is one of the ways of controlling population growth; however, contraceptive methods used today are not available to many individuals due to sociological, financial, or educational limitations. Immunocontraception may provide alternative to the current contraceptive methods due to its reversibility and the fact that most of the developing countries where population growth is highest have service infrastructure for delivery of disease vaccines into which contraceptive vaccines could be incorporated [2].

13.1.2 Harmful and Dangerous Early Clinical Trials

Since the editorials of the *JAMA* [3] suggested that sperm could cause the formation of specific antibodies which are capable of preventing fertilization, leading to sterility of the prostitute, harmful and dangerous clinical trials were carried out.

Rosenfeld [4] suggested that a woman repeatedly injected with human semen became infertile. He attempted to immunize women with multiple subcutaneous injections of semen and found serological evidence of immunity to sperm. Mild local reactions were noted but no systemic disturbance occurred.

Following this report, Baskin [5] also reported that active immunization with human semen induced temporary sterility in female patients. In 1937, Baskin has issued the US patent for a spermotoxic vaccine, which produced reversible sterilization in fertile women. However, despite 12 claims made in the patent, there is no record in the literature whether the patented product achieved the goals set forth [6]. This may have been due to anaphylactic and hypersensitivity reactions seen following immunization with whole semen.

Such experiments were turned down because ethical regulations, such as the Nuremberg Code in 1947, the Geneva Declaration in 1948, and the Helsinki Declaration in 1964, for the human research were advocated following the World War II. Moreover, Otani et al. [7] demonstrated that immunization with human seminal plasma proteins or diluted semen resulted in anaphylactic responses in several women (usually after the seventh to tenth injection), whereas inoculation with sperm proteins did not. Hypersensitivity reactions to seminal plasma antigens have been similarly reported [8]. Obviously, an immunocontraceptive based on the injection of fresh semen samples is unacceptable, but these experiments in humans indicate that immunization with sperm antigens may be an acceptable form of contraception and focus attention on the identification of the optimal sperm antigens.

13.1.3 Lessons from Female Animals Immunized with Sperm or Testis

In the early 1900s, there were some studies of immunological contraception in women [9]. In some of these reports, infertility and humoral antibody formation were achieved by repeated injections of semen without the use of immunological adjuvants. However, the results obtained were conflicting, with some showing positive relationship, while others negative relationship.

In the era, there were also controversial reports in animal experiments. McCartney [10] reported a positive data on ASA and infertility in rats, while Henle et al. [11] suggested a negative data on ASA and infertility in guinea pigs. Animal studies have yet to reach a conclusion in these days.

It was dissolved by the introduction of Freund's adjuvant for enhancing immunization that encouraged researchers to reinvestigate this approach [12]. The purpose of their study was to give an account of an effort to purify the antigenic material, and of observations on anaphylactic systemic and cutaneous sensitizations to homologous testicular extracts.

Guinea pig testicles were extracted with acetic acid, and the extract was purified by removing material in consecutive precipitations with 30% saturated ammonium sulfate, trichloroacetic acid, and chloroform. The solution so purified, when administered with complete adjuvants, was highly active in inducing impairment of spermatogenesis in guinea pigs. Anaphylactic shock and cutaneous reaction to the purified homologous extract occurred in guinea pigs sensitized by the extract combined with adjuvants. For the production of aspermatogenesis, it was essential to incorporate killed mycobacteria into the water-in-oil emulsion containing the antigen, but anaphylactic sensitization did not require the presence of mycobacteria.

In 1959, Isojima et al. [13] clearly demonstrated sterility in adult virgin female guinea pigs induced by injection with an emulsion of homologous adult testis and Freund's adjuvant before exposure to males [see Chap. 1]. Katsh [14] also obtained a similar result by immunizing female guinea pigs with sperm using complete Freund's adjuvant. Later, Otani et al. [15] re-evaluated and proved that female guinea pigs immunized with homologous testis or sperm using complete Freund's adjuvant reduced pregnancy rates in immunized animals. In these days, hormonal contraceptives were developed. However, it was available only in some countries. Moreover, lethal side effects, such as thrombosis, were reported in some women prescribed.

On the other hand, two investigators reported the opposite conclusions in mice. In 1964, McLaren [16] showed the reduction of fertilization in females received three intraperitoneal injections of live homologous sperm a week for 7 weeks. However, Edwards [17] showed that immunization of mice with sperm and adjuvant has not led to drastic reductions in infertility. The incidence of fertilization was higher and more independent of the amount of circulating antibody than was the incidence of infertility found by McLaren [16]. Therefore, it was suggested that there might exist variability of the immune response among the immunized individuals as well as species.

13.1.4 Focus Attention on the Identification of the Optimal Sperm Antigens for the Development of Immunocontraceptives in Female

Immunization of female mice or humans with either sperm or their extracts led to the production of ASA and infertility. Though these studies suggested that immunization with sperm can lead to infertility, they had a drawback of the presence of several proteins that were shared by other somatic cells. It is critical that antibodies developed by contraceptive vaccine based on sperm-specific proteins should not react with any other somatic cells [18]. Candidate immunogens should have a fertility-related function that can either be blocked by antibody or be located on the sperm surface for ASA to affect sperm motility. They should also be shown to inhibit fertility in animal models and reversibility of the contraceptive effect is desirable [6].

The goal of immunizations against sperm components is to develop a contraceptive vaccine which will induce ASA in the female reproductive tract at sufficient levels to block fertilization. It is envisioned that antibodies developed by this vaccine will act to agglutinate, immobilize, or block fertilization by the sperm in the oviduct, uterus, cervix, or vagina. Thus, Herr [19] suggested that the inhibition of prefertilization events is desired because a contraceptive that blocks implantation or induces embryonic mortality is unacceptable to many individuals.

ASA impair fertility by inhibiting sperm transport [20–22] and/or gamete interaction [23–39]. Bivalent antibodies can agglutinate sperm to form immobile networks of cells in the reproductive tract [40]. ASA bound to the sperm surface can alter swimming patterns, impair sperm penetration of cervical mucus, immobilize sperm, or invoke the complement cascade resulting in sperm lysis [41, 42].

Some approaches have been taken to identify candidate sperm antigens for the development of immunocontraceptives. That includes the study of clinical ASA-mediated infertility as "experiments of nature" in fertility reduction, the direct identification of sperm surface antigens, the identification of antigens with mAbs or polyclonal antisera generated against human sperm, and the assessment of known sperm antigens [6].

Multiple approaches have been taken to identify candidate sperm antigens for immunocontraceptive development. The following discussion describes these approaches and representative antigens thus identified.

13.1.4.1 Investigation of the Experiments of Nature

To develop birth control immunization strategies based on antigens involved in reproduction, the World Health Organization (WHO)'s Task Force on Immunological Methods for Fertility Regulation was established in 1973. Hjort and Griffin [43] investigate the natural immunities to reproductive antigens which led to infertility to outline the "experiments of nature." To accomplish this goal, a library of serum samples were collected from 329 infertile individuals. Numerous investigators have reported the identification of specific antigens using sera from the WHO reference bank or from independent collections [44–54]. However, characterization of sperm antigens was limited to SDS-PAGE in most studies. Furthermore, results varied widely due to the methods and reagents employed [6]. Moreover, Koyama et al. [55] suggested that when they identified the corresponding antigen against the sera from refractory infertile women, carbohydrate antigens were often identified, and it was difficult to mass-produce specific carbohydrate antigens using advanced science and technology.

An alternate strategy has combined recombinant DNA technology with the experimental approach described by Hjort and Griffin [43]. Sera from ASA-positive infertile patients and vasectomized men identified testis antigens expressed by recombinant bacteriophage in human testis libraries [56–58]. Isolation of relevant cDNAs permits the direct assessment of tissue specificity and expression of recombinant proteins for subsequent characterization. Furthermore, future selection of peptide epitopes for immunocontraceptive development is facilitated by the absence of carbohydrate epitopes on the recombinant antigens [59].

13.1.4.2 Direct Identification of Sperm Surface Antigens by Vectorial Labeling and 2-D Electrophoresis

To identify the repertoire of proteins exposed on the surface of ejaculated human sperm, Naaby-Hansen et al. [60] developed high-resolution two-dimensional (2-D) gel systems for separation of human sperm and seminal plasma proteins. 1397 sperm proteins with a molecular mass between 5 and 160 kDa and isoelectric points (pI) from 4 to 11 were catalogued from silver-stained gels loaded with human sperm extracts harvested by Percoll density gradient centrifugation, and 1191 proteins were resolved following extraction. Sperm surface proteins accessible to vectorial labeling with ^{125}I or N-hydroxysuccinimide biotin were identified; 181 protein spots were radiolabeled with ^{125}I, while 228 protein spots were biotinylated, including several groups of protein isoforms. Ninety-eight sperm surface proteins were labeled by both iodine and biotin, and 22 sperm surface proteins, representing five groups of protein isoforms, were shown to contain phosphotyrosine. A composite computer image showing the position of the dually vectorially labeled sperm surface proteins was constructed.

Our group have identified human sperm antigens recognized by the sera from infertile women having sperm-immobilizing antibodies that has been well known as most closely related with infertility in women, using the high-resolution 2-D gel electrophoresis (Fig. 13.2) [61]. Sperm proteins were transferred to the nitrocellulose membranes and immunoblotted with seven sera from infertile women with high titers of antibodies and six sera from those without sperm-immobilizing antibodies. The blots were compared to the 2-D composite image of human sperm proteins

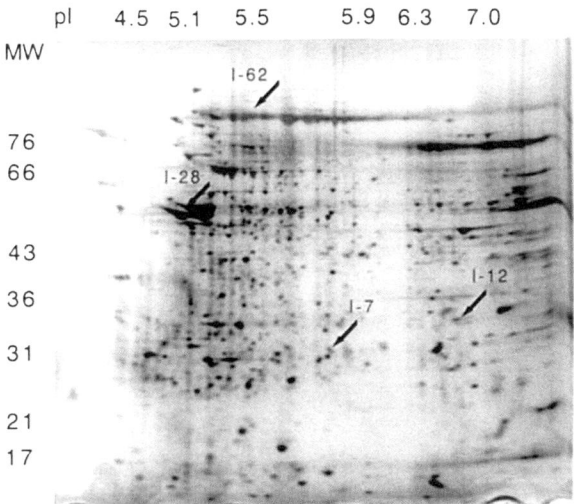

Fig. 13.2 The coordinates of prominent immunoreactive sperm proteins corresponded to sperm-immobilizing antibodies in the sera of infertile women [61]. Human sperm antigens recognized by the sera from infertile women having sperm-immobilizing antibodies were identified using the high-resolution 2-D gel electrophoresis. A subset of sperm surface proteins which were unique to the sperm-immobilizing antibodies were identified by the following criteria: the sperm protein spots which were highly reactive with the infertile sera containing sperm-immobilizing antibodies but not reactive with any of the antibody-negative infertile sera. The coordinates of four prominent immunoreactive sperm proteins were as follows: (MW, pI) = (32.6, 5.9), (35.6, 6.6), (58.6, 4.9), and (93.8, 5.5). These sperm surface antigens were considered as possibly relevant to antibody-mediated female infertility

[Sperm Protein Encyclopedia], and sperm surface index and the sperm surface proteins recognized by infertile sera were identified. Fifty-two human sperm surface proteins reacted with sera containing sperm-immobilizing antibodies, while 35 of these were reactive with the antibody-negative control sera. The average numbers of protein spots reacted with test and control sera were 24.6 and 15.0, respectively. A subset of sperm surface proteins which were unique to the sperm-immobilizing antibodies were identified by the following criteria: the sperm protein spots which were highly reactive with the infertile sera containing sperm-immobilizing antibodies but not reactive with any of the antibody-negative infertile sera. As a result, the coordinates of four prominent immunoreactive sperm proteins were identified as follows: $(MW, pI) = (32.6, 5.9), (35.6, 6.6), (58.6, 4.9)$, and $(93.8, 5.5)$. These sperm surface antigens were considered as possibly relevant to antibody-mediated female infertility [61].

Shetty et al. [62, 63] have also identified immunodominant sperm antigens recognized by ASA detected by the immunobead test (IBT) in serum samples from infertile women and men. High-resolution 2-D gel electrophoresis was employed to separate human sperm proteins followed by PAGE and Western blotting. In their early study [62], they identified a subset of 6 iso- and autoantigens as follows: $(MW, pI) = (34.0, 4.2), (38.0, 4.3), (60.0, 6.2), (60.0, 6.4), (88.2, 4.0)$, and $(140.0, 4.2)$. These sperm surface antigens were possibly relevant to antibody-mediated infertility. In their later study [63], serum samples from five infertile women and five infertile men that contained ASA were analyzed to identify the basic sperm antigens reactive to the sera. Serum samples from five fertile women and five fertile men that were ASA-negative were used as controls. Five sperm isoantigenic basic spots were recognized by infertile women but not from fertile subjects. Five alloantigenic basic protein spots were also recognized by sera from infertile men but not from fertile subjects. The differences of the corresponding antigens between these studies using circulating ASA might be due to the reason that ASA detected by means of IBT may not be associated with infertility and could differ from sperm-immobilizing antibodies [61–63]. Indeed, Shetty et al. [63] described that a high percentage of intra-acrosomal proteins were reactive to ASA, which calls a modification of the IBT to test infertile sera with populations of acrosome-reacted sperm.

To examine these proteins as potential candidates for the future development of immunocontraceptives, microsequence selected surface proteins by tandem mass spectrometry, cloning and sequencing the proteins, and performing tissue specificity studies were required. Several alloantigenic protein spots identified in the 2-D sperm proteome are expressed post-meiotically and have been cloned and characterized [63]. Examples are equatorial segment protein (ESP) [64], sperm acrosomal membrane-associated protein (SAMP) 2 [65], SAMP14 [66], A-kinase anchor protein (AKAP)3 [67], calcium-binding tyrosine phosphorylation-regulated protein (CABYR) [68], and radial spoke protein (RSP) 44, a radial spoke protein present in the axonemes of both sperm tail and cilia [69]. Among these antigens, ESP, SAMP32, and SAMP14 are intra-acrosomal proteins and become exposed on the sperm surface following the acrosome reaction. ESP is localized to the equatorial segment of the acrosome. Both SAMP32 and SAMP14 are acrosomal membrane proteins. AKAP3 and CABYR, in contrast, are localized to the fibrous sheath of the principal piece of the sperm tail [67, 68].

13.1.4.3 Monoclonal and Polyclonal Anti-sperm Antibodies (ASA)

Many investigators have raised anti-sperm monoclonal and polyclonal antibodies that inhibit sperm motility and gamete interaction [70, 71]. Their cognate antigens represent excellent targets for immunocontraception. A variety of sperm-specific proteins have been identified and characterized and their potential to inhibit fertility evaluated in suitable animal models. The characteristics of some of these proteins and their efficacy to inhibit fertility are described below (Table 13.2).

13.1.4.3.1 CD52

Isojima et al. [72] established a stable human-mouse heterohybridoma, H6-3C4, producing a human mAb with high titers of complement-dependent sperm-immobilizing activity by using peripheral B lymphocytes from an infertile woman. They showed that the H6-3C4 mAb and related murine mAbs were subsequently shown to recognize N-linked oligosaccharide epitopes on a sperm surface glycoprotein identified as a series of 15 to 25 kDa immunoreactive bands on Western blots

Table 13.2 Candidate sperm antigens for the development of contraceptive vaccines in female

Antigen	Localization	Molecular weight (kDa)	Biological activity
CD52	Entire surface	15–25	Sperm immobilization and agglutination, suppress complement activation
RSA	Cytoplasm of the head throughout the tail	14–18	Dose-dependent reduction of fertility in mice
SP-10	Inner and outer acrosomal membranes	18–34	Partial reduction of fertility in baboon
PH-20	Inner acrosomal membrane	64	Hyaluronidase activity, complete inhibition of fertility in guinea pigs
PH-30	Post-acrosomal membrane	60 (α chain) 44 (β chain)	Inhibition of sperm-egg fusion in guinea pigs Partial reduction of fertility in rabbits
LDH-C4	Head to tail	35	Partial reduction of fertilization
FA-1	Head to tail	23	Reduction of fertility in sperm-zona binding
YLP12	Acrosomal region	72	Binding to ZP3-primary sperm receptor Reduction of fertility
Izumo	Inner acrosomal membrane	56.4 (mouse) 37.2 (human)	Incapable of fusing with eggs

[73]. Tsuji et al. [74] demonstrated that the H6-3C4 mAb reacted with a carbohydrate epitope containing repetitive N-acetyllactosamine. Subsequently, it was suggested that mAb H6-3C4 recognized the male reproductive tract CD52 (mrtCD52) molecule [75, 76]. CD52 was originally identified as an antigen for an antibody called campath-1 that was produced using a human spleen cell antigen [77]. Human CD52 is expressed in virtually all lymphocytes, mature sperm, and seminal plasma [78].

Diekman et al. [79] described sperm agglutination antigen-1 (SAGA-1), a polymorphic (15–25 kDa), highly acidic hydrophobic glycoprotein that is localized over the entire surface of the human sperm. SAGA-1 was characterized with S19, a murine mAb generated by the immunization of mice with human sperm homogenates that exhibited sperm inhibitory actions in vitro including agglutination, inhibition of cervical mucus penetration, and inhibition of human sperm-zona pellucida binding [71, 79, 80]. Furthermore, Diekman et al. [79] demonstrated that the S19 mAb recognized a carbohydrate epitope on the SAGA-1 glycoprotein. Competitive inhibition studies, as well as similarities in reported mass, suggested that SAGA-1 and the H6-3C4 antigen represent the same sperm glycoproteins [73, 79, 81].

Hasegawa et al. [82] analyzed the epitope on mrtCD52 for mAb H6-3C4 and found that it was polymorphic in Western blot analysis and disappeared after enzymatic removal of the N-linked carbohydrate moiety. Two other mAbs (1G12, campath-1) with sperm-immobilizing activity recognized mrtCD52 in a polymorphic manner similar to mAb H6-3C4. Further analysis showed that 1G12 recognized a structure formed by the peptide and/or a glycosylphosphatidylinositol (GPI) anchor portion as does campath-1. Results of a lectin binding assay suggested the presence of O-linked carbohydrates on mrtCD52. They also indicated that the peptide portion of CD52 could serve as an epitope for sperm-immobilizing antibodies. They concluded that the epitope of mAb H6-3C4 is similar to, but distinct from, those of 1G12 and campath-1 and that mrtCD52 contains different antigenic epitopes.

Interestingly, by using the hemizona assay (HZA) [83], the zona pellucida penetration assay (ZPA) [84], and the sperm penetration assay (SPA) [85], Shibahara et al. [86] reported that mAb 1G12 showed strong inhibitory effects in all the assays (HZA, ZPA, and SPA); however, mAb (H6-3C4) seemed to have no inhibitory effects on fertilization. These findings suggest that ASA block fertilization at specific stages. Some of them may inhibit sperm capacitation and thus prevent all processes of fertilization that follow. Some other antibodies may not affect capacitation and sperm binding to zona pellucida but inhibit the acrosome reaction, followed by the blocking of sperm penetration through zona pellucida and ooplasm.

To develop a contraceptive vaccine using CD52, Hasegawa et al. [87] investigated the immunogenicity of CD52 by intranasal immunization. They prepared synthetic peptide corresponding to CD52 core peptide (GQNDTSQTSSPS) and conjugated with diphtheria toxoid as a carrier protein. The immunogen was given to mice with DOTAP/cholesterol liposome adjuvant intranasally. As a result, the CD52 core peptide elicited IgA class as well as Ig antibodies both in the sera and vaginal secretions after intranasal immunization. An additional nasal inoculation after decrease of the antibody titer raised the antibody level to its highest level during the experiment. They suggested that the common mucosal immune system (CMIS) could induce Ig and IgA class antibodies reactive to CD52 core peptide in the female genital tract. Intranasal immunization of a sperm-specific antigen would be a promising regimen for a safe and easy contraceptive vaccine.

Later, they found that mrt-CD52 played a role in suppressing complement activation [88]. Purified mrt-CD52 from human seminal plasma interfered with the classical pathway, but not lectin binding or alternative pathways of the complement system [89]. They also found that mrt-CD52 might bind to C1q via N-linked carbohydrate moiety and suppress complement activation to avoid sperm impairment if an antigen-antibody complex forms on the sperm surface [90]. Taken together, they suggested that mrt-CD52 protects sperm function from complement attack if ASA is generated in the female genital tract.

CD52 is recently applied as multipurpose prevention technologies [91]. More details of this topics will be introduced later.

13.1.4.3.2 Rabbit Sperm Autoantigens (RSA)/Sp17

O'Rand et al. [92] identified the rabbit sperm autoantigens (RSA) using an anti-sperm polyclonal antiserum. RSA-1, RSA-2, RSA-3, and RSA-4 are a family of low molecular weight membrane glycoproteins (14, 16, 17, 18 kD) that function as zona pellucida binding proteins [93]. RSA-3 cDNA clones, designated Sp17, were isolated from rabbit, mouse, and human testis libraries, and the Sp17 mRNA was shown to be testis-specific in these species [94–96]. Rabbit Sp17 was expressed in COS cells and was shown to bind rabbit zona pellucida glycoproteins [97]. O'Rand et al. [98] used a peptide pin-block method to identify an immunodominant B-cell epitope of rabbit Sp17, and this epitope was also identified in mouse Sp17. Localization studies revealed that the presence of Sp17 was in the cytoplasm of the head region of the sperm and throughout the tail region [99]. Immunization of female BALB/c mice with synthetic peptide encompassing dominant B-cell epitope of rabbit Sp17 (aa residues 58–66; AEWGAKVDD) and promiscuous T-cell epitope of bovine RNase (aa residues 94–104) led to a dose-dependent reduction in fertility [100]. However, the infertility mediated by this chimeric peptide was mouse strain-specific as no effect on fertility was observed in the immunized B6AF1 mice. It was further demonstrated that induction of peptide-specific T-cell responses and cytokines was a major factor in mediating infertility [100].

13.1.4.3.3 SP-10

Herr and colleagues identified the testis-/sperm-specific, intra-acrosomal sperm antigen SP-10 using MHS-10, a mAb generated against whole human sperm [101]. SP-10 was identified by immunoblot analysis as a series of protein bands (18–34 kDa), the polymorphism of which was attributed to alternative splicing and endoproteolytic cleavage [102, 103]. Ultrastructural and biochemical studies indicated that SP-10 is a hydrophilic protein localized to the luminal aspect of the inner and outer acrosomal membranes [101, 104]. Using the bovine IVF model, both monoclonal and polyclonal antibodies to SP-10 were found to inhibit secondary binding of capacitated sperm to the zona pellucida [105]. This finding gave further support to the hypothesis that antibodies to SP-10 act when SP-10 becomes accessible after the acrosome reaction is initiated and fusion pores form [104]. As a model sperm antigen for the development of an oral contraceptive, female mice were fed avirulent *Salmonella* that express SP-10- and SP-10-specific secretory immune responses that were observed in the vaginal washes of these mice. It was the first demonstration of a secretory immune response in the reproductive tract to a genetically engineered sperm antigen administered via an oral vector [19, 106]. Later, it was shown that immunization of male mice with recombinant SP-10 resulted in sterility [107].

13.1.4.3.4   PH-20

PH-20 is a very promising contraceptive immunogen candidate based on its fertility-related function, its localization, and the effects of PH-20 immunization. Primakoff and their colleagues identified PH-20 on the sperm surface with three mAbs generated against membrane preparations of guinea pig sperm [108]. Guinea pig PH-20 is a sperm surface glycoprotein (64 kD) involved in zona pellucida adhesion [109]. In acrosome-intact sperm, PH-20 is located on the inner acrosomal membrane and on the sperm surface in the posterior head region [110]. This latter population migrates to the inner acrosomal membrane after the acrosome reaction. The PH-20 protein was shown to exhibit hyaluronidase activity [111]. Furthermore, they demonstrated that this hyaluronidase activity enables sperm to penetrate the cumulus cell layer surrounding the ovum [111]. Primakoff et al. [112] reported a complete inhibition of fertility after immunization of male or female guinea pigs with PH-20, and this contraceptive effect was reversible. Human and cynomolgus monkey PH-20 cDNAs have been isolated to express recombinant proteins for testing contraceptive effects in primates [113].

13.1.4.3.5   Fertilin (PH-30)

The PH-30 antigen was a second post-acrosomal plasma membrane antigen identified with the series of anti-guinea pig sperm mAbs [108]. Unlike PH-20, the PH-30 mAb immunolocalized exclusively to the post-acrosomal plasma membrane of acrosome-intact and acrosome-reacted guinea pig sperm [108, 114]. Myles et al. [115] designated the PH-30 antigen as fertilin based on its apparent role during fertilization. Fertilin is composed of two subunits which are both integral membrane glycoproteins. The α subunit (60 kDa) contains a peptide typical of viral fusion proteins, while the β subunit (44 kDa) is an integrin ligand (disintegrin)

[116, 117]. Furthermore, the PH-30 mAb inhibited guinea pig sperm-egg fusion in a concentration-dependent manner [114]. Blobel et al. [116] proposed that the fertilin β subunit functions as a ligand for integrins on the egg surface, while fertilin α is responsible for gamete membrane fusion. Myles et al. [115] employed peptide competition assays to demonstrate that the binding of sperm and the oocyte plasma membrane through the disintegrin domain of fertilin β is required for gamete membrane fusion. Immunization of male guinea pigs with purified guinea pig fertilin resulted in complete infertility, whereas immunization of female guinea pigs resulted only in partial infertility [118]. As a preliminary step toward testing in primate models, cDNA clones encoding the monkey [118] and human [119] fertilin β subunits have been isolated and characterized. On the contrary, immunization of male and female rabbits with *E. coli*-expressed recombinant α or β-fertilin led to partial inhibition of fertility [120]. Only 4 out of 33 immunized female rabbits failed to conceive.

13.1.4.4 Contraceptive Potential of Other Sperm Antigens

In addition to the identification of potential contraceptive immunogens by the immunological and surface localization methods described above, sperm proteins were selected for fertility testing based on their known testis/sperm specificity or function in fertilization. The examination of testis-specific proteins identified an isoform of lactate dehydrogenase as a promising contraceptive immunogen candidate.

13.1.4.4.1 LDH-C$_4$

The testis-specific isozyme lactate dehydrogenase-C$_4$ (LDH-C$_4$) described by Goldberg et al. [121] is perhaps the most extensively characterized sperm antigen [6]. This is formulated as a DNA-based vaccine and used to control many animal populations where applicable [122]. LDH-C$_4$ null male mice showed normal spermatogenesis and testes development but reduced fertility [123]. Sperm from these mice had lower motility and reduced fertilization capacity. Female mice immunized with mouse LDH-C$_4$ by intrauterine route developed antibodies against LDH-C$_4$ and showed subfertility [124]. Male mice immunized systemically with LDH-C$_4$ also showed reduction in fecundity [125].

Immunogenicity and contraceptive efficacy of LDH-C4 peptides (human LDH-C4 aa residues 1–20 and 9–20 and baboon LDH-C4 aa residues 5–19) either alone or as a chimeric peptide incorporating promiscuous T-cell epitope of TT have been evaluated [126]. The chimeric peptide (baboon LDH-C4 aa residues 5–19) was more immunogenic and reduced fertility by 62% in the immunized baboons [126]. However, subsequent contraceptive efficacy studies with the same chimeric peptide in female cynomolgus macaque did not yield the desirable contraceptive efficacy [127].

13.1.4.4.2 Fertilization Antigen (FA)-1

Fertilization antigen-1 (FA-1) is a germ cell-specific glycoprotein present on the surface of mouse and human sperm. Antibodies to FA-1 block sperm-egg binding, and reduce sperm capacitation and acrosome reaction by inhibiting phosphorylation at tyrosine, serine, and threonine residues [128]. FA-1, expressed in post capacitation, reduced female fertility rates by 70% in B6 mice [129–132]. FA-1 is shown to interact with zona pellucida glycoprotein-3 (ZP3) in vitro in mice [133]. It is localized on post-acrosomal region of the sperm, and the mAbs against human FA-1 were shown to inhibit IVF in bovine [134]. Immunization of mice with recombinant FA-1 led to generation of the sperm-/testis-specific antibody response resulting in reversible inhibition of fertility by affecting sperm-zona binding and the following fertilization process [135].

13.1.4.4.3 YLP12

Naz et al. [136] identified a sperm protein and isolated only 12 amino acids of peptide YLP12 (novel dodecamer peptide sequence). YLP12 was identified that binds to the ZP3-primary sperm receptor [136]. It was localized in the acrosomal region of the human sperm. In the Western blot procedure involving ten different human tissue extracts, the anti-YLP12 Fab antibodies recognized a protein band of ~72 ± 2 kDa only in the testis lane [136]. Antibodies to synthetic YLP12 peptide inhibit sperm-egg binding in mouse and human. Active immunization studies with this peptide conjugated with the binding subunit of recombinant cholera toxin led to block in fertility, which was reversible as antibody titers declined [137]. Immunizing against antigen YLP12 was found to reduce fertility [138]. Particularly, this vaccine may be rendered as a recombinant DNA-based vaccine to reduce fertility over a prolonged period in mouse models [122].

13.1.4.4.4 Izumo

Izumo is an immunoglobulin superfamily protein and plays an important role in fusion of sperm membrane with oolemma. It is located on the inner acrosomal membrane of sperm and is accessible to the antibodies only after acrosome reaction. Izumo knockout male mice are sterile but otherwise healthy [139]. Mouse Izumo was shown to be a testis (sperm)-specific 56.4 kDa antigen with a polyclonal antibody raised against recombinant mouse Izumo. Izumo was also detectable as a 37.2 kDa protein of human sperm with anti-human Izumo antibody [139]. Sperm from these animals bind to the zona pellucida and penetrate normally but were incapable of fusing with the eggs [139]. Active immunization studies in mice with the recombinant Ig-like domain of Izumo showed reduction in fertility of immunized animals [140]. In addition to Ig-like domain of Izumo, synthetic peptides corresponding to Izumo (aa residues 166–182, 308–323, 341–359, 371–385) conjugated with different carrier proteins also showed their contraceptive potential during active immunization studies in mice [128].

Xue et al. [141] identified linear B-cell epitopes (BCE) in Izumo using biosynthetic peptides and used to immunize female mice. Five Izumo BCE were identified: DLVLDCL177-183, YSFYRV196-201 (named BCE-2), YLT217-219, SMVGPED221-227, and DAGNY228-232. Active immunization with the BCE-2 vaccine sharply decreased the fertility rate in female mice in a safe and reversible manner. IVF showed that the BCE-2 vaccine interferes with and blocks the fusion of the sperm and the ovum. They suggested that B-cell epitopes-2 may be a new candidate for the development of contraceptive vaccine due to its effectiveness, safety, and reversibility.

13.1.5 Construction of Multi-antigen Peptides (MAPs) as Immunocontraceptives

Many sperm antigens identified so far have potential; however, no single protein fulfills all the criteria necessary for a human vaccine. Therefore, some researchers have opted for construction of multi-antigen peptide (MAP) vaccines as a mechanism to overcome the limitations of a single molecule (Table 13.3). Kurth et al. [142] described that the vaccine comprised of multiple acrosomal antigens that have the potential for interrupting the fertilization process is that a multivalent vaccine may evoke a greater antifertility effect in females than immunization with a single sperm antigen. Although titers to a given epitope may wane with antibody catabolism, the overall number of different antibodies targeted to the sperm surface is predicted to be greater. Further, in an outbreed population such as humans, variability in host responsiveness to any single epitope is likely to be present. With administration of multiple antigens, there is a greater possibility of activating the host individual's immune system to produce a range of antibodies to surface exposed acrosomal epitopes.

Table 13.3 Multi-antigen peptides (MAPs) developed as contraceptive vaccines

Authors [Ref. no.]	Formulation	Species	Contraceptive effects
Kurth et al. [142]	ESP, SLLP-1, SAMP 32, SP10, SAMP 14	Cynomolgus monkeys	All monkeys developed both IgG and IgA response
Naz [128]	Izumo, FA-1, YLP12, SP56	Mice	73% reduction in fertility
Hardy et al. [143]	SP56, ZP3, ZP2, ZP1	Mice	60–80% reduction in fertility
			No apparent ovarian pathology
Choudhury et al. [144]	YLP12, ZP3	Mice	50% reduction in fertility

Kurth et al. [142] used a MAPs vaccine formulation comprised of five recombinant human intra-acrosomal sperm proteins (ESP, SLLP-1, SAMP 32, SP-10, and SAMP 14) to immunize female cynomolgus monkeys. Five female cynomolgus monkeys were inoculated intramuscularly three times at monthly intervals. All five monkeys developed both IgG and IgA serum responses to each recombinant immunogen. For antigens that induced an IgA response, the duration of the IgA response was longer than the IgG response to the same antigens. They suggested that a multivalent contraceptive vaccine may be administered to female primates evoking both peripheral (IgG) and mucosal (IgA) responses to each component immunogen following an intramuscular route of inoculation with a mild adjuvant, aluminum hydroxide, approved for human use.

Naz [128] studied the immunocontraceptive effect by using synthetic MAPs based upon four sperm proteins, namely, Izumo, FA-1, YLP12, and SP56, that are involved in various steps of the fertilization cascade. Female mice were immunized with various peptide vaccines, and each booster injection was given with the peptide conjugated to a different carrier protein. The contraceptive efficacy of synthetic peptide corresponding to Izumo was enhanced when the animals were co-immunized with synthetic peptides corresponding to FA-1, SP56, and YLP12 conjugated with various carrier proteins, resulting in an overall 73.33% reduction in fertility. When the antibodies against the peptides disappeared after >9–10 months from circulation and genital tract, all the animals regained fertility. They suggested that the immunization with Izumo and other sperm peptides induces antibodies in serum and genital tract that cause a reversible long-term contraceptive effect in female mice.

Hardy et al. [143] administered multi-antigen recombinant polypeptides comprising the mouse reproductive antigens SP56, ZP3, ZP2, and ZP1 to female mice. The MAPs contained three tandem repeats each of ZP2 and ZP3 peptide epitopes and a single copy of a ZP1 peptide sequence all of which had previously been demonstrated to individually have immunodominant or contraceptive effects. Female BALB/c mice actively immunized with these antigens in Freund's adjuvants produced variable serum antibody responses to the component peptides. Fertility rates for animals immunized were significantly reduced (20–40%) compared to the control mice (90%), but the reduction in fertility did not correlate with peptide-specific serum antibody levels. They suggested that high levels of immunocontraception in mice can be achieved without apparent ovarian pathology, using species-specific immunogens that include repeated peptides from proteins involved in fertilization.

Choudhury et al. [144] also immunized female mice with viruslike particles (VLPs) derived from Johnsongrass mosaic virus coat protein expressing YLP12 as a fusion peptide with ZP3 peptide or physical mixture of VLPs presenting either YLP12 or ZP3 epitope led to the curtailment of fertility. For safer and effective immunocontraception, zona (ZP3)- and sperm-specific (YLP12) peptides have been presented on VLPs. Immunization of FvB/cJ female mice with VLPs presenting YLP12-ZP3 fusion peptide and a physical mixture of VLPs presenting either YLP12 or ZP3 epitope led to generation of specific antibody responses and a significant reduction in litters born per mice ($p < 0.005$). Significant curtailment (50%) of fertility was also observed in animals immunized with adjuvnated ZP3 and YLP12 synthetic peptides. They suggested that VLPs can be used to present gamete epitopes for immunocontraception.

13.1.6   Monoclonal Antibody-Based Contraception and Multipurpose Prevention Technology (MPT)

Sexually transmitted infections are highly prevalent, and over 40% of pregnancies are unplanned. New mAb-based multipurpose prevention technology (MPT) products have been engineered to address these problems and fill an unmet need in female reproductive health [91]. Anderson et al. [91] used a *Nicotiana* platform to manufacture mAbs against two prevalent sexually transmitted pathogens, HIV-1 and HSV-2, and incorporated them into a vaginal film (MB66). Preclinical studies indicated that the antibodies in MB66 remain active under low pH genital tract conditions and effectively prevent macaques from vaginal SHIV transmission [145]. They also conducted a Phase I clinical trial to assess the safety, pharmacokinetics (PK), and ex vivo efficacy of single and repeated doses of MB66 when used intravaginally by enrolling healthy reproductive-aged, sexually abstinent women [146]. The product was generally safe and well tolerated, with no serious adverse events. Acceptability and willingness to use the product were judged to be high by post-use surveys. Concentrations of VRC01-N and HSV8-N in vaginal secretions were assessed over time. Antibody levels peaked 1 h postdosing with active film and remained significantly elevated at 24 h post first and seventh film. Cervicovaginal lavage samples collected 24 h after MB66 insertion significantly neutralized both HIV-1 and HSV-2 ex vivo.

They added a monoclonal ASA to MB66 to provide a contraception option for women [147]. The human IgM ASA, H6-3C4, originally isolated by Isojima et al. [72] from the blood of an infertile woman, recognizes a carbohydrate epitope on CD52, a glycosylphosphatidylinositol-anchored glycoprotein found in abundance on the surface of human sperm. They engineered the antibody for production in *Nicotiana* and called the new antibody as "human contraception antibody (HCA)," which effectively agglutinates sperm and maintains activity. In brief, heavy and light chain variable region DNA sequences of H6-3C4 were inserted with human IgG$_1$ constant region sequences into an agrobacterium and transfected into *Nicotiana benthamiana*. The product, an IgG$_1$ mAb (HCA), was purified on Protein A columns. HCA was obtained at concentrations ranging from 0.4 to 4 mg/mL and consisted of >90% IgG monomers. The mAb specifically reacted with a glycan epitope on CD52, a glycoprotein produced in the male reproductive tract and found in abundance on sperm. HCA potently agglutinated sperm under a variety of relevant physiological conditions at concentrations $\geqq 6.25$ µg/mL, and mediated complement-dependent sperm immobilization at concentrations $\geqq 1$ µg/mL. HCA and its immune complexes did not induce inflammation in a human vaginal tissue (EpiVaginal™) model. They suggested that HCA, an IgG1 mAb with potent sperm agglutination and immobilization activity and a good safety profile, is a promising candidate for female contraception.

References

1. Tuckwell HC, Koziol JA. World population. Nature. 1992;359:200.
2. Adoyo PA, Koyama K. Prospects for sperm immunocontraceptive vaccines. Reprod Immunol Biol. 2003;18:7–18.
3. Editorial. Have spermatozoa functions or effects other than fertilization? JAMA. 1921;77:42–3.
4. Rosenfeld SS. Semen injection with serologic studies: a preliminary report. Am J Obstet Gynecol. 1926;12:385–7.
5. Baskin MJ. Temporary sterilization by the injection of human spermatozoa. A preliminary report. Am J Obstet Gynecol. 1932;24:892–7.
6. Diekman AB, Herr JC. Sperm antigens and their use in the development of an immunocontraceptive. Am J Reprod Immunol. 1997;37:111–7.
7. Otani Y, Ino H, Inoue S, Kagami T. Immunization of human female with human sperm and semen. Int J Fertil. 1971;16:19–23.
8. Bemstein L, Englander B, Gallagher J, Nathan P, Marcus Z. Localized and systemic hypersensitivity reactions to human seminal fluid. Ann Intern Med. 1981;94:459–65.
9. Joel CA. Historical survey of research on spermatozoa from antiquity to the present. In: Joel CA, editor. Fertility disturbances in men and women. Basel: Karger; 1971. p. 3–13.
10. McCartney LJ. Am J Phys. 1923;63:207.
11. Henle W, Henle G. J Immunol. 1940;38:105.
12. Freund J, Thompson GE, Lipton MM. Aspermatogenesis, anaphylaxis, and cutaneous sensitization induced in the guinea pig by homologous testicular extract. J Exp Med. 1955;101:591–604.
13. Isojima S, Graham RM, Graham JB. Sterility in female guinea pigs induced by injection with testis. Science. 1959;129:44.
14. Katsh S. Infertility in female guinea pigs induced by injection of homologous sperm. Am J Obstet Gynecol. 1959;78:276–8.
15. Otani Y, Behrman SJ, Porter CW, Nakayama M. Reduction of fertility in immunized guinea pigs. Int J Fertil. 1963;8:835–9.
16. McLaren A. Immunological controls of fertility in female mice. Nature. 1964;201:582–5.
17. Edwards RG. Immunological control of fertility in female mice. Nature. 1964;203:50–3.
18. Gupta SK, Shrestha A, Minhas V. Milestones in contraceptive vaccines development and hurdles in their application. Hum Vaccin Immunother. 2014;10:911–25.
19. Herr JC. Update on the center for recombinant gamete contraceptive vaccinogens. Am J Reprod Immunol. 1996;35:184–9.
20. Koyama K, Ikuma K, Kubota K, Isojima S. Effects of antisperm antibodies on sperm migration through cervical mucus. Excertta Med Int Congr Ser. 1980;512:705–8.
21. Shibahara H, Shiraishi Y, Hirano Y, Kasumi H, Koyama K, Suzuki M. Relationship between level of serum sperm immobilizing antibody and its inhibitory effect on sperm migration through cervical mucus in immunologically infertile women. Am J Reprod Immunol. 2007;57:142–6.
22. Shibahara H, Shigeta M, Toji H, Koyama K. Sperm immobilizing antibodies interfere with sperm migration from the uterine cavity through the fallopian tubes. Am J Reprod Immunol. 1995;34:120–4.

23. Nip MM, Taylor PV, Rutherford AJ, Hanecock KW. Autoantibodies and antisperm antibodies in sera and follicular fluids of infertile patients; relation to reproductive outcome after in-vitro fertilization. Hum Reprod. 1995;10:2564–9.
24. Marin-Briggiler CI, Vazquez-Levin MH, Gonzalez-Echeverria F, Blaquier JA, Miranda PV, Tezon JG. Effect of antisperm antibodies present in human follicular fluid upon the acrosome reaction and sperm-zona pellucida interaction. Am J Reprod Immunol. 2003;50:209–19.
25. Vujisic S, Lepej SZ, Jerkovic L, Emedi I, Sokolic B. Antisperm antibodies in semen, sera and follicular fluids of infertile patients: relation to reproductive outcome after in vitro fertilization. Am J Reprod Immunol. 2005;54:13–20.
26. Shibahara H, Hirano Y, Shiraishi Y, et al. Effects of in vivo exposure to eggs with sperm-immobilizing antibodies in follicular fluid on subsequent fertilization and embryo development in vitro. Reprod Med Biol. 2006;5:137–43.
27. Taneichi A, Shibahara H, Hirano Y, Suzuki T, Obara H, Fujiwara H, Takamizawa S, Sato I. Sperm immobilizing antibodies in the sera of infertile women cause low fertilization rates and poor embryo quality in vitro. Am J Reprod Immunol. 2002;47:46–51.
28. Benoff S, Cooper GW, Huriey I, et al. Antisperm antibody binding to human sperm inhibits capacitation induced changes in the levels of plasma membrane sterols. Am J Reprod Immunol. 1993;30:113–30.
29. Coddington CC, Alexander NJ, Fulgham D, et al. Hemizona assay (HZA) demonstrates effects of characterized mouse antihuman sperm antibodies on sperm zona binding. Andrologia. 1992;24:271–7.
30. Zouari R, De Almeida M, Feneux D. Effect of sperm-associated antibodies on the dynamics of sperm movement and on the acrosome reaction of human spermatozoa. J Reprod Immunol. 1992;22:59–72.
31. Bandoh R, Yamano S, Kamada M, Daitoh T, Aono T. Effect of sperm-immobilizing antibodies on the acrosome reaction of human spermatozoa. Fertil Steril. 1992;57:387–92.
32. Bronson RA, Cooper GW, Rosenfeld DL. Sperm-specific isoantibodies and autoantibodies inhibit the binding of human sperm to the human zona pellucida. Fertil Steril. 1982;38:724–9.
33. Alexander NJ. Antibodies to human spermatozoa impede sperm penetration of cervical mucus or hamster eggs. Fertil Steril. 1984;41:433–9.
34. Kamada M, Daitoh T, Hasebe H, Irahara M, Yamano S, Mori T. Blocking of human fertilization in vitro by sera with sperm-immobilizing antibodies. Am J Obstet Gynecol. 1985;153:328–31.
35. Mahony MC, Blackmore PF, Bronson RA, et al. Inhibition of human sperm-zona pellucida tight binding in the presence of antisperm antibody positive polyclonal patient sera. J Reprod Immunol. 1991;19:287–301.
36. Tsukui S, Noda Y, Fukuda A, et al. Blocking effect of sperm immobilizing antibodies on sperm penetration of human zonae pellucida. J In Vitro Fert Embryo Transf. 1988;5:123–8.
37. Shibahara H, Shigeta M, Koyama K, Burkman LJ, Alexander NJ, Isojima S. Inhibition of sperm-zona pellucida tight binding by sperm immobilizing antibodies as assessed by the hemizona assay (HZA). Acta Obstet Gynaecol Jpn. 1991;43:237–8.
38. Shibahara H, Burkman LJ, Isojima S, Alexander NJ. Effects of sperm-immobilizing antibodies on sperm-zona pellucida tight binding. Fertil Steril. 1993;60:533–9.
39. Mahony MC, Fulgham DL, Blackmore PF, Alexander NJ. Evaluation of human sperm-zona pellucida tight binding by presence of monoclonal antibodies to sperm antigens. J Reprod Immunol. 1991;19:269–85.
40. Tung KSK, Okada A, Yanagimachi R. Sperm autoantigens and fertilization. I. Effects of antisperm autoantibodies on rouleaux formation, viability, and acrosome reaction of guinea pig spermatozoa. Biol Reprod. 1980;23:171–83.
41. Bronson RG, Cooper G, Rosenfeld D. Sperm antibodies: their role in infertility. Fertil Steril. 1984;42:171–83.

42. D'Cruz OJ, Haas GG, Wang B, DeBault LE. Activation of human complement by IgG antisperm antibody and the demonstration of C3 and C5b-p-mediated immune injury to human sperm. J Immunol. 1991;146:611–20.
43. Hjort T, Griffin PD. The identification of candidate antigens for the development of birth control vaccines. J Reprod Immunol. 1985;8:271–8.
44. Aitken RJ, Parslow JM, Hargreave TB, Hendry WE. Influence of antisperm antibodies on human sperm function. Br J Urol. 1988;62:367–73.
45. D'Cruz OJ, Haas GG, Lambert H. Heterogeneity of human sperm surface antigens identified by indirect immunoprecipitation of antisperm antibody bound to biotinylated sperm. J Immunol. 1993;151:1062–74.
46. Lee C-YG, Lum V, Wong E, Menge A, Huang Y-S. Identification of human sperm antigens to antisperm antibodies. Am J Reprod Immunol. 1983;3:183–7.
47. Lehmann D, Temminck B, Rugna DD, Leibundgut B, Miiller H. Blot-immunoblotting test for the detection of anti-sperm antibodies. J Reprod Immunol. 1985;8:329–36.
48. Naaby-Hansen S, Bjerrum OJ. Auto- and iso-antigens of human spermatozoa detected by immunoblotting with human sera after SDS-PAGE. J Reprod Immunol. 1985;7:41–57.
49. Naz RK, Gateva E, Morte C. Autoantigenicity of human sperm: molecular identities of sperm antigens recognized by sera of immunoinfertile men in the immunoprecipitation procedure. Arch Androl. 1995;35:225–31.
50. Primakoff P, Lathrop W, Bronson R. Identification of human sperm surface glycoproteins recognized by autoantisera from immune infertile men, women, and vasectomized men. Biol Reprod. 1990;42:929–42.
51. Saji F, Ohashi K, Kamiura S, Negoro T, Tanizawa O. Identification and characterization of a human sperm antigen corresponding to sperm-immobilizing antibodies. Am J Reprod Immunol. 1988;17:128–33.
52. Shai S, Naot Y. Identification of human sperm antigens reacting with antisperm antibodies from sera and genital tract secretions. Fertil Steril. 1992;58:593–8.
53. Snow K, Ball GD. Characterization of human sperm antigens and antisperm antibodies in infertile patients. Fertil Steril. 1992;58:1011–9.
54. Wingate EW, Patrick RT, Mathur S. Antigens in capacitated spermatozoa eliciting autoimmune responses. J Urol. 1993;149:1331–7.
55. Koyama K, Kameda K, Nakamura N, Kubota K, Shigeta M, Isojima S. Recognition of carbohydrate antigen epitopes by sperm-immobilizing antibodies in sera of infertile women. Fertil Steril. 1991;56:954–9.
56. Diekman AB, Goldberg E. Characterization of a human antigen with sera from infertile patients. Biol Reprod. 1984;50:1087–93.
57. Liang ZG, O'Hem PA, Yavetz H, Yavetz B, Goldberg E. Human testis cDNAs identified by sera from infertile patients: a molecular biological approach to immunocontraceptive development. Reprod Fertil Dev. 1994;6:297–305.
58. Liu QY, Wang LF, Miao SY, Catterall JF. Expression and characterization of a novel human sperm membrane protein. Biol Reprod. 1996;54:323–30.
59. Young RA, Davis RW. Efficient isolation of genes by using antibody probes. Proc Natl Acad Sci U S A. 1983;80:1194–8.
60. Naaby-Hansen S, Flickinger CJ, Herr JC. Two-dimensional gel electrophoretic analysis of vectorially labeled surface proteins of human spermatozoa. Biol Reprod. 1997;56:771–87.
61. Shibahara H, Sato I, Shetty J, Naaby-Hansen S, Herr JC, Wakimoto E, Koyama K. Two-dimensional electrophoretic analysis of sperm antigens recognized by sperm immobilizing antibodies detected in infertile women. J Reprod Immunol. 2002;53:1–12.
62. Shetty J, Naaby-Hansen S, Shibahara H, Bronson R, Flickinger CJ, Herr JC. Human sperm proteome: immunodominant sperm surface antigens identified with sera from infertile men and women. Biol Reprod. 1999;61:61–9.

63. Shetty J, Bronson RA, Herr JC. Human sperm protein encyclopedia and alloantigen index: mining novel allo-antigens using sera from ASA-positive infertile patients and vasectomized men. J Reprod Immunol. 2008;77:23–31.

64. Wolkowicz MJ, Shetty J, Westbrook A, Klotz K, Jayes F, Mandal A, Flickinger CJ, Herr JC. Equatorial segment protein defines a discrete acrosomal subcompartment persisting throughout acrosomal biogenesis. Biol Reprod. 2003;69:735–45.

65. Hao Z, Wolkowicz MJ, Shetty J, Klotz K, Bolling L, Sen B, Westbrook VA, Coonrod S, Flickinger CJ, Herr JC. SAMP32, a testis-specific, isoantigenic sperm acrosomal membrane-associated protein. Biol Reprod. 2002;66:735–44.

66. Shetty J, Wolkowicz MJ, Digilio LC, Klotz KL, Jayes FL, Diekman AB, Westbrook VA, Farris EM, Hao Z, Coonrod SA, Flickinger CJ, Herr JC. SAMP14, a novel, acrosomal membrane-associated, glycosylphosphatidylinositol-anchored member of the Ly-6/urokinase-type plasminogen activator receptor superfamily with a role in sperm–egg interaction. J Biol Chem. 2003;278:30506–15.

67. Mandal A, Naaby-Hansen S, Wolkowicz MJ, Klotz K, Shetty J, Retief JD, Coonrod SA, Kinter M, Sherman N, Cesar F, Flickinger CJ, Herr JC. FSP95, a testis-specific 95-kilodalton fibrous sheath antigen that undergoes tyrosine phosphorylation in capacitated human spermatozoa. Biol Reprod. 1999;61:1184–97.

68. Naaby-Hansen S, Mandal A, Wolkowicz MJ, Sen B, Westbrook VA, Shetty J, Coonrod SA, Klotz KL, Kim YH, Bush LA, Flickinger CJ, Herr JC. CABYR, a novel calcium-binding tyrosine phosphorylation-regulated fibrous sheath protein involved in capacitation. Dev Biol. 2002;242:236–54.

69. Shetty J, Klotz KL, Wolkowicz MJ, Flickinger CJ, Herr JC. Radial spoke protein 44 (human meichroacidin) is an axonemal alloantigen of sperm and cilia. Gene. 2007;396:93–107.

70. Lee C-YG, Wong E, Richter DE, Menge AC. Monoclonal antibodies to human sperm antigens II. J Reprod Immunol. 1984;6:227–38.

71. Anderson DJ, Johnson PM, Alexander NJ, Jones WR, Griffin PD. Monoclonal antibodies to human trophoblast and sperm antigens: report of two WHO-sponsored workshops, June 30, 1986-Toronto, Canada. J Reprod Immunol. 1987;10:231–57.

72. Isojima S, Kameda K, Tsuji Y, Shigeta M, Ikeda Y, Koyama K. Establishment and characterization of a human hybridoma secreting monoclonal antibody with high titers of sperm immobilizing and agglutinating activities against human seminal plasma. J Reprod Immunol. 1987;10:67–78.

73. Kameda K, Tsuji Y, Koyama K, Isojima S. Comparative studies of the antigens recognized by sperm-immobilizing monoclonal antibodies. Biol Reprod. 1992;46:349–57.

74. Tsuji Y, Clausen H, Nudelman E, Kaizu T, Hakomori S, Isojima S. Human sperm carbohydrate antigens defined by an antisperm human monoclonal antibody derived from an infertile woman bearing antisperm antibodies in her serum. J Exp Med. 1988;168:343–56.

75. Diekman AB, Norton EJ, Klotz KL, Westbrook VA, Shibahara H, Naaby-Hansen S, Flickinger CJ, Herr JC. N-linked glycan of a sperm CD52 glycoform associated with human infertility. FASEB J. 1999;13:1303–13.

76. Kirchhoff C, Schroter S. New insights into the origin, structure and role of CD52: a major component of the mammalian sperm glycocalyx. Cells Tissues Organs. 2001;168:93–104.

77. Hale G, Xia MQ, Tighe HP, Dyer MJ, Waldmann H. The CAMPATH-1 antigen (CDw52). Tissue Antigens. 1990;35:118–27.

78. Hale G, Rye PD, Warford A, Lauder I, Brito-Babapulle A. The glycosylphosphatidylinositol-anchored lymphocyte antigen CDw52 is associated with the mrt maturation of human spermatozoa. J Reprod Immunol. 1993;23:189–205.

79. Diekman AB, Westbrook-Case VA, Naaby-Hansen S, Klotz KL, Flickinger CJ, Herr JC. Biochemical characterization of sperm agglutination antigen-1, a human sperm antigen implicated in gamete interactions. Biol Reprod. 1997;57:1136–45.

80. Mahony MC, Fulgham DL, Blackmore PF, Alexander NJ. Evaluation of human sperm-zona pellucida tight binding by presence of monoclonal antibodies to sperm antigens. J Reprod Immunol. 1991;1:269–85.
81. Isojima S, Tsuji Y, Kameda K, Shigeta M, Hakomori S. Coating antigens of human seminal plasma involved in fertilization. In: Alexander NJ, Griffin D, Spieler JM, Waites GMH, editors. Gamete interaction: prospects for immunocontraception. New York: Wiley-Liss, Inc.; 1990. p. 63–74.
82. Hasegawa A, Fu Y, Tsubamoto H, Tsuji Y, Sawai H, Komori S, Koyama K. Epitope analysis for human sperm-immobilizing monoclonal antibodies, MAb H6-3C4, 1G12 and campath-1. Mol Hum Reprod. 2003;9:337–43.
83. Burkman LJ, Coddmgton CC, Franken DR, Krugen TF, Rosenwaks Z, Hogen GD. The hemizona assay (HZA)- development of a diagnostic test for the binding of human spermatozoa to the human hemizona pellucida to predict fertilization potential. Fertil Steril. 1988;49: 688–97.
84. Yanagimachi R, Yanagimach H, Rogers BJ. The use of zona-free animal ova as a test-system for the assessment of the fertilizing capacity of human spermatozoa. Biol Reprod. 1976;15: 471–6.
85. Yanagimachi R, Lopata A, Odom CB, Bronson RA, Mahi CA. Penetration of biologic characteristics of zona pellucida in highly concentrated salt solution the use of salt stored eggs for assessing the fertilizing capacity of spermatozoa. Fertil Steril. 1979;35:562–74.
86. Shibahara H, Shigeta M, Inoue K, Hasegawa A, Koyama K, Alexander NJ, Isojima S. Diversity of the blocking effects of antisperm antibodies on fertilization in human and mouse. Hum Reprod. 1996;11:2595–9.
87. Hasegawa A, Fu Y, Koyama K. Nasal immunization with diphtheria toxoid conjugated-CD52 core peptide induced specific antibody production in genital tract of female mice. Am J Reprod Immunol. 2002;48:305–11.
88. Hasegawa A, Koyama K. Antigenic epitope for sperm—immobilizing antibody detected in infertile women. J Reprod Immunol. 2005;67:77–86.
89. Koyama K, Hasegawa A, Komori S. Functional aspects of CD52 in reproduction. J Reprod Immunol. 2009;83:56–9.
90. Hardiyanto L, Hasegawa A, Komori S. The N-linked carbohydrate moiety of male reproductive tract CD52 (mrt-CD52) interferes with the complement system via binding to C1q. J Reprod Immunol. 2012;94:142–50.
91. Anderson DJ, Politch JA, Cone RA, Zeitlin L, Lai SK, Santangelo PJ, Moench TR, Whaley KJ. Engineering monoclonal antibody-based contraception and multipurpose prevention technologies. Biol Reprod. 2020;103:275–85.
92. O'Rand MG, Romrell LJ. Appearance of cell surface auto and isoantigens during spermatogenesis in the rabbit. Dev Biol. 1977;55:347–58.
93. Richardson RT, Widgren EE, O'Rand MG. Molecular comparison of two sperm autoantigens. In: Dondero F, Johnson PM, editors. Serono symposia publications: reproductive immunology. New York: Raven Press; 1993. p. 41–6.
94. Richardson RT, Yamasaki N, O'Rand MG. Sequence of a rabbit sperm zona pellucida binding protein and localization during the acrosome reaction. Dev Biol. 1994;165:688–701.
95. Kong M, Richardson RT, Widgren EE, O'Rand MG. Sequence and localization of the mouse sperm autoantigenic protein, Sp17. Biol Reprod. 1995;53:579–90.
96. Lea I, Richardson RT, Widgren EE, O'Rand MG. Cloning and sequencing of human Sp17, a sperm zona binding protein. Mol Biol Cell. 1993;4:248a.
97. Yamasaki N, Richardson RT, O'Rand MG. Expression of the rabbit sperm protein Sp17 in COS cells and interaction of recombinant Sp17 with the rabbit zona pellucida. Mol Reprod Dev. 1995;40:48–55.
98. O'Rand MG, Widgren EE. Identification of sperm antigen targets for immunocontraception: B-cell epitope analysis of Sp17. Reprod Fertil Dev. 1994;6:289–96.

99. Frayne J, Hall L. A re-evaluation of sperm protein 17 (Sp17) indicates a regulatory role in an A-kinase anchoring protein complex, rather than a unique role in sperm-zona pellucida binding. Reproduction. 2002;124:767–74.

100. Lea IA, van Lierop MJ, Widgren EE, Grootenhuis A, Wen Y, van Duin M, O'Rand MG. A chimeric sperm peptide induces antibodies and strain-specific reversible infertility in mice. Biol Reprod. 1998;59:527–36.

101. Herr JC, Flickinger CJ, Homyk M, Klotz K, John E. Biochemical and morphological characterization of the intraacrosomal antigen SP-10 from human sperm. Biol Reprod. 1990;42:181–93.

102. Wright RM, Sun AK, Komreich B, Flickinger CJ, Herr JC. Cloning and characterization of the gene coding for the human acrosomal protein SP-10. Biol Reprod. 1993;49:316–25.

103. Herr JC, Klotz K, Shannon J, Wright RM, Flickinger CJ. Purification and microsequencing of the intra-acrosomal protein SP-10. Evidence that SP- 10 heterogeneity results from endoproteolytic processes. Biol Reprod. 1992;47:11–20.

104. Foster JA, Herr JC. Interactions of human sperm acrosomal protein SP- 10 with the acrosomal membranes. Biol Reprod. 1992;46:981–90.

105. Coonrod SA, Westusin ME, Herr JC. Inhibition of bovine fertilization in vitro by antibodies to SP-10. J Reprod Fertil. 1996;107:287–97.

106. Srinivasan J, Tinge S, Wright R, Herr JC, Curtiss R III. Oral immunization with attenuated *Salmonella* expressing human sperm antigen induces antibodies in serum and the reproductive tract. Biol Reprod. 1995;53:462–571.

107. Goyal S, Manivannan B, Kumraj GR, Ansari AS, Lohiya NK. Evaluation of efficacy and safety of recombinant sperm-specific contraceptive vaccine in albino mice. Am J Reprod Immunol. 2013;69:495–508.

108. Primakoff P, Myles DG. A map of the guinea pig sperm surface constructed with monoclonal antibodies. Dev Biol. 1983;98:417–28.

109. Primakoff P, Hyatt H, Myles DG. A role for the migrating sperm surface antigen PH-20 in guinea pig sperm binding to the egg zona pellucida. J Cell Biol. 1985;101:2239–44.

110. Myles DG, Primakoff P. Sperm proteins that serve as receptors for the zona pellucida and their post-testicular modification. Ann N Y Acad Sci. 1991;637:486–91.

111. Lin Y, Mahan K, Lathrop WF, Myles DG, Primakoff P. A hyaluronidase activity of the sperm plasma membrane protein PH-20 enables sperm to penetrate the cumulus cell layer surrounding the egg. J Cell Biol. 1994;125:1157–63.

112. Primakoff P, Lathrop W, Woolman L, Cowan A, Myles D. Fully effective contraception in male and female guinea pigs immunized with the sperm protein PH-20. Nature. 1998;335:543–6.

113. Lin Y, Kimmel LH, Myles DG, Primakoff P. Molecular cloning of the human and monkey sperm surface protein PH-20. Proc Natl Acad Sci U S A. 1993;90:10071–5.

114. Primakoff P, Hyatt H, Tredick-Kline J. Identification and purification of a sperm surface protein with a potential role in sperm-egg membrane fusion. J Cell Biol. 1987;104:141–9.

115. Myles DG, Kimmel LH, Blobel CP, White JM, Primakoff P. Identification of a binding site in the disintegrin domain of fertilin required for sperm-egg fusion. Proc Natl Acad Sci U S A. 1994;91:4195–8.

116. Blobel CP, Wolfsberg TG, Turck CW, Myles DG, Primakoff P, White JM. A potential fusion peptide and an integrin ligand domain in a protein active in sperm-egg fusion. Nature. 1992;356:248–52.

117. Wolfsberg TG, Bazan JF, Blobel CP, Myles DG, Primakoff P, White JM. The precursor region of a protein active in sperm-egg fusion contains a metalloprotease and a disintegrin domain: structural, functional, and evolutionary implications. Proc Natl Acad Sci U S A. 1993;90:10738–87.

118. Ramarao CS, Myles DG, White JM, Primakoff P. Initial evaluation of fertilin as an immunocontraceptive antigen and molecular cloning of the cynomolgus monkey fertilin β subunit. Mol Reprod Dev. 1996;43:70–5.

119. Vidaeus CM, Kap-Herr C, Golden W, Eddy RL, Shows TB, Herr JC. Cloning and characterization of a candidate contraceptive vaccinogen. Am J Reprod Immunol. 1996;35:466.
120. Hardy CM, Clarke HG, Nixon B, Grigg JA, Hinds LA, Holland MK. Examination of the immunocontraceptive potential of recombinant rabbit fertilin subunits in rabbit. Biol Reprod. 1997;57:879–86.
121. Goldberg E. Lactic and malic dehydrogenases in human spermatozoa. Science. 1963;139:602–3.
122. Vickram AS, Dhama K, Chakraborty S, Samad HA, Latheef SK, Sharun K, Khurana SK, Archana K, Tiwari R, Bhatt P, Vyshali K, Chaicumpa W. Role of antisperm antibodies in infertility, pregnancy, and potential for contraceptive and antifertility vaccine designs: research Progress and pioneering vision. Vaccines (Basel). 2019;7:116. https://doi.org/10.3390/vaccines7030116.
123. Odet F, Duan C, Willis WD, Goulding EH, Kung A, Eddy EM, Goldberg E. Expression of the gene for mouse lactate dehydrogenase C (Ldhc) is required for male fertility. Biol Reprod. 2008;79:26–34.
124. Shelton JA, Goldberg E. Local reproductive tract immunity to sperm-specific lactate dehydrogenase-C4. Biol Reprod. 1986;35:873–6.
125. Mahi-Brown CA, VandeVoort CA, McGuinness RP, Overstreet JW, O'Hern P, Goldberg E. Immunization of male but not female mice with the sperm-specific isozyme of lactate dehydrogenase (LDH-C4) impairs fertilization in vivo. Am J Reprod Immunol. 1990;24:1–8.
126. O'Hern PA, Liang Z-G, Bambra CS, Goldberg E. Colinear synthesis of an antigen-specific B-cell epitope with a 'promiscuous' tetanus toxin T-cell epitope: a synthetic peptide immunocontraceptive. Vaccine. 1997;15:1761–6.
127. Tollner TL, Overstreet JW, Branciforte D, Primakoff PD. Immunization of female cynomolgus macaques with a synthetic epitope of sperm-specific lactate dehydrogenase results in high antibody titers but does not reduce fertility. Mol Reprod Dev. 2002;62:257–64.
128. Naz RK. Immunocontraceptive effect of Izumo and enhancement by combination vaccination. Mol Reprod Dev. 2008;75:336–44.
129. Naz RK, Wolf DP. Antibodies to sperm-specific human FA-1 inhibit in vitro fertilization in rhesus monkeys: development of a simian model for testing of anti-FA-1 contraceptive vaccine. J Reprod Immunol. 1994;27:111–21.
130. Kadam AL, Fateh M, Naz RK. Fertilization antigen (FA-1) completely blocks human sperm binding to human zona pellucida: FA-1 antigen may be a sperm receptor for zona pellucida in humans. J Reprod Immunol. 1995;29:19–30.
131. Naz RK. Effect of sperm DNA vaccine on fertility of female mice. Mol Reprod Dev. 2006;73:918–28.
132. Naz RK. Effect of fertilization antigen (FA-1) DNA vaccine on fertility of female mice. Mol Reprod Dev. 2006;73:1473–9.
133. Zhu X, Naz RK. Fertilization antigen-1: cDNA cloning, testis-specific expression, and immunocontraceptive effects. Proc Natl Acad Sci U S A. 1997;94:4704–9.
134. Coonrod SA, Westhusin ME, Naz RK. Monoclonal antibody to human fertilization antigen-1 (FA-1) inhibits bovine fertilization in vitro: application in immunocontraception. Biol Reprod. 1994;51:14–23.
135. Naz RK, Zhu X. Recombinant fertilization antigen-1 causes a contraceptive effect in actively immunized mice. Biol Reprod. 1998;59:1095–100.
136. Naz RK, Zhu X, Kadam AL. Identification of human sperm peptide sequence involved in egg binding for immunocontraception. Biol Reprod. 2000;62:318–24.
137. Naz RK, Chauhan SC. Human sperm-specific peptide vaccine that causes long-term reversible contraception. Biol Reprod. 2002;67:674–80.
138. Naz RK, Chauhan SC, Trivedi RN. Monoclonal antibody against human sperm-specific YLP 12 peptide sequence involved in oocyte binding. Arch Androl. 2002;48:169–75.
139. Inoue N, Ikawa M, Isotani A, Okabe M. The immunoglobulin superfamily protein Izumo is required for sperm to fuse with eggs. Nature. 2005;434:234–8.

140. Wang M, Lv Z, Shi J, Hu Y, Xu C. Immunocontraceptive potential of the Ig-like domain of Izumo. Mol Reprod Dev. 2009;76:794–801.

141. Xue F, Wang L, Liu Y, Tang H, Xu W, Xu C. Vaccination with an epitope peptide of IZUMO1 to induce contraception in female mice. Am J Reprod Immunol. 2016;75:474–85.

142. Kurth BE, Digilio L, Snow P, Bush LA, Wolkowicz M, Shetty J, Mandal A, Hao Z, Reddi PP, Flickinger CJ, Herr JC. Immunogenicity of a multi-component recombinant human acrosomal protein vaccine in female Macaca fascicularis. J Reprod Immunol. 2008;77:126–41.

143. Hardy CM, Beaton S, Hinds LA. Immunocontraception in mice using repeated, multi-antigen peptides: immunization with purified recombinant antigens. Mol Reprod Dev. 2008;75:126–35.

144. Choudhury S, Kakkar V, Suman P, Chakrabarti K, Vrati S, Gupta SK. Immunogenicity of zona pellucida glycoprotein-3 and spermatozoa YLP(12) peptides presented on Johnson grass mosaic virus-like particles. Vaccine. 2009;27:2948–53.

145. Anderson DJ, Politch JA, Zeitlin L, Hiatt A, Kadasia K, Mayer KH, Ruprecht RM, Villinger F, Whaley KJ. Systemic and topical use of monoclonal antibodies to prevent the sexual transmission of HIV. AIDS. 2017;31:1505–17.

146. Politch JA, Cu-Uvin S, Moench TR, Tashima KT, Marathe JG, Guthrie KM, Cabral H, Nyhuis T, Brennan M, Zeitlin L, Spiegel HML, Mayer KH, Whaley KJ, Anderson DJ. Safety, acceptability, and pharmacokinetics of a monoclonal antibody-based vaginal multipurpose prevention film (MB66): A Phase I randomized trial. PLoS Med. 2021;18(2): e1003495. https://doi.org/10.1371/journal.pmed.1003495.

147. Baldeon-Vaca G, Marathe JG, Politch JA, Mausser E, Pudney J, Doud J, Nador E, Zeitlin L, Pauly M, Moench TR, Brennan M, Whaley KJ, Anderson DJ. Production and characterization of a human antisperm monoclonal antibody against CD52g for topical contraception in women. EBioMedicine. 2021;69:103478. https://doi.org/10.1016/j.ebiom.2021.103478.

Chapter 14
Development of Immunocontraceptives in Male

Hiroaki Shibahara

Abstract The currently applicable two forms of male contraception are condoms and the vasectomies. Despite ongoing efforts to develop male contraceptive options, to date, highly effective and reversible male contraceptive methods are not available.

Some sperm antigens, which have been regarded as useful in the development of contraceptive vaccines on the female side, are also expected to be contraceptive vaccines on the male side. Besides them, there are also some proteins identified for male contraception, which we will introduce here. These proteins play important biological roles, such as sperm motility and various processes at fertilization. A variety of sperm-specific proteins have been identified, and the characteristics of these proteins and their efficacy to inhibit fertility have been investigated in suitable animal models.

So far, however, no contraceptive vaccine based on sperm-specific proteins has gone through preclinical safety evaluation in animal models and thus not entered into the clinical trials in humans. Further scientific inputs and rigorous investigations are required to propose a candidate contraceptive vaccine based on sperm-specific antigens for human use.

14.1 Introduction

Unintended pregnancy is a global public health problem. The two currently available forms of male contraception are condoms and the vasectomies. Condoms have a high failure rate, 13% with regular use [1], while vasectomies are invasive and are not reliably reversible [2]. Despite ongoing efforts to develop male contraceptive options, to date, highly effective and reversible male contraceptive methods are unavailable.

H. Shibahara (✉)
Department of Obstetrics and Gynecology, School of Medicine, Hyogo Medical University, Nishinomiya, Hyogo, Japan
e-mail: sibahara@hyo-med.ac.jp

© The Editor(s) (if applicable) and The Author(s), under exclusive license to
Springer Nature Singapore Pte Ltd. 2022
H. Shibahara, A. Hasegawa (eds.), *Gamete Immunology*,
https://doi.org/10.1007/978-981-16-9625-1_14

Nonhormonal male contraceptive approaches include physically blocking sperm passage through the male reproductive tract (vaso-occlusion), altering sperm motility, and interrupting intratesticular sperm maturation, among others [3]. Some sperm antigens, which have been regarded as useful in the development of contraceptive vaccines on the female side, are also expected to be contraceptive vaccines on the male side. Here, some candidate proteins for the development of contraceptive vaccine in male are introduced. For the details of the characteristics of some identical proteins, please refer to Chap. 16.

14.2   Potential Targets for a Contraceptive Vaccine in Male

To develop immunocontraceptives for male, a variety of sperm-specific proteins have been identified and characterized, and their potential to inhibit fertility was evaluated in suitable animal models. The characteristics of some of these proteins and their efficacy to inhibit fertility are described below (Table 14.1).

Table 14.1 Candidate sperm antigens for the development of contraceptive vaccines in male

Antigen	Localization	Molecular weight (kDa)	Biological activity/immunization effect
SP-10	Inner and outer acrosomal membranes	18–34	Secondary binding of capacitated sperm to ZP Partial reduction of fertility in baboon
PH-20	Inner acrosomal membrane	64	Hyaluronidase activity, complete inhibition of fertility in guinea pigs
PH-30	Post-acrosomal membrane	60 (α chain) and 44 (β chain)	Sperm-egg fusion in guinea pigs Partial reduction of fertility in rabbits
LDH-C4	Head to tail	35	Impairment of fertilization Partial reduction of fertility in mice
CatSper1	Principal piece of the tail	80	Hyperactivated motility of sperm Partial reduction of fertility in mice
Eppin	Post-acrosomal membrane	18 and 11	Liquefaction of the ejaculate 78% reduction of fertility in bonnet monkeys
80 kDa HSA	Head	80	Complete reduction of fertility in rats Partial reduction of fertility in rabbits and marmosets
Proacrosin	Acrosome	53	Binding to ZP glycoproteins, release of acrosome content Limited proteolysis of the ZP
CRISP1	Acrosome	32	Fusion with the egg 75% reduction of fertility in mice

14.2.1 SP-10

The testis-/sperm-specific, intra-acrosomal sperm antigen SP-10, with a series of protein bands (18–34 kDa), was identified by Herr and colleagues [4]. Both monoclonal and polyclonal antibodies to SP-10 were found to inhibit secondary binding of capacitated sperm to the zona pellucida (ZP) [5].

It was shown that immunization of male mice with recombinant SP-10 resulted in sterility [6]. In brief, Goyal et al. [6] investigated the efficacy and safety of recombinant SP-10 protein for male contraception. Adult male mice were divided into five groups. Group I was placebo-treated control, while Groups II–V were immunized with recombinant SP-10 protein on day 0 and 21 with different doses (range 25–100 μg). Group I showed normal fertility. Groups II–V showed dose-dependent reduction in fertility. In contrast, at higher dose (75 and 100 μg), all animals were sterile for 3 months. Further, histological and hematological parameters, sperm characteristics, serum clinical biochemistry, and testosterone levels in experimental groups were comparable to those of control animals. They suggested that SP-10 was promising candidate for contraceptive vaccine development.

14.2.2 PH-20

Primakoff et al. [7, 8] identified PH-20, with a sperm surface glycoprotein (64 kD), on the sperm surface with three mAbs generated against membrane preparations of guinea pig sperm. They reported a complete inhibition of fertility after immunization of male or female guinea pigs with PH-20, and this contraceptive effect was reversible [9]. Antisera from immunized females had high titers, specifically recognized PH-20 in sperm extracts, and blocked sperm adhesion to the egg ZP in vitro.

In their initial experiment to assess contraceptive effectiveness of PH-20 immunization in males, six male guinea pigs were immunized with 2.5–50 μg PH-20; seven control males received injections without PH-20. About 2 months after the initial injection, each male was housed with two females for mating. None of the 12 females mated with the PH-20-immunized males carried fetuses; 14 out of 14 of the females mated with control males did. Four of the immunized six males regained fertility by 7 months after the initial injection and thus did not show the irreversible sterility associated with autoimmune orchitis. They suggested that a human functional analogue of PH-20 would be a candidate for an effective contraceptive immunogen [9].

14.2.3   Fertilin (PH-30)

The PH-30 antigen was a second post-acrosomal plasma membrane antigen [8]. Myles et al. [10] designated the PH-30 antigen as fertilin based on its apparent role during fertilization. Fertilin is composed of two subunits which are both integral membrane glycoproteins. The α subunit (60 kDa) contains a peptide typical of viral fusion proteins, while the β subunit (44 kDa) is an integrin ligand (disintegrin) [11, 12].

The PH-30 mAb inhibited guinea pig sperm-egg fusion in a concentration-dependent manner [13]. Blobel et al. [11] proposed that the fertilin β subunit functions as a ligand for integrins on the egg surface, while fertilin α is responsible for gamete membrane fusion. Immunization of male guinea pigs with purified guinea pig fertilin resulted in complete infertility [14]. On the contrary, immunization of male and female rabbits with E. coli-expressed recombinant α or β-fertilin led to partial inhibition of fertility [15].

14.2.4   LDH-C$_4$

Goldberg et al. [16] extensively characterized the testis-specific isozyme lactate dehydrogenase-C$_4$ (LDH-C$_4$). LDH-C$_4$ genetically null male mice showed normal spermatogenesis and testes development but reduced fertility [17]. Sperm from these mice had lower motility and reduced fertilization capacity. Mahi-Brown et al. [18] immunized both male and female mice with LDH-C$_4$ in two systemic doses in Freund's adjuvants. The presence of anti-LDH-C$_4$ antibodies in uterine fluid was confirmed. Oviducts were viewed directly 1 h after mating for the presence of sperm. No significant effect of immunization on sperm transport to the oviduct could be demonstrated. Fertilization was evaluated 4 h after mating. It was found that immunization of males, but not females, impaired fertilization (19.6% versus 50.8% of the oocytes penetrated in 17 and 19 females, respectively). Orchitis was found histologically in 43.8% of the immunized males and 10% of the control males. They suggested that immunization of male but not female mice with LDH-C$_4$ impaired fertilization and that there is no obvious synergistic effect.

14.2.5   CatSper1

There are four members in Cation channels of sperm (CatSper) family, which form heterotetrameric channels. Mice deficient in any member of the family are completely sterile. These channels are located in the principal piece of the sperm tail and are required for its hyperactivation and motility.

To evaluate the contraceptive ability of two B-cell epitopes in CatSper1, Li et al. [19] synthesized predicted two B-cell epitopes in the extracellular part of transmembrane domains and pore region of CatSper1 to immunize to male mice. Significant reduction of fertility was observed in mating trial with no evident systemic illness or abnormal mating behavior. Epididymal sperm of immunized males exhibited impaired ability to fertilize eggs in vitro, and showed sperm agglutination in some animals, while presenting no changes in sperm viability or progressive motility. High titer of antibodies was induced in the sera, and the binding of the antibodies to epididymal sperm of immunized males was observed. They suggested that CatSper members could be the effective and viable targets for immunocontraception. These two epitopes in CatSper1 share high identity between mouse and human and may be effective for fertility regulation in humans.

Nazari et al. [20] have explored targeting of the extracellular loop as an approach for the generation of antibodies with the potential ability to block the ion channel and applicable method to the next generation of nonhormonal contraceptive. A small extracellular fragment of CatSper1 channel was cloned in pET-32a and pEGFP-N1 plasmids. The results showed that vaccination of the experimental group with DNA vaccine caused to produce antibody with ($p < 0.05$) unlike the control group. This antibody extracted from Balb/c serum could recognize the antigen, and it may be used potentially as a male contraception to prevent sperm motility. They suggested that the antibody can be used in male for blocking CatSper channel and it has the potential ability to use as a contraceptive.

Yu et al. [21] aimed to develop DNA vaccine targeting CatSper1 for male contraception; the whole open reading frame of mouse CatSper1 was cloned into the plasmid pEGFP-N1 to obtain a DNA vaccine pEGFP-N1-CatSper1. The vaccine was confirmed to be transcribed and translated in mouse N2a cell in vitro and mouse muscle tissue in vivo. Intramuscular injection with the vaccine on male mice induced specific immune reaction and caused significant inhibition on sperm hyperactivated motility and progressive motility ($P < 0.001$ for both), and consequently reduced male fertility. The fertility rate of experimental group was 40.9%, which was significantly lower ($P = 0.012$) than control group (81.8%). No significant change in mating behavior, sperm production, and histology of testis/epididymis was observed. CatSper1 exhibits a high degree of homology among different species. They suggested that CatSper1 DNA vaccine might be a good strategy for developing an immunocontraceptive vaccine for human and animal use.

From a clinical point of view, Williams et al. [22] reported that specific loss of CatSper function is sufficient to compromise fertilizing capacity of human sperm. In brief, a total of 101/102 (99%) IVF patients and 22/23 (96%) donors exhibited a normal Ca^{2+} response. The mean normalized peak response did not differ between the two groups. In recall patients, 9/10 (90%) showed a normal Ca^{2+} response. Three men were initially identified with a defective Ca^{2+} influx. However, only one, who had normal semen characteristics but showed a total fertilization failure in IVF, had a defective response in repeat semen samples. These data add significantly to the understanding of the role of CatSper in human sperm function and its impact on male fertility. They suggested that CatSper is a suitable and specific target for human male contraception.

14.2.6 Eppin

Eppin stands for epididymal protease inhibitor, which is a serine protease inhibitor present on the surface of entire sperm with a stronger reactivity to a band at 18 kDa and weaker reactivity to a band at 11 kDa [23]. It is 134 aa long protein in mice and 133 aa long in humans [24]. Eppin functions in liquefaction of the ejaculate, the absence of which severely impairs sperm motility [25].

O'Rand et al. [26] reported that immunization of male bonnet monkeys (*Macaca radiata*) with Eppin led to the generation of high antibody titers with concomitant infertility in approximately 78% of the immunized animals. Block in fertility was reversible as five out of seven high-titer monkeys regained fertility, subsequent to decline in the antibody titers. They suggested that this method of immunocontraception may be extended to humans.

They worked on developing small molecules that inhibit Eppin binding to the protein semenogelin, which is a necessary step in sperm liquefaction [27]. Recently, they demonstrated that intravenous administration of the small molecule EP055 reduced sperm motility by 80% in male macaques [28]. In brief, the macaques were infused with a low dose (75 ± 80 mg/kg) followed by a recovery period and a subsequent high dose (125 ± 130 mg/kg) of EP055. After high-dose administration, sperm motility fell to approximately 20% of pretreatment levels within 6 h post-infusion; no normal motility was observed at 30 h post-infusion. Recovery of sperm motility was obvious by 78 h post-infusion, with full recovery in all animals by 18 days post-infusion. They suggested that EP055 has the potential to be a male contraceptive that would provide a reversible, short-lived pharmacological alternative to condoms or vasectomy.

Thirumalai et al. [29] mentioned that they are now working on the development of potent, oral compounds in animal studies. Hopefully, continued work on this approach will result in a pill that can effectively reduce sperm motility as a male contraceptive.

14.2.7 80 kDa HSA

An 80 kDa human sperm antigen (HSA) was initially identified from human sperm extract as an antigen responsible for inducing immunological infertility and later demonstrated to be the promising candidate for development of antifertility vaccine [30]. Protein band of 80 kDa from human sperm extract reacted specifically with the serum of an immunoinfertile woman; 80 kDa HSA was purified from human sperm extract. Active immunization with 10 μg of purified 80 kDa HSA elicited gradual increase in antibody titer and induced infertility in male and female rats. Immunized animals regained fertility with decline in antibody titer [31].

Immunization with the synthetic peptides corresponding to 80 kDa HSA conjugated to keyhole limpet hemocyanin (KLH) led to reversible block of fertility in male rabbits and male marmosets [32].

14.2.8 Proacrosin

Acrosin is a trypsin-like protease localized to the sperm acrosome as an inactive zymogen (proacrosin) that is converted to the active enzyme and released during the acrosomal exocytosis [33]. Activation of proacrosin (53 kDa) leads to the 34 kDa enzymatically active form, β-acrosin, through proteolytic cleavage at both the N- and C-protein termini [34]. Several functions have been attributed to the proacrosin/acrosin system, mainly binding to ZP glycoproteins, participation in the release of acrosomal content, and limited proteolysis of the ZP [35].

Active immunization of male mice with DNA vaccine encoding proacrosin led to decrease in fertility of the immunized animals [36]. A decrease in litter size was also observed as compared with the control group. The antibodies thus elicited by immunization with DNA vaccine also led to the inhibition of sperm-ZP binding and Ca^{2+}-ionophore-induced acrosome reaction [36].

14.2.9 CRISP1

Cysteine-rich secretory protein 1 (CRISP1) is an epididymal protein which belongs to CRISP family [37]. The molecular weight of CRISP1 was confirmed as 32 kDa [38]. Its size in humans is 249 aa and in mice it is 244 aa. *CRISP1*-deficient mice showed normal fecundity, but sperm showed less capability to fuse with egg during in vitro fertilization [39]. Mouse CRISP1 has been localized in the dorsal region of the acrosome and human CRISP1 in the post-acrosomal region of the sperm head. A part of the protein is lost during the capacitation, while remaining part migrates from the dorsal side of sperm to the equatorial region after capacitation and acrosome reaction [39].

Luo et al. [40] carried out immunization of mice with the DNA vaccine encoding mouse CRISP1 (mCRISP1) to examine the immunocontraceptive properties of the plasmid pcDNA-mCRISP1 and compare them to the corresponding recombinant mCRISP1 (r-mCRISP1). Three groups of mice received three injections of r-mCRISP1, pcDNA-mCRISP1, or pcDNA vector, respectively. The titers of anti-mCRISP1 antibodies from DNA-immunized mice were significantly lower than that of r-mCRISP1-immunized mice, but it lasted relatively longer. Male and female pcDNA-mCRISP1-injected animals presented a statistically significant reduction in their fertility with no signs of immunopathologic effects. They suggested that the feasibility of generating an immune response to mCRISP1 protein by DNA vaccine and pcDNA-mCRISP1 plasmid causes significant antifertility potential.

Later, Luo et al. [41] also evaluated the effectiveness and security of a contraceptive vaccine using pcDNA-mCRISP1 as antigen and chitosan-pcDNA3.1-mCRISP1 nanoparticles (CS-DNA NPs) as the carrier. Four groups of mice received three injections of 0.9% normal saline, pcDNA3.1 vector, pcDNA3.1-CRISP1, or CS-DNA NPs, respectively. CS-DNA NPs showed high DNAse resistance and good transfection efficiency without cell toxicity. The titers of anti-mCRISP1 antibodies from CS-DNA NP-immunized mice were significantly higher than that of pcDNA3.1-CRISP1 group. Male and female CS-DNA NP-immunized animals were recognized with a statistically significant reduction in their fertility compared with pcDNA3.1-CRISP1-immunized mice. They suggested that using chitosan-DNA nanoparticles as the carrier can improve the immunogenicity of mCRISP1 DNA contraceptive vaccine with good security.

14.3   Construction of Multi-antigen Peptides (MAPs) as Immunocontraceptives

As discussed in Chap. 13, several immunocontraceptives using multi-antigen peptides (MAPs) for female have been constructed to overcome the limitations of a single molecule. Mortazavi et al. [42] also designed a recombinant protein containing three main sperm epitopes (IZUMO1, SACA3, and PH-20) and evaluated as contraceptive vaccine to control fertility in male mice. Both IZUMO1 and PH-20 were described in detail in Chap. 13. SACA3 is another surface sperm protein identified in acrosome. It belongs to lysozyme-like proteins, which is encoded by Spaca3 gene [43]. This antigen mediates adhesion and fusion of sperm-oocyte and participates at early embryo development in hamster, marine, and cattle [44, 45]. A research has shown the activation of SACA3 antisera in infertile couples had been suggested as an important candidate for the development of contraceptive vaccine [46].

Twenty inbred male BALB/c mice (6–8 weeks) were immunized by 100 μg purified protein [42]. The results revealed that the designed chimeric protein stimulated the immune system of the mice effectively. The level of IgG antibody in the sera was significantly higher in vaccinated mouse rather than control mouse. Eighty percent of the vaccinated mice became infertile, and in the remaining ones, the number of children decreased to 4–6 offspring instead of 10–12 in normal mice (Table 14.2). Histopathological studies showed that no organs including the heart,

Table 14.2 Multi-antigen peptides (MAPs) developed as contraceptive vaccines in male

Author [Ref. no.]	Formulation	Species	Contraceptive effects
Mortazavi et al. [42]	Izumo, SACA3, PH-20	Mice	80% of the vaccinated mice became infertile

brain, lung, liver, kidney, and intestine were damaged. However, normal spermato-genesis has been disrupted and necrotic spermatogonia cells were reported in seminiferous tubules. They suggested that the designed chimeric protein containing IZUMO1, SACA3, and PH-20 epitopes can stimulate the immune system and cause male contraception without any side effects.

14.4 Direct Identification of Sperm Surface Antigens by Vectorial Labeling and 2-D Electrophoresis

To identify the repertoire of proteins exposed on the surface of ejaculated human sperm, Naaby-Hansen et al. [47] developed high-resolution two-dimensional (2-D) gel systems for separation of human sperm and seminal plasma proteins. Shetty et al. [48] employed sera from infertile men containing ASA were on 2-D immunoblots of human sperm proteins to identify novel sperm alloantigens relevant to immune infertility. An immunoreactive protein spot (MW, 44 kDa; pI, 4.5) was microsequenced, and the related cDNA was cloned yielding a 309 amino acid sequence corresponding to TSGA2 homolog (mouse) to signify "testis-specific gene A2," which was named human meichroacidin (hMCA). In human testis, hMCA localized to the tails of developing spermatids and did not localize to the nucleus of either spermatocytes or spermatids. Electron microscopic immunocyto-chemistry showed that hMCA localized within the radial spokes of the axonemal complex of the sperm flagellum, and immunofluorescence studies revealed hMCA in the cilia of epithelial cells in the trachea and ependyma. Bioinformatic identification of orthologues of MCA in several lower organisms including ciliates and flagellates suggest the protein plays a role in flagellar motility across phyla. Therefore, they proposed the term radial spoke protein 44 (RSP44) as an accurate designation, preferable to hMCA because it denotes the restricted localization of the protein to the radial spokes of the axonemes of both sperm and cilia.

Shiraishi et al. [49] have identified human sperm antigens recognized by the sera from infertile men with sperm-immobilizing antibodies using the high-resolution 2-D gel electrophoresis. To depict the target antigen of the sperm-immobilizing antibody, 2-D electrophoresis was utilized so that one possible target antigen with the trait of (MW, pI) = (43.1, 6.2) was discovered. It was later analyzed using matrix-assisted laser desorption/ionization-time of flight mass spectrometry (MALDI-TOF MS) and found the molecule to be glyceraldehyde 3-phosphate dehydrogenase, testis-specific (G3PT), which is an enzyme in glycolysis [50].

References

1. Sundaram A, Vaughan B, Kost K, Bankole A, Finer L, Singh S, Trussell J. Contraceptive failure in the United States: estimates from the 2006-2010 National Survey of Family Growth. Perspect Sex Reprod Health. 2017;49:7–16.
2. Shih G, Turok DK, Parker WJ. Vasectomy: the other (better) form of sterilization. Contraception. 2011;83:310–5.
3. Abbe CR, Page ST, Thirumalai A. Male contraception. Yale J Biol Med. 2020;93:603–13.
4. Herr JC, Flickinger CJ, Homyk M, Klotz K, John E. Biochemical and morphological characterization of the intraacrosomal antigen SP-10 from human sperm. Biol Reprod. 1990;42:181–93.
5. Coonrod SA, Westusin ME, Herr JC. Inhibition of bovine fertilization in vitro by antibodies to SP-10. J Reprod Fertil. 1996;107:287–97.
6. Goyal S, Manivannan B, Kumraj GR, Ansari AS, Lohiya NK. Evaluation of efficacy and safety of recombinant sperm-specific contraceptive vaccine in albino mice. Am J Reprod Immunol. 2013;69:495–508.
7. Primakoff P, Myles DG. A map of the guinea pig sperm surface constructed with monoclonal antibodies. Dev Biol. 1983;98:417–28.
8. Primakoff P, Hyatt H, Myles DG. A role for the migrating sperm surface antigen PH-20 in guinea pig sperm binding to the egg zona pellucida. J Cell Biol. 1985;101:2239–44.
9. Primakoff P, Lathrop W, Woolman L, Cowan A, Myles D. Fully effective contraception in male and female guinea pigs immunized with the sperm protein PH-20. Nature. 1998;335:543–6.
10. Myles DG, Kimmel LH, Blobel CP, White JM, Primakoff P. Identification of a binding site in the disintegrin domain of fertilin required for sperm-egg fusion. Proc Natl Acad Sci U S A. 1994;91:4195–8.
11. Blobel CP, Wolfsberg TG, Turck CW, Myles DG, Primakoff P, White JM. A potential fusion peptide and an integrin ligand domain in a protein active in sperm-egg fusion. Nature. 1992;356:248–52.
12. Wolfsberg TG, Bazan JF, Blobel CP, Myles DG, Primakoff P, White JM. The precursor region of a protein active in sperm-egg fusion contains a metalloprotease and a disintegrin domain: structural, functional, and evolutionary implications. Proc Natl Acad Sci U S A. 1993;90: 10738–87.
13. Primakoff P, Hyatt H, Tredick-Kline J. Identification and purification of a sperm surface protein with a potential role in sperm-egg membrane fusion. J Cell Biol. 1987;104:141–9.
14. Ramarao CS, Myles DG, White JM, Primakoff P. Initial evaluation of fertilin as an immunocontraceptive antigen and molecular cloning of the cynomolgus monkey fertilin β subunit. Mol Reprod Dev. 1996;43:70–5.
15. Hardy CM, Clarke HG, Nixon B, Grigg JA, Hinds LA, Holland MK. Examination of the immunocontraceptive potential of recombinant rabbit fertilin subunits in rabbit. Biol Reprod. 1997;57:879–86.
16. Goldberg E. Lactic and malic dehydrogenases in human spermatozoa. Science. 1963;139:602–3.
17. Odet F, Duan C, Willis WD, Goulding EH, Kung A, Eddy EM, Goldberg E. Expression of the gene for mouse lactate dehydrogenase C (Ldhc) is required for male fertility. Biol Reprod. 2008;79:26–34.
18. Mahi-Brown CA, VandeVoort CA, McGuinness RP, Overstreet JW, O'Hern P, Goldberg E. Immunization of male but not female mice with the sperm-specific isozyme of lactate dehydrogenase (LDH-C4) impairs fertilization in vivo. Am J Reprod Immunol. 1990;24:1–8.
19. Li H, Ding X, Guo C, Guan H, Xiong C. Immunization of male mice with B-cell epitopes in transmembrane domains of CatSper1 inhibits fertility. Fertil Steril. 2012;97:445–52.
20. Nazari M, Mirshahi M, Mowla SJ, Bamdad T, Sarikhani S. Investigation in vitro expression of CatSper sub fragment followed by production of polyclonal antibody: potential candidate for the next generation of non hormonal contraceptive. Cell J. 2012;14:215–24.

21. Yu Q, Mei XQ, Ding XF, Dong TT, Dong WW, Li HG. Construction of a catsper1 DNA vaccine and its antifertility effect on male mice. PLoS One. 2015;10(5):e0127508. https://doi.org/10.1371/journal.pone.0127508.
22. Williams HL, Mansell S, Alasmari W, Brown SG, Wilson SM, Sutton KA, Miller MR, Lishko PV, Barratteven CLR, Publicover SJ, da Silva SM. Specific loss of CatSper function is sufficient to compromise fertilizing capacity of human spermatozoa. Hum Reprod. 2015;30:2737–46.
23. Sivashanmugam P, Hall SH, Hamil KG, French FS, O'Rand MG, Richardson RT. Characterization of mouse *Eppin* and a gene cluster of similar protease inhibitors on mouse chromosome 2. Gene. 2003;312:125–34.
24. Silva EJ, Patrão MT, Tsuruta JK, O'Rand MG, Avellar MC. Epididymal protease inhibitor (EPPIN) is differentially expressed in the male rat reproductive tract and immunolocalized in maturing spermatozoa. Mol Reprod Dev. 2012;79:832–42.
25. O'Rand MG, Widgren EE, Hamil KG, Silva EJ, Richardson RT. Functional studies of Eppin. Biochem Soc Trans. 2011;39:1447–9.
26. O'Rand MG, Widgren EE, Sivashanmugam P, Richardson RT, Hall SH, French FS, VandeVoort CA, Ramachandra SG, Ramesh V, Jagannadha RA. Reversible immunocontraception in male monkeys immunized with Eppin. Science. 2004;306:1189–90.
27. O'Rand MG, Silva EJ, Hamil KG. Non-hormonal male contraception: a review and development of an eppin based contraceptive. Pharmacol Ther. 2016;157:105–11.
28. O'Rand MG, Hamil KG, Adevai T, Zelinski M. Inhibition of sperm motility in male macaques with EP055, a potential non-hormonal male contraceptive. PLoS One. 2018;13:e0195953.
29. Thirumalai A, Amory J. Emerging approaches to male contraception. Fertil Steril. 2021;115: 1369–76.
30. Bandivdekar AH. Development of antifertility vaccine using sperm specific proteins. Indian J Med Res. 2014;140(Suppl):73–7.
31. Bandivdekar AH, Gopalkrishnan K, Garde SV, Fernandez PX, Moodbidri SB, Sheth AR, Koide SS. Antifertility effect in rats actively immunized with 80 kDa human semen glycoprotein. Indian J Exp Biol. 1992;30:1017–23.
32. Khobarekar BG, Vernekar V, Raghavan V, Kamada M, Maegawa M, Bandivdekar AH. Evaluation of the potential of synthetic peptides of 80 kDa human sperm antigen (80 kDa HSA) for the development of contraceptive vaccine for male. Vaccine. 2008;26: 3711–8.
33. Tesarik J, Drahorad J, Testart J, Mendoza C. Acrosin activation follows its surface exposure and precedes membrane fusion in human sperm acrosome reaction. Development. 1990;110:391–400.
34. Zahn A, Furlong LI, Biancotti JC, Ghiringhelli PD, Marin-Briggiler CI, Vazquez-Levin MH. Evaluation of the proacrosin/acrosin system and its mechanism of activation in human sperm extracts. J Reprod Immunol. 2002;54:43–63.
35. Furlong LI, Harris JD, Vazquez-Levin MH. Binding of recombinant human proacrosin/acrosin to zona pellucida (ZP) glycoproteins. I. Studies with recombinant human ZPA, ZPB, and ZPC. Fertil Steril. 2005;83:1780–90.
36. García L, Veiga MF, Lustig L, Vazquez-Levin MH, Veaute C. DNA immunization against proacrosin impairs fertility in male mice. Am J Reprod Immunol. 2012;68:56–67.
37. Cohen DJ, Maldera JA, Weigel Muñoz M, Ernesto JI, Vasen G, Cuasnicu PS. Cysteine rich secretory proteins (CRISP) and their role in mammalian fertilization. Biol Res. 2011;44:135–8.
38. Maldera JA, Vasen G, Ernesto J, Weigel-Munoz M, Cohen DJ, Cuasnicu PS. Evidence for the involvement of zinc in the association of CRISP1 with rat sperm during Epididymal maturation. Biol Reprod. 2011;85:503–10.
39. Cohen DJ, Maldera JA, Vasen G, Ernesto JI, Muñoz MW, Battistone MA, Cuasnicú PS. Epididymal protein CRISP1 plays different roles during the fertilization process. J Androl. 2011;32:672–8.

40. Luo J, Yang J, Cheng Y, Li W, Yin TL, Xu WM, Zou YJ. Immunogenicity study of plasmid DNA encoding mouse cysteine-rich secretory protein-1 (mCRISP1) as a contraceptive vaccine. Am J Reprod Immunol. 2012;68:47–55.
41. Luo J, Liu XL, Zhang Y, Wang YQ, Xu WM, Yang J. The immunogenicity of CRISP1 plasmid-based contraceptive vaccine can be improved when using chitosan nanoparticles as the carrier. Am J Reprod Immunol. 2016;75:643–53.
42. Mortazavi B, Fard NA, Karkhane AS, Shokrpoor S, Heidari F. Evaluation of multi-epitope recombinant protein as a candidate for a contraceptive vaccine. J Reprod Immunol. 2021;145: 103325.
43. Mandal A, Klotz KL, Shetty J, Jayes FL, Wolkowicz MJ, Bolling LC, Coonrod SA, Black MB, Diekman AB, Haystead TA. SLLP1, a unique, intra-acrosomal, non-bacteriolytic, c lysozyme-like protein of human spermatozoa. Biol Reprod. 2003;68:1525–37.
44. Chiu W, Erikson E, Sole C, Shelling A, Chamley L. SPRASA, a novel sperm protein involved in immune-mediated infertility. Hum Reprod. 2004;19:243–9.
45. Herrero MB, Mandal A, Digilio LC, Coonrod SA, Maier B, Herr JC. Mouse SLLP1, a sperm lysozyme-like protein involved in sperm–egg binding and fertilization. Dev Biol. 2005;284: 126–42.
46. Wagner A, Holland OJ, Tong M, Shelling AN, Chamley LW. The role of SPRASA in female fertility. Reprod Sci. 2015;22:452–61.
47. Naaby-Hansen S, Flickinger CJ, Herr JC. Two-dimensional gel electrophoretic analysis of vectorially labeled surface proteins of human spermatozoa. Biol Reprod. 1997;56:771–87.
48. Shetty J, Klotz KL, Wolkowicz MJ, Flickinger CJ, Herr JC. Radial spoke protein 44 (human meichroacidin) is an axonemal alloantigen of sperm and cilia. Gene. 2007;396:93–107.
49. Shiraishi Y, Shibahara H, Koriyama J, Kikuchi K, Hirano Y, Suzuki M. Clinical and biological characteristics of immunologically infertile men with sperm immobilizing antibody. 64th American Society for Reproductive Medicine (ASRM) Scientific Congress. (abstract); 2008.
50. Shiraishi Y, Maekawa M, Shibahara H, Toshimori K. Proteomic analysis of the antigen protein on the sperm surface detected by the sperm immobilizing antibody among the immunologically infertile men. (manuscript in prep.)

Chapter 15
Wildlife Overpopulation Control

Hiroaki Shibahara

Abstract There is uncontrolled increase in the population of some animal species that is leading to an increasing conflict for habitation between humans and wildlife species. Efforts at controlling wildlife overpopulation by means of contraception began as early as the 1950s. Steroids or nonsteroids were attempted, which ended in failure, because of several reasons written in detail in this text.

Fertility control by immunocontraception is suggested to offer a long-term, effective, and human approach for reproduction control in captive animals as well as to reduce free-ranging wildlife populations. The most widely tested immunocontraceptive vaccine for wildlife species is based on developing antibodies to zona pellucida. In addition, contraceptive vaccines based on neutralization of gonadotropin-releasing hormone have also shown promising contraceptive efficacy in various animal models.

Another candidate protein for the production of contraceptive vaccines is sperm antigens. A variety of proteins derived from sperm have been experimentally assessed for their contraceptive potential in wildlife species. However, none of them had been used yet for contraception in exotic species.

Here, the present state of the development of the contraceptive vaccine using the sperm antigen for the adjustment of wild animals, such as fox in Australia and sika deer in Japan, will be discussed.

15.1 Introduction

In addition to growing human population, there is uncontrolled increase in the population of some animal species. For example, increasing population of elephants in Africa and kangaroos in Australia is leading to an increasing conflict for habitation

H. Shibahara (✉)
Department of Obstetrics and Gynecology, School of Medicine, Hyogo Medical University, Nishinomiya, Hyogo, Japan
e-mail: sibahara@hyo-med.ac.jp

H. Shibahara, A. Hasegawa (eds.), *Gamete Immunology*,
https://doi.org/10.1007/978-981-16-9625-1_15

between humans and wildlife species [1]. The environment suffers due to the strain from the natural activities of the overpopulated species. The results can be devastating as animals scrape for food and water into unnatural habitats in search of something to eat.

Moreover, wild animals may act as vectors or reservoirs for various diseases of zoonoses, which has been accounted for approximately 60% of all infectious disease pathogens. For example, dogs are one of the main vectors harboring rabies virus, which is transmitted to humans by rabid dog bite. Surgical sterilization such as spaying of female dogs and castration of male dogs have failed to control population of the wild dogs in several developing countries. As a result, rabies infection is prevalent in these countries with significant human mortality [1].

To minimize human-animal conflicts for habitation and to reduce the burden of various zoonotic diseases, it is imperative to develop new tools for wildlife population management [2]. Efforts at controlling wildlife overpopulation by means of contraception began as early as the 1950s, and most research involved natural and synthetic steroids, or a variety of nonsteroidal compounds [3–5]. Virtually, all of these early attempts, while pharmacologically successful, ended in failure, for a variety of reasons including toxicity, passage through the food chain, adverse effects on social behaviors, high cost of application, difficulty in delivering the compounds remotely, health risks in pregnant animals, and a host of regulatory issues [6].

Fertility control by immunocontraception is suggested to offer a long-term, effective, and human approach for reproduction control in captive animals as well as to reduce free-ranging wildlife populations [7]. The most widely tested immunocontraceptive vaccine for wildlife species is based on developing antibodies to zona pellucida (ZP). Contraceptive vaccines based on native porcine ZP glycoproteins have been successfully used to manage population of feral horses [8] and white-tailed deer [9]. In addition, contraceptive vaccines based on neutralization of gonadotropin-releasing hormone (GnRH) have also shown promising contraceptive efficacy in various animal models [10–12]. Both ZP- and GnRH-based contraceptive vaccines are commercially available, and their usefulness was reported by monitoring ovarian activity in mares [13].

Another candidate protein for the production of contraceptive vaccines is sperm antigens. A variety of proteins derived from sperm have been experimentally assessed for their contraceptive potential in wildlife species. Some of the sperm proteins already proved to be promising antigens for contraceptive vaccine include lactate dehydrogenase [14], protein hyaluronidase (PH-20) [15], Eppin [16], or CatSper [17]. After identification of the specific sperm fusion protein, Izumo [18], is discussed as a potential contraceptive antigen as well [19]. Despite sperm-egg receptors, many other biological active components of reproduction are considered for contraceptive vaccines [20], but none of them had been used yet for contraception in exotic species.

This chapter describes the present state of the development of the contraceptive vaccine using the sperm antigen for the adjustment of wild animals, such as fox in Australia and sika deer in Japan.

15.2 Development of an Immunocontraceptive Vaccine by Using Sperm Antigen to Control Fox Populations in Australia

Bradley [21] mentioned that the development of an immunocontraceptive vaccine to control fox populations in Australia would confer considerable advantages in controlling the long-term impact of this predator on native and endangered species. It is proposed that such a vaccine would be delivered orally in a bait, thereby stimulating a mucosal immune response to the foreign antigens. Attenuated *Salmonella typhimurium* strains are potential "safe" delivery vectors of an oral immunocontraceptive vaccine for the European red fox (*Vulpes vulpes*).

His group performed a study in which model bacterial (*Escherichia coli* heat-labile enterotoxin B subunit, LTB) and fox sperm (fSP10) antigens were expressed in *S. typhimurium* SL3261 (delta aroA) under the control of the trc promoter [22]. Adult female foxes were given three oral immunizations with SL3261 containing either LTB (SL3261/pLTB), fSP10 (SL3261/pFSP10), or a control plasmid (pKK233-2 or pTrc99A). All foxes raised serum (IgG) and vaginal (IgG and IgA) antibodies against *S. typhimurium* lipopolysaccharide (LPS). Each fox that received SL3261/pLTB raised high-titer LTB-specific serum and vaginal IgG antibodies. However, only one of four foxes immunized with SL3261/pFSP10 raised an anti-fSP10 immune response, in the form of low-titer serum and vaginal IgG antibodies. No vaginal IgA antibodies were raised against either LTB or fSP10 in these experiments. The immune responses against recombinant LTB and fSP10 resulted chiefly from the initial dose of antigen in the inocula and were minimally influenced by continued in vivo antigen expression. They suggested that the study demonstrated for the first time in the female red fox that oral *Salmonella* can elicit specific systemic and reproductive tract antibodies against heterologous, recombinant proteins.

15.3 Development of an Immunocontraceptive Vaccine by Using Sperm to Control Sika Deer Populations in Japan

The impact of deer overabundance is a worldwide problem. In Japan, the amount of agricultural damage caused by wildlife has been estimated to be approximately 20 billion yen (approx. 175.4 million USD) per year during the latest decade, with deer accounting for more than one third of the total. Along with habitat expansion and population increase, damage by sika deer to the forest ecosystem and agriculture has become a serious issue. Deer also transmit a number of diseases and parasites to humans and livestock. The overabundance of deer is a result of their strong fecundity [23].

Castration is a way to make males infertile. Castrated males become indifferent to females. Since male deer form harems during the mating season, the remaining normal males will mate with the females even if castrated males are produced, and the female pregnancy rate will not decrease. Therefore, it is important to be a "male that mates but does not allow females to become pregnant." Therefore, the present situation should be tackled by experts in reproductive biology [23].

Recently, Noguchi et al. [24] developed a lipopolysaccharide (LPS)-supplemented adjuvant for the purpose of stimulating both humoral and cell-mediated immune responses. The practically used adjuvant AdjuVac® was developed by modifying a mycoplasma vaccine, Mycoper®, and was proven not to cause inflammation at the injection site [25]. However, its use in Japan is probably impermissible, since AdjuVac contains killed bacteria (*Mycobacterium* spp.), thus contravening the domestic Policy for Control of Infectious Diseases in Domestic Animals. Freund's complete adjuvant (FCA) is a strong adjuvant containing *Mycoplasma* spp., but it was proven to cause inflammation at the injection site [25], thus making it suitable only for experimental use in terms of animal welfare. With the long-term aim of achieving effective immunocontraception for density control of sika deer in Japan, they developed an alternative adjuvant that would overcome the two problems described above, allowing its registration as a vaccine adjuvant for field use. To achieve efficacy and gain public acceptance, they investigated the effects of adding nonpathogenic *Escherichia coli* LPS to montanide ISA 71VG®, a mineral oil-based water-in-oil-type veterinary vaccine adjuvant.

Supplementation with LPS from nonpathogenic *Escherichia coli* was found to enhance the adjuvant effects of a veterinary vaccine adjuvant (ISA 71VG®). Sperm immunization using 71VG as an adjuvant in the immature period induced infertility in 25% of male rats, whereas this increased to 62.5% after immunization with 71VG + LPS or FCA. Mean testicular weight of non-sterile males in the 71VG + LPS group was significantly lower than that in the 71VG or FCA group. Histological examination of testicular tissue from sterile males demonstrated severe impairment of spermatogenesis due to experimental autoimmune orchitis, a cell-mediated auto-immune condition. The serum anti-sperm antibody (ASA) titer was elevated in the three sperm-immunized groups relative to male rats treated with adjuvant alone, but the titer was higher in the 71VG + LPS and FCA groups than in the 71VG group. They suggested that the LPS-supplemented adjuvant stimulates both humoral and cell-mediated immune responses to an extent comparable to FCA.

It was clarified that male rats can be inoculated with rat sperm to induce autoimmune epididymitis and become "males that can be mated but females cannot be pregnant." According to another their reports [26], it was similarly confirmed that goats could be inoculated with goat sperm to induce autoimmune epididymitis. They have developed an immunostimulatory agent that can be used in Japan. They succeeded in inducing autoimmune epididymitis using a single inoculation and heterologous sperm, which are issues for practical use.

References

1. Gupta SK, Shrestha A, Minhas V. Milestones in contraceptive vaccines development and hurdles in their application. Hum Vaccin Immunother. 2014;10:911–25.
2. Minhas V, Kumar R, Moitra T, Singh R, Panda AK, Gupta SK. Immunogenicity and contraceptive efficacy of recombinant fusion protein encompassing Sp17 spermatozoa-specific protein and GnRH: relevance of adjuvants and microparticles based delivery to minimize number of injections. Am J Reprod Immunol. 2020;83:e13218.
3. Kirkpatrick JF, Turner JW. Chemical fertility control and wildlife management. Bioscience. 1985;35:485–91.
4. Kirkpatrick JF, Turner JW. Reversible fertility control in nondomestic animals. J Zoo Wildl Med. 1991;22:392–408.
5. Seal US. Fertility control as a tool for regulating captive and freeranging wildlife populations. J Zoo Wildl Med. 1991;22:1–5.
6. Kirkpatrick JF, Lyda RO, Frank KM. Contraceptive vaccines for wildlife: a review. Am J Reprod Immunol. 2011;66:40–50.
7. Jewgenow K. Immune contraception in wildlife animals. In: Krause WKH, Naz RK, editors. Immune infertility. Cham: Springer Int Pub; 2017. p. 263–80.
8. Kirkpatrick JF, Liu IKM, Turner JW Jr. Remotely-delivered immunocontraception in feral horses. Wildl Soc Bull. 1990;18:326–30.
9. Naugle RE, Rutberg AT, Underwood HB, Turner JW Jr, Liu IK. Field testing of immunocontraception on white-tailed deer (Odocoileus virginianus) on Fire Island National Seashore, New York, USA. Reprod Suppl. 2002;60:143–53.
10. Yoder CA, Miller LA. Effect of GonaCon™ vaccine on black-tailed prairie dogs: immune response and health effects. Vaccine. 2010;29:233–9.
11. Brunius C, Zamaratskaia G, Andersson K, Chen G, Norrby M, Lundstrom K. Early immunocastration of male pigs with Improvac® -effect on boar taint, hormones and reproductive organs. Vaccine. 2011;29:9514–20.
12. Levy JK, Friary JA, Miller LA, Tucker SJ, Fagerstone KA. Long-term fertility control in female cats with GonaCon™, a GnRH immunocontraceptive. Theriogenology. 2011;76:1517–25.
13. Nolan MB, Bertschinger HJ, Roth R, Crampton M, Martins IS, Fosgate GT, Stout TA, Schulman ML. Ovarian function following immunocontraceptive vaccination of mares using native porcine and recombinant zona pellucida vaccines formulated with a non-Freund's adjuvant and anti-GnRH vaccines. Theriogenology. 2018;120:111–6.
14. Goldberg E. Lactic and malic dehydrogenases in human spermatozoa. Science. 1963;139:602–3.
15. Primakoff P, Lathrop W, Woolman L, Cowan A, Myles D. Fully effective contraception in male and female guinea pigs immunized with the sperm protein PH-20. Nature. 1998;335:543–6.
16. O'Rand MG, Widgren EE, Sivashanmugam P, Richardson RT, Hall SH, French FS, VandeVoort CA, Ramachandra SG, Ramesh V, Jagannadha Rao A. Reversible immunocontraception in male monkeys immunized with Eppin. Science. 2004;306:1189–90.
17. Yu Q, Mei XQ, Ding XF, Dong IT, Dong WW, Li HG. Construction of a Catsper1 DNA vaccine and its antifertility effect on male mice. PLoS One. 2015;10:e0127508.
18. Inoue N, Ikawa M, Isotani A, Okabe M. The immunoglobulin superfamily protein Izumo is required for sperm to fuse with eggs. Nature. 2005;434(7030):234–8.
19. Naz RK. Vaccine for human contraception targeting sperm Izumo protein and YLPI2 dodecamer peptide. Protein Sci. 2014;23(7):857–68.
20. Sun LL, Li JT, Wu YZ, Ni B, Long L, Xiang YL, et al. Screening and identification of dominant functional fragments of human epididymal protease inhibitor. Vaccine. 2010;28(7):1847–53.
21. Bradley MP. Experimental strategies for the development of an immunocontraceptive vaccine for the European red fox, Vulpes vulpes. Reprod Fertil Dev. 1994;6:307–17.

22. de Jersey J, Bird PH, Verma NK, Bradley MP. Antigen-specific systemic and reproductive tract antibodies in foxes immunized with Salmonella typhimurium expressing bacterial and sperm proteins. Reprod Fertil Dev. 1999;11:219–28.
23. Noguchi J. Overabundance of sika deer and immunocontraception. J Reprod Dev. 2017;63:13–6.
24. Noguchi J, Watanabe S, Nguyen TQD, Kikuchi K, Kaneko H. Development of a lipopolysaccharide (LPS)-supplemented adjuvant and its effects on cell-mediated and humoral immune responses in male rats immunized against sperm. J Reprod Dev. 2017;63:111–5.
25. Powers JG, Nash PB, Rhyan JC, Yoder CA, Miller LA. Comparison of immune and adverse effects induced by AdjuVac and Freunds complete adjuvant in New Zealand white rabbits (Oryctolagus cuniculus). Lab Anim (NY). 2007;36:51–8.
26. Noguchi J. Development of male contraceptive method for wildlife population regulation. In: Ministry of Agriculture, Forestry and Fisheries introduces food industry science and technology research promotion project. (in Japanese); 2016.

Part IV
Anti-Zona Pellucida Antibody (AZPA): Zona Pellucida (ZP) in Basic Aspect

Chapter 16
Structure of ZP

Akiko Hasegawa

Abstract The zona pellucida (ZP) or egg coat is an architecture that surrounds animal oocytes from fish to mammals. ZP is composed of three to four glycoproteins and plays several roles in reproduction. The ancestor gene arose about five million years ago and underwent mutation and variation over time in various animal species. The classical nomenclature of ZP proteins was derived from apparent molecular weight by migration in SDS-PAGE. The current nomenclature based on ZP genes is accepted for all vertebrate animals including mammals, birds, amphibians, and fish. Crystal structure analysis revealed that there is a common region (named ZP module) consisting of ~260 amino acids. ZP module is conserved in most animal species. *ZP3*, the smallest among all *ZP* genes, contains a single ZP module comprising the ZP-N and ZP-C regions. Other ZP genes contain several ZP-N regions and a single ZP-C region. ZP module is presumed to be a structural component of ZP. Sperm-ZP interaction is believed to have several steps involving different ZP proteins. According to a classical model, sperm initially just attach to ZP via ZP3 molecule and induce acrosome reaction. Subsequently, sperm tightly bind to ZP via ZP2 and penetrate through the ZP.

16.1 Evolution

It is paradoxical that sexual reproduction has developed, maintaining both species diversity and sperm-oocyte compatibility. This chapter describes sperm-oocyte recognition from a viewpoint of the extracellular matrix.

In both vertebrates and invertebrates, oocytes are surrounded by an extracellular matrix named the egg coat. This egg coat is also termed zona pellucida (in mammals) or vitelline envelope (in non-mammals). The evolution of the egg coat is supposed to

A. Hasegawa (✉)
Department of Obstetrics and Gynecology, School of Medicine, Hyogo Medical University, Nishinomiya, Hyogo, Japan
e-mail: zonapel@hyo-med.ac.jp

© The Editor(s) (if applicable) and The Author(s), under exclusive license to Springer Nature Singapore Pte Ltd. 2022
H. Shibahara, A. Hasegawa (eds.), *Gamete Immunology*,
https://doi.org/10.1007/978-981-16-9625-1_16

211

have started with whole genome duplication (WGD), which has now been firmly established in all major eukaryotic kingdoms [1]. All vertebrates descend from at least two rounds of WGDs, which occurred in their ancestors about 300–500 million years ago.

The egg coat in vertebrates has been a popular target for biological and biochemical research for more than 50 years. It acts as a barrier to spermatozoa and finely regulates penetration of a single sperm. The mechanisms underlying these processes and related molecules differ among animal species [2–4].

In classical studies, Hedrick performed comparative analysis of egg coat and ZP between *Xenopus* and mammals. The vitelline envelope from *Xenopus laevis* eggs (an extracellular matrix structure equivalent to the mammalian zona pellucida) is composed of glycoproteins that are homologous to pig ZP glycoproteins [5]. He predicted structural and antigenic similarities of ZP glycoproteins across evolution before the advance of genetic analysis [6]. Afterward, although genomic analysis made the relations between ZP proteins and coding genes clear, complications paradoxically increased.

Figure 16.1 shows a schematic illustration of egg coat gene profiles and the evolutionary process during vertebrate evolution [7]. ZP evolution is predicted to have occurred in the common ancestral gene, which produced two lines by WGD. One line descended to the present ZP3 gene, and the other generated the five current ZP genes. At this moment in time, the ZP3 genes exist universally in most animal species. The primitive ZP3 gene is putatively regarded as the ancestral gene for all current ZP genes.

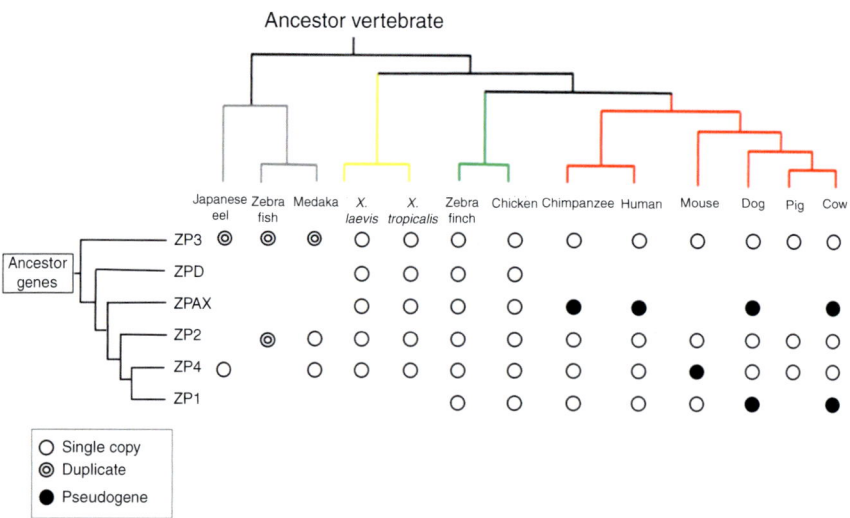

Fig. 16.1 ZP genes in present-day vertebrate animals. All the species have ZP genes and their protein products. Mammalian species are shown in red. Humans express four ZP proteins except for a pseudogene. Pigs have three ZP genes and the three corresponding proteins (modified from Molecular Ecology 24:4052 2015)

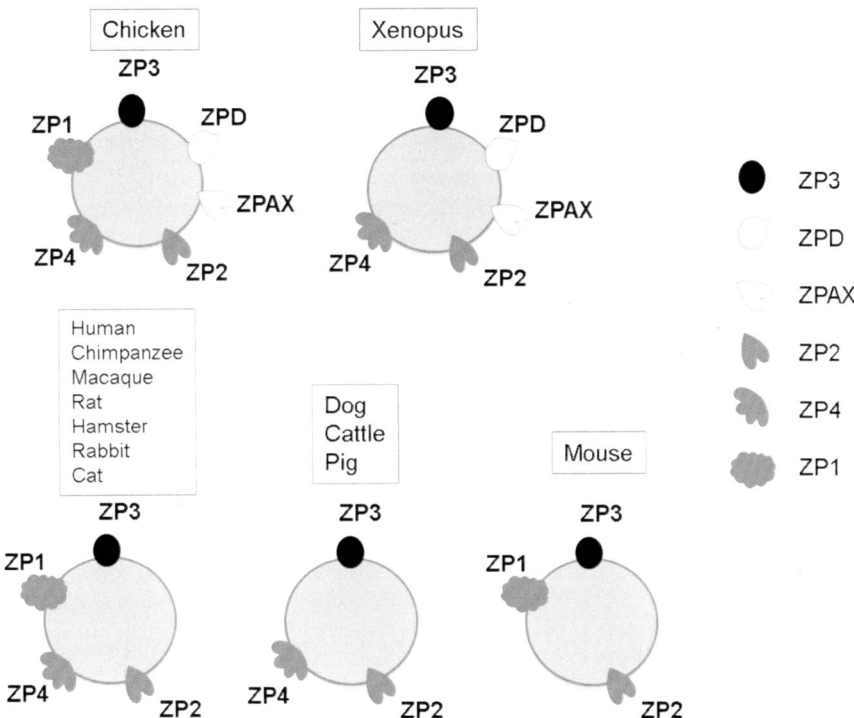

Fig. 16.2 Compositions of ZP in vertebrates except fish. In the chicken, ZP is composed of six proteins. In *Xenopus*, ZP is composed of five proteins. In mammals, there are three groups. In one group (human, chimpanzee, macaque, rat, rabbit, and cat), ZP is composed of four proteins. In the second group, ZP is composed of three proteins. In the third group (only mouse), ZP is composed of three proteins but different proteins are present from the second group (modified from Biol Reprod 78:796 2008)

On the other hand, the second line underwent multiple duplications and mutations to produce ZP gene diversities (Fig. 16.1). Moreover, mutations including DNA deletion, duplication, and transposing of DNA segments occurred repeatedly in this line from ancient times down to the present [8].

Using the most common laboratory animals, mice, ZP has been shown to be composed of three proteins, ZP1, ZP2, and ZP3, from genomic and proteomic studies. The data of mouse ZP did not fit in all mammals. Actually, in many mammalian species, the ZP has been clarified to be composed of four proteins, not three.

The complexity of the ZP proteins increased across the evolutionary tree [9, 10]. ZP genes evolved in various animals during evolution. Precious analysis of ZP genes and proteins concludes three to six ZP proteins are present in all vertebrates except fish (Fig. 16.2). In birds and amphibians, ZP tends to have relatively higher numbers of ZP proteins. This is probably because oocytes and eggs are directly exposed to more severe environments than during gestation. In primates including humans, four ZP proteins have been detected [11]. Rat and hamster ZP were also known to be composed of four ZP proteins named ZP1, ZP2, ZP3, and ZP4 [12].

In 2012, a study made clear that ZP1 existed in rabbits and showed high identity with other species: 70% identity with human and horse ZP1, and 67% identity with mouse and rat ZP1. In addition, a molecular phylogenetic analysis of ZP1 showed that pseudogenization of this gene had occurred at least four times during the evolution of mammals [13]. Furthermore, ZP4-deficient rabbits recently created by the CRISPR-Cas9 method showed lower fecundity compared to wild-type [14].

In domestic cats (*Felis catus*), before 2015 ZP was regarded as composed of three glycoproteins, ZP2, ZP3, and ZP4. The presence of a fourth protein ZP1 was not known. Recently, a study using proteomic analysis and cDNA amplification by reverse-transcribed polymerase chain reaction revealed that cat also possesses ZP1 in ZP [15].

More surprisingly, ZPAX orthologue (which had been thought not to exist in mammals) was detected in a marsupial species [16]. Marsupials are classified between amphibians and mammals in terms of development.

16.2 Nomenclature

The classical nomenclature of ZP proteins was based on electrophoretic analysis in which the apparent molecular weight of mouse ZP proteins was determined from migration in SDS-PAGE. The mouse ZP proteins were named ZP1, ZP2, and ZP3 from higher to lower molecular weights [17]. Classical stoichiometric analysis and electron microscopic observations provided a hypothetical structure of mouse ZP (Fig. 16.3). In this schematic structure, ZP2 and ZP3 are linked to each other to form

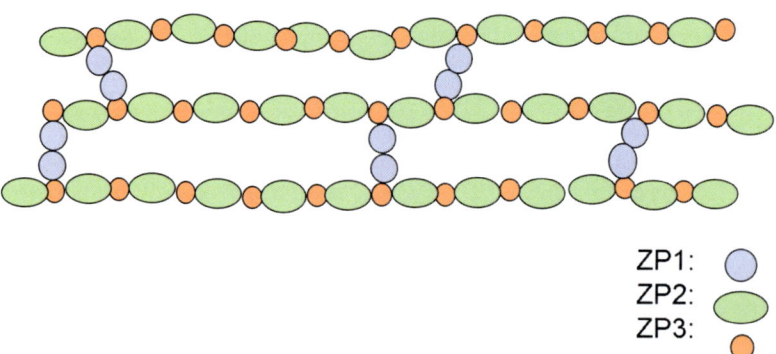

Fig. 16.3 Classical and simple schematic drawing of mouse ZP structure. This diagram is based on that by Wassarman et al. (mZP2 and mZP3 are linked as heterodimers and form fibrous structures). The fibers are connected to each other with mZP1 homodimers. mZP3 is essential for mZP1 connection but mZP1 is not essential for ZP formation, although the ZP structure is fragile

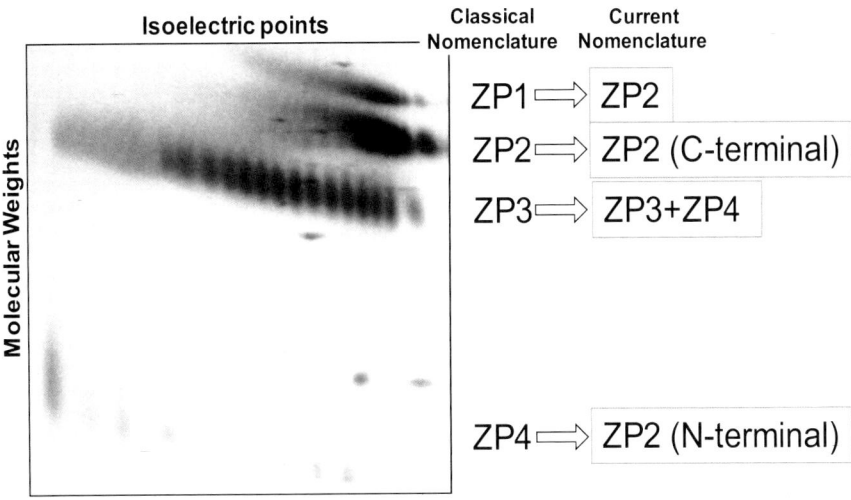

Fig. 16.4 Two-dimensional electrophoresis of pig ZP. In protein analysis, pig ZP was apparently separated into four glycoprotein families with different molecular weights. In earlier studies, the protein families were named ZP1, ZP2, ZP3, and ZP4. Later, it was found that ZP3 included two protein families coded by different genes. Additionally, ZP1 is shown to cleave into two regions. From electrophoretic analysis, the four protein families correspond to ZP2, ZP2 (C-terminal), ZP3 + ZP4, and ZP2 (N-terminal) by the new nomenclature based on gene sequencing. Remarkably, ZP1 is absent in pigs. Each family contains multiple members with slightly different isoelectric points. This diversity is due to posttranscriptional modification, i.e., characteristic properties of carbohydrate moieties (modified from J Reprod Fertil 105:295 1995)

a heterodimer and make filament structures [18]. ZP1 connects the filament structures mediated with ZP3 to form a layer that accumulates as multiple layers.

The space in the structure holds water and some essential factors for oocyte growth and fertilization. Nowadays, more precise analyses have revealed that ZP domains located in each ZP protein serve important roles to maintain the three-dimensional structure of ZP, as described in the next section. In pigs, however, complicated electrophoretic migration patterns appeared. Four ZP proteins were detected in two-dimensional SDS-PAGE (Fig. 16.4), and named ZP1, ZP2, ZP3, and ZP4 in descending molecular weight [19–21].

Later, it was found that ZP3 contained two ZP subfamilies of proteins coded by two independent genes, named ZP3α and ZP3β [22]. Additionally, classical ZP1 was revealed to cleave into two regions, classical ZP2 and classical ZP4. The classical ZP4 as shown in Fig. 16.4 occurs from classical ZP1 by the action of proteolytic enzymes after fertilization, or spontaneous rupture of cortical granules that included ZP cleavage proteinase [23].

Collectively, the four protein families, ZP1, ZP2, ZP3, and ZP4, named in the classical nomenclature from electrophoresis correspond to intact ZP2, fragment ZP2 (C-terminal), ZP3 + ZP4, and fragment ZP2 (N-terminal) from the modern nomenclature based on gene sequencing as shown in Fig. 16.4. More complicatedly, however, as *zp1* gene is a pseudogene, ZP1 protein is absent in pigs. Each protein family contains multiple members with slightly different isoelectric points. This diversity is due to posttranscriptional modifications that are characteristic properties of carbohydrate moieties. In general, egg coats include carbohydrate moieties. Pig ZP especially shows tremendous diversities (Fig. 16.4).

Different research groups used different nomenclatures for the same proteins and increased the confusion. To unify nomenclatures among mammalian ZP glycoproteins, Harris proposed a nomenclature based on genes [24], named *ZPA, ZPB,* and *ZPC*, corresponding to gene names, but this was still insufficient because they used the alphabet—*ZPA* is the longest and *ZPC* the shortest.

Extensive phylogenetic analyses showed a dual nomenclature system was no longer feasible. Spargo and Hope classified major paralogous ZP proteins because non-mammalian vertebrates contain higher numbers of ZP proteins than mammals [9, 25].

To remove the ambiguities, Conner et al. proposed a simplified nomenclature limited to mammalian animals, in which the ZP genes were called modern ZP1, ZP2, ZP3, and ZP4, which are not classical nomenclatures [10]. They also analyzed other species across the evolutionary tree and defined pathology to cover all mammalian species, as shown in Fig. 16.1. Moreover, this nomenclature can cover other vertebrates as well as mammals. ZPD/ZPAX proteins present in birds and reptiles have not been detected in mammals. In birds and reptiles, the names ZPD/ZPAX are still used in this field (Figs. 16.1 and 16.2).

DNA technology has increased comprehensive understanding of egg coat structure in the evolutional processes. The presence of interspecific variations of egg coat proteins became clear and valid. Besides mammalian ZP, the proteins coded by *ZPD/ZPAX* appear in reptiles and birds. The *ZPAX* gene is also present in the mammalian genome, albeit as a non-transcriptional pseudogene. *ZPD* gene seems to be completely absent in present-day mammals. There are higher numbers of *ZP* genes in reptiles and birds than in mammalians. Since oviparous animals should have avoided harmful environment for embryo growth, additional ZP products were beneficial for species survival. Mammalian animals might have lost *ZPD/ZPAX* genes due to the development of gestation systems.

Differing from other mammalian animals, ZP4 is a pseudogene in mice. As shown in Figs. 16.1 and 16.2, primates such as chimpanzees and humans have four ZP genes that are translated into the corresponding four proteins, whereas mice, dogs, and cows have only three ZP proteins. In dogs and cows, *ZPAX* and *ZP1* are pseudogenes. In mice, *ZP4* is a pseudogene (Fig. 16.1). Additionally, *ZP1* gene is lacking in pigs, as mentioned previously. Recently, it was found that rabbit and cat also express four ZP proteins as shown in Fig. 16.2 [15]. The ZP in the present animal species seems to have developed with different evolutional processes.

16.3 Conservation of ZP Module (or ZP Domain)

The primary structures of mouse [26–28] and pig [29, 30] ZP proteins were deduced by cDNA sequences that were prolificated from ovarian tissues. In mice, a region named the zona domain (consisting of 260 amino acids) had been identified in all ZP proteins. The domain had been implicated in the polymerization of extracellular matrices [31–33]. The connections of each ZP protein, as shown in Fig. 16.3, are mediated with the ZP domains. Furthermore, this domain contained conserved cysteine residues serving a structural function to form disulfide bonds. It is likely that ZP domain maintains common structures in various animal egg coats.

In 2003, each mouse ZP sequence was directly determined using mass spectrometry of isolated natural ZP proteins [31]. The authors found that the N-termini of ZP1 and ZP3, but not ZP2, were blocked by cyclization of glutamine to pyroglutamate and that consensus sequences of protein cleavage sites lie near the C-termini of ZP1, ZP2, and ZP3.

Thereafter, ZP studies were carried out by artificial DNA mutagenesis, i.e., gene manipulation. First, the technology focused on the hydrophobic transmembrane domain of ZP3 at the C-terminus. ES cell lines transformed by modified *ZP3* gene could not secrete recombinant ZP3 protein or assemble to ZP formation [32, 33]. The region coding the hydrophobic transmembrane domain of ZP3 was essential for intact ZP assembly. Coincidently, it was demonstrated that the C-terminal transmembrane domain and short cytoplasmic tail of ZP2 and ZP3 are essential for assembly of these ZP molecules [34]. It indicated that the C-terminal region is conserved in ZP proteins and serves for polymerization of ZP. Later, Jovine et al. demonstrated by crystallographic study that the region, named ZP module, constituted the three-dimensional structure of ZP [35].

As the most prominent study, Jovine's group used an X-ray crystallization method to determine the sperm binding site of ZP protein. They succeeded in the crystallization of ZP2 and ZP3 proteins. They made clear that there are ZP domains possessing sequence homology in these ZP proteins in mice. The module is composed of N-terminal (ZP-N) and C-terminal (ZP-C) domains (comprising ~260 amino acids), and the ZP-N domains were repeated in ZP2 protein. Furthermore, the N-terminal domain located in the N-terminal of ZP2 named ZP-N1 formed a sperm binding groove. The current consensus domain structure of ZP1-ZP4 proteins is proposed as shown in Fig. 16.4.

ZP3 and other ZP proteins include a common region (named ZP module) consisting of ~260 amino acids as shown in Fig. 16.5. ZP module is conserved in egg coat proteins in most animal species [35].

Protein crystallization analysis revealed that ZP module was composed of the N-terminal (ZP-N) and C-terminal (ZP-C) regions. As *ZP3* is the smallest among all *ZP* genes, it contains a single ZP module comprising the ZP-N and ZP-C regions. The other ZP genes were found to contain several ZP-N regions and a single ZP-C region in ZP module. The extension of ZP-N in ZP2 is thought to form sperm binding structures in mice.

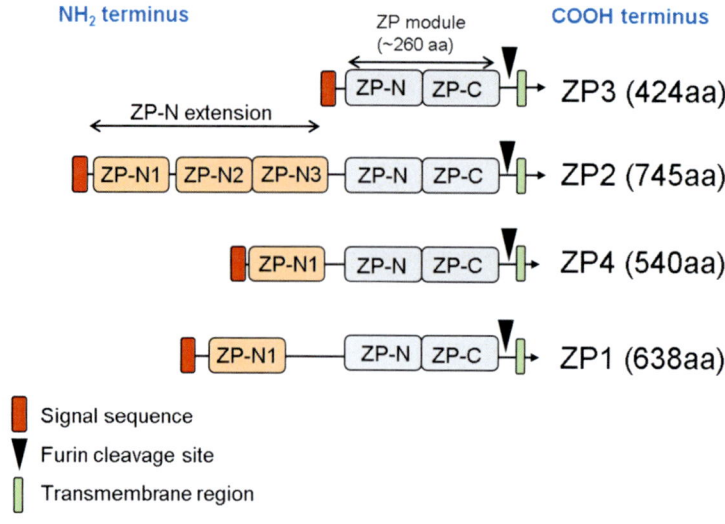

Fig. 16.5 Schematic diagram of ZP1, ZP2, ZP3, and ZP4 based on human homologs. "ZP modules" are included in all ZP proteins as connected forms of ZP-N and ZP-C. The homologous regions of ZP-N, named ZP-N and ZP-N1-4, are probably derived from the original ZP-N located in the ZP module during evolution (modified from Bioinformatics 23:1871 2007)

This structural element, present in extracellular proteins from a wide variety of organisms from nematodes to mammals, consists of ~260 amino acids. The domain had been identified in not only ZP but also TGF-beta Type III receptor, uromodulin, and major zymogen granule membrane protein (GP-2). It is suggested that the domain has a similar tertiary structure and function to ZP. In addition, these domains conserve eight cysteine (Cys) residues and are located close to the C-terminus of the polypeptide [36]. Inter- and intra-disulfide bonds in the ZP module are significantly folded ZP structures in chickens [37], rats [38], and pigs [39]. Furthermore, it was demonstrated that invertebrate egg coat protein VERL interact with sperm protein lysin. VERL repeats have essential function for sperm binding, similar to N-terminal repeat of mammalian sperm receptor ZP2 [40]. In addition, other conservative regions such as the trefoil region and consensus region for carbohydrate moieties should also contribute to ZP structure and functions [41, 42]. Details are issues in the future.

Investigations of database analysis revealed that ZP-N domains in ZP motif extend to the N-terminus [37]. Conclusively, they showed that there are one, two, two, and four ZP-N domains in ZP3, ZP1, ZP4, and ZP2, respectively (Fig. 16.5). Additionally, ZPAX, which is present in *Xenopus* but not in mammals, contained six ZP-N domains [43].

Recently, crystallographic analysis revealed that the segment of ZP-N repeats in egg coat is the lysin binding site. Lysin is a protein released from abalone (a species of mollusk) sperm at fertilization and egg coat lysis. The segment, therefore, was proposed to be a sperm binding site in egg coat of abalone [44, 45].

It also suggested that there is a common site for sperm recognition folding between invertebrates (abalone) and vertebrates (mice). Surprisingly, although sequence homology is just 10%, the region forms similar sperm recognition structures such as grooves. They claimed that this segment has been conserved evolutionally from abalone to mammals as a fertilization-related site.

Furthermore, it was documented that the N-terminal region of ZP2 served an important role in species-specific binding of sperm. The researchers proposed a species-specific sperm binding site on ZP2 using a mouse gene manipulation method [46]. The gene-manipulated mouse ZP in which mouse ZP2 was replaced with human ZP2 could bind to human sperm. The sperm binding site corresponded to the region (51–149) of human ZP2. The region was the sperm binding site that had been evolutionally conserved.

16.4   The Classical Paradigm of Sperm-ZP Interaction

It was observed that sperm needed to spend some time in the female reproductive tract before penetrating the ZP [47, 48]. This was referred to as sperm "capacitation." Later, it was found that in both in vitro and in vivo fertilization, the acrosome reaction of sperm was absolutely necessary to penetrate the ZP [49, 50]. Capacitation and acrosome reaction are thus essential sperm fertilization processes in various mammalian species. Drastic biochemical changes occur in sperm through acrosome reaction [51]. The natural trigger of acrosome reaction induction is still unknown, however, although ZP, albumin, cholesterol, and heparin have been candidated.

In the past, mouse ZP morphologies were investigated using microscopic observation and stoichiometric analysis [17, 52]. Wassarman and his colleagues hypothesized that ZP2 and ZP3 formed heterodimers to fibrous structures and ZP1 connected the fibers (Fig. 16.3).

Around the same time, sperm-ZP interaction was intensively studied. The results showed that there were several steps in sperm-ZP interaction. First, sperm attached to ZP, then it bound tightly, and then it started to penetrate. Each step showed a different molecular interaction between sperm and ZP. Biochemical studies indicated that carbohydrate portions were located in ZP and played critical roles for fertilization [53, 54].

As mentioned previously, spermatozoa must undergo acrosome reaction by attaching to ZP. The classical paradigm of sperm-ZP interaction was established by PM Wassarman and his colleagues [55, 56]. They have been analyzing sperm-ZP binding mechanisms for more than 40 years. In the 1980s, Bleil et al. analyzed isolated mouse ZP proteins and revealed ZP composition and apparent molecular weights by SDS-PAGE [57]. Eventually mouse ZP protein terminology was determined as ZP1, ZP2, and ZP3. They also reported ZP2 cleavage after fertilization and suggested that ZP2 contributes to polyspermy blocking [58, 59]. The study provided the principal evidence to date. The research group proposed a concept of sperm receptors in ZP. According to the concept, the O-linked carbohydrates of ZP3 (named the primary receptor) induce acrosome reaction in acrosome-intact sperm [60, 61], and the acrosome-reacted sperm tightly binds to ZP via ZP2 (named the secondary receptor) which triggers sperm penetration of ZP [62]. This multistep recognition has been widely accepted for many years.

Their research on sperm-ZP interaction progressed to sub-molecular levels. They showed the O-linked carbohydrate in ZP3 to be the primary sperm receptor by biochemical and in vitro fertilization techniques. · It was assumed that acrosome-intact sperm bound to the sperm receptor on ZP3 and induced acrosome reaction, and then acrosome-reacted sperm bound tightly to the second sperm receptor on ZP2. The secondary binding was proposed to be a protein-protein interaction which triggered sperm penetration of ZP.

In that same period, on the sperm side, intriguing molecular systems were hypothesized as a sperm-ZP binding interaction by Shur's group [63, 64]. They discovered the presence of β1,4-galactotransferase (GalTase) on mouse sperm acrosome surfaces. This enzyme recognized N-acetylglucosamine residue to transfer galactose to the carbohydrate terminal [65].

The enzyme was thought to be the primary binding ligand of sperm-ZP at a molecular level. The finding matched the hypothesis that the O-linked carbohydrate nonreducing terminal galactose was essential for sperm binding to ZP. Although this hypothesis was very attractive, it was later weakened by gene-targeting technology. GalTase gene knockout (KO) mice were in fact found to maintain fertility, as sperm could still bind and penetrate ZP [66]. The minor phenotype of GalTase KO sperm showed an inability to bind epididymal glycoconjugates. The researchers claimed that function of GalTase is to normally maintain sperm in an "uncapacitated" state to prevent premature capacitation. In GalTase gene knockout (KO) mice, sperm capacitation easily occurred followed by acrosome reaction and fertilization [67].

Shur reassessed mouse sperm-ZP binding, looking over his past research, and concluded that sperm-ZP binding is a result of multiple and distinct events, not a single molecular mechanism like ligand-receptor interaction. Biochemical and gene-targeting approaches are now the focus of researchers' interest. Thus, the real mechanism of sperm-ZP binding has not been deciphered, because it is unknown how gamete growth, storage, maturation, and transport are related to the targeted molecule [68].

Gene-targeting technology has also been applied to the ZP3 side. Genetically, KO mice lacking the ZP3 region for O-linked carbohydrate were produced. The ZP from gene-modified mice could not bind sperm. The female mice were infertile as expected. This result was evidence that the O-linked carbohydrates on ZP3 do contribute to sperm recognition or binding as a primary receptor [69]. However, controversial observations were also reported, in which O-linked carbohydrate moieties were not required for sperm-ZP binding [70].

Sperm and egg coat recognition is thought to be relatively species-specific, but in vitro it depends on the combination of mammalian species, as we showed (Fig. 16.6). Human sperm do not attach or bind to ZP from different animal species such as farm animals (pig, etc.) in vitro except for rabbits. Rabbit sperm, however, attach to the ZP from various foreign species (as shown in red). Also rabbit ZP

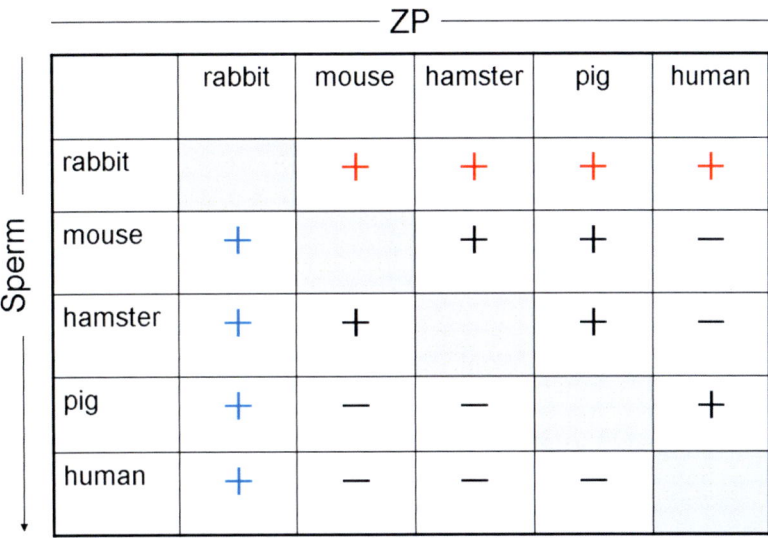

Fig. 16.6 Interspecies sperm and ZP binding. Rabbit sperm binds to the ZP from foreign species (as shown in red) and rabbit ZP also allows foreign sperm to bind and penetrate (as shown in blue). Mouse and hamster sperm binds to the ZP from three foreign species. Pig sperm binds to the ZP from two foreign species. Human sperm does not bind to the ZP from foreign species except rabbit. Greyed cells indicate homologous species combinations of sperm and ZP

allows foreign sperm to bind and penetrate (as shown in blue). Mouse and hamster sperm bind to the ZP from three foreign species. Pig sperm bind to the ZP from two foreign species.

Similar observations were reported in a previous paper [71]. Sperm can associate with ZP in a nonspecific as well as specific manner. In addition, decades ago, it was reported that human spermatozoa would not attach to the zona surface of sub-hominoid primates (baboon, rhesus monkey, squirrel monkey) [72].

The enzymatic affinity of acrosin, a sperm-specific protease required for ZP penetration, may reflect the species-specific manner of ZP penetration by sperm. Sperm binding to and penetration of ZP are possibly based on different molecular mechanisms in mammalian species. Indeed, in other species such as pigs and bovines, sperm-ZP binding has been shown as having quite different mechanisms than in mice. Domestic animals such as pigs and bovines showed that the neutral N-linked carbohydrate chains are sperm binding sites [73, 74].

Figure 16.7 shows sperm-ZP binding patterns in homologous and heterologous combinations. In homologous combinations, mouse and pig sperm bind exclusively to homologous ZP. In heterologous combinations, mouse sperm bind to pig ZP, but not the reverse. Mouse sperm probably bind to pig ZP in a lectin-like manner. It is assumed that pig sperm do not have a lectin-like molecule that recognizes carbohydrates in mouse ZP. The binding is not a system for species recognition. The results of these studies suggest that spermatozoa can associate with zona in a nonspecific as well as a specific manner.

Homologous combinations of sperm and ZP

Mouse sperm x Mouse ZP   Pig sperm x Pig ZP

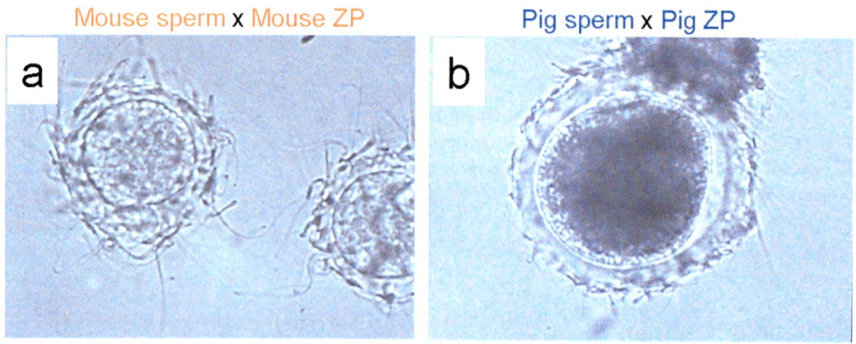

Heterologous combinations of sperm and ZP

Pig sperm x Mouse ZP   Mouse sperm x Pig ZP

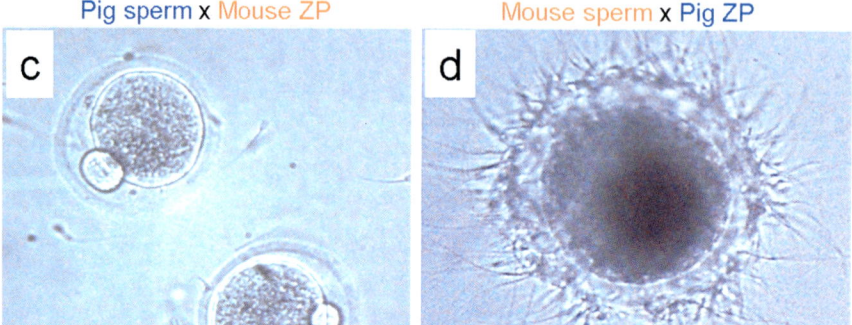

Fig. 16.7 Comparison of ZP-sperm binding in homologous and heterologous species combinations. It is natural that numerous sperms bind to ZP both in mice (**a**) and pigs (**b**) under homologous conditions. Under heterologous conditions, however, pig sperm does not bind to mouse ZP (**c**) but mouse sperm binds to pig ZP (**d**). The patterns of sperm binding differ according to the species combination

It was reported that the sperm acrosome reaction did not necessarily occur on the ZP surface after attachment to the ZP [75]. In the results, acrosome-reacted sperm were observed before attachment which penetrated ZP. Moreover, Inoue et al. reported that motile sperm having penetrated the ZP into the perivitelline space could penetrate the ZP of pre-fertilized oocytes [76]. Similar observation has been reported in rabbits [77].

These results indicated that acrosome reaction need not take place on the surface of ZP. Acrosome reaction is required to penetrate ZP, but different molecules could possibly become triggers in vitro. On the other hand, Baibakov et al. documented that sperm binding to the ZP is not sufficient to induce acrosome reaction [78]. Taken together, what is the physiological trigger for acrosome reaction? Current complicated state may suggest a paradigm shift in the classical fertilization model.

Gene manipulation technology has brought significant progress in sperm-ZP recognition, but the molecular mechanisms of sperm-ZP interaction are not sufficiently understood. Comprehensive studies of testis-specific gene expression are going on to determine sperm function at fertilization. These results indicate that what has been observed in in vitro fertilization (IVF) differs significantly from what occurs during "physiological" fertilization [79]. The most classically anticipated sperm molecules did not contribute to ZP binding or penetration according to a recent review [80].

References

1. Ohno S, Wolf U, Atkin NB. Evolution from fish to mammals by gene duplication. Hereditas. 1968;59:169–87.
2. Grey RD, Don PW, Hedrick JL. Formation and structure of the fertilization envelope in Xenopus laevis. Dev Biol. 1974;36:44–61.
3. Grey RD, Working PK, Hedrick JL. Alteration of structure and penetrability of the vitelline envelope after passage of eggs from coelom to oviduct in Xenopus laevis. J Exp Zool. 1977;201 (1):73–83.
4. Gerton GL, Hedrick JL. The vitelline envelope to fertilization envelope conversion in eggs of Xenopus laevis. Dev Biol. 1986;116(1):1–7.
5. Hedrick JL. Comparative structural and antigenic properties of zona pellucida glycoproteins. J Reprod Fertil. 1996;50:9–17.
6. Hedrick JL. Anuran and pig egg zona pellucida glycoproteins in fertilization and early development. Int J Dev Biol. 2008;52:683–701.
7. Shu L, Suter MJ, Räsänen K. Evolution of egg coats: linking molecular biology and ecology. Mol Ecol. 2015;24:4052–73.
8. Goudet G, Mugnier S, Callebaut I, Monget P. Phylogenetic analysis and identification of pseudogenes reveal a progressive loss of zona pellucida genes during evolution of vertebrates. Biol Reprod. 2008;78(5):796–806.
9. Spargo SC, Hope RM. Evolution and nomenclature of the zona pellucida gene family. Biol Reprod. 2003;68:358–62.
10. Conner SJ, Lefièvre L, Hughes DC, Barratt CL. Cracking the egg: increased complexity in the zona pellucida. Hum Reprod. 2005;20(5):1148–52.
11. Ganguly A, Sharma RK, Gupta SK. Bonnet monkey (Macaca Radiata) ovaries, like human oocytes, express four zona pellucida glycoproteins. Mol Reprod Dev. 2008;75(1):156–66.
12. Izquierdo-Rico MJ, Jimenez-Movilla M, Llop E, Perez-Oliva AB, Ballesta J, Gutierrez-Gallego R, Jimenez-Cervantes C, Aviles M. Hamster zona pellucida is formed by four glycoproteins: ZP1, ZP2, ZP3, and ZP4. J Proteome Res. 2009;8(2):926–41.
13. Stetson I, Izquierdo-Rico MJ, Moros C, Chevret P, Lorenzo PL, Ballesta J, Rebollar PG, Gutiérrez-Gallego R, Avilés M. Rabbit zona pellucida composition: a molecular, proteomic and phylogenetic approach. J Proteome. 2012;75(18):5920–35.
14. Lamas-Toranzo I, Fonseca Balvís N, Querejeta-Fernández A, Izquierdo-Rico MJ, González-Brusi L, Lorenzo PL, García-Rebollar P, Avilés M, Bermejo-Álvarez P. ZP4 confers structural properties to the zona pellucida essential for embryo development. Elife. 2019;8:e48904.
15. Stetson I, Avilés M, Moros C, García-Vázquez FA, Gimeno L, Torrecillas A, Aliaga C, Bernardo-Pisa MV, Ballesta J, Izquierdo-Rico MJ. Four glycoproteins are expressed in the cat zona pellucida. Theriogenology. 2015;83(7):1162–73.
16. Frankenberg F, Renfree MB. Conceptus coats of marsupials and monotremes. Curr Top Dev Biol. 2018;130:357–77.

17. Bleil JD, Wassarman PM. Structure and function of the zona pellucida: identification and characterization of the proteins of the mouse oocyte's zona pellucida. Dev Biol. 1980;76: 185–202.

18. Greve JM, Wassarman PM. Mouse egg extracellular coat is a matrix of interconnected filaments possessing a structural repeat. J Mol Biol. 1985;181(2):253–64.

19. Hedrick JL, Wardrip NJ. Isolation of the zona pellucida and purification of its glycoprotein families from pig oocytes. Anal Biochem. 1986;157(15):63–70.

20. Yurewiczs EC, Sacco AG, Subramanian MG. Structural characterization of porcine oocyte zona pellucida the Mr 55,000 antigen (ZP3) of porcine oocyte. J Biol Chem. 1987;262(2):564–71.

21. Koyama K, Hasegawa A, Inoue M, Isojima S. Blocking of human sperm-zona interaction by monoclonal antibodies to a glycoprotein family (ZP4) of porcine zona pellucida. Biol Reprod. 1991;45(5):727–35.

22. Sacco AG, Yurewicz EC, Subramanian MG, Matzat PD. Porcine zona pellucida: association of sperm receptor activity with the alpha-glycoprotein component of the Mr = 55,000 family. Biol Reprod. 1989;41(3):523–32.

23. Hedrick JL, Wardrip NJ. On the macromolecular composition of the zona pellucida from porcine oocytes. Dev Biol. 1987;121(2):478–88.

24. Harris JD, Hibler DW, Fontenot GK, Hsu KT, Yurewicz EC, Sacco AG. Cloning and characterization of zona pellucida genes and cDNAs from a variety of mammalian species: the ZPA, ZPB and ZPC gene families. DNA Seq. 1994;4(6):361–93.

25. Lefièvre L, Conner SJ, Salpekar A, Olufowobi O, Ashton P, Pavlovic B, Lenton W, Afnan M, Brewis IA, Monk M, Hughes DC, Barratt CL. Four zona pellucida glycoproteins are expressed in the human. Hum Reprod. 2004;19(7):1580–6.

26. Ringuette MJ, Chamberlin ME, Baur AW, Sobieski DA, Dean J. Molecular analysis of cDNA coding for ZP3, a sperm binding protein of the mouse zona pellucida. Dev Biol. 1988;127:287–95.

27. Liang LF, Chamow SM, Dean J. Oocyte-specific expression of mouse Zp-2: developmental regulation of the zona pellucida genes. Mol Cell Biol. 1990;10:1507–15.

28. Epifano O, Liang LF, Familari M, Moos MC Jr, Dean J. Coordinate expression of the three zona pellucida genes during mouse oogenesis. Development. 1995;121:1947–56.

29. Hasegawa A, Koyama K, Okazaki Y, Sugimoto M, Isojima S. Amino acid sequence of a porcine zona pellucida glycoprotein ZP4 determined by peptide mapping and cDNA cloning. J Reprod Fertil. 1994;100(1):245–55.

30. Taya T, Yamasaki N, Tsubamoto H, Hasegawa A, Koyama K. Cloning of a cDNA coding for porcine zona pellucida glycoprotein ZP1 and its genomic organization. Biochem Biophys Res Commun. 1995;207(2):790–9.

31. Boja ES, Hoodbhoy T, Fales HM, Dean J. Structural characterization of native mouse zona pellucida proteins using mass spectrometry. J Biol Chem. 2003;278(36):34189–202.

32. Litscher ES, Qi H, Wassarman PM. Mouse zona pellucida glycoproteins mZP2 and mZP3 undergo carboxy-terminal proteolytic processing in growing oocytes. Biochemistry. 1999;38:2280–7.

33. Zhao M, Gold L, Dorward H, Liang L, Hoodbhoy T, Boja E, Fales HM, Dean J. Mutation of a conserved hydrophobic patch prevents incorporation of ZP3 into the zona pellucida surrounding mouse eggs. Mol Cell Biol. 2003;23(24):8982–91.

34. Bork P, Sander C. A large domain common to sperm receptors (Zp2 and Zp3) and TGF-β type III receptor. FEBS Lett. 1992;300:237–40.

35. Jovine L, Qi H, Williams Z, Litscher E, Wassarman PM. The ZP domain is a conserved module for polymerization of extracellular proteins. Nat Cell Biol. 2002;4(6):457–61.

36. Jovine L, Qi H, Williams Z, Litscher ES, Wassarman PM. A duplicated motif controls assembly of zona pellucida domain proteins. Proc Natl Acad Sci U S A. 2004;101(16):5922–7.

37. Jovine L, Darie CC, Litscher ES, Wassarman PM. Zona pellucida domain proteins. Annu Rev Biochem. 2005;74:83–114.

38. Killingbeck EE, Swanson WJ. Egg coat proteins across metazoan evolution. Curr Top Dev Biol. 2018;130:443–88.
39. Han L, Monné M, Okumura H, Schwend T, Cherry AL, Flot D, Matsuda T, Jovine L. Insights into egg coat assembly and egg-sperm interaction from the x-ray structure of full-length ZP3. Cell. 2010;143:404–15.
40. Boja ES, Hoodbhoy T, Garfield M, Fales HM. Structural conservation of mouse and rat zona pellucida glycoproteins. Probing the native rat zona pellucida proteome by mass spectrometry. Biochemistry. 2005;44:16445–60.
41. Kanai S, Kitayama T, Yonezawa N, Sawano Y, Tanokura M, Nakano M. Disulfide linkage patterns of pig zona pellucida glycoproteins ZP3 and ZP4. Mol Reprod Dev. 2008;75:847–56.
42. Raj I, Hosseini HSAI, Dioguardi E, Nishimura K, Han L. Structural basis of egg coat-sperm recognition at fertilization. Cell. 2017;169(7):1315–26.
43. Monné M, Han L, Schwend T, Burendahl S, Jovine L. Crystal structure of the ZP-N domain of ZP3 reveals the core fold of animal egg coats. Nature. 2008;456(7222):653–7.
44. Monné M, Jovine L. A structural view of egg coat architecture and function in fertilization. Biol Reprod. 2011;85(4):661–9.
45. Callebaut I, Mornon JP, Monget P. Isolated ZP-N domains constitute the N-terminal extensions of zona pellucida proteins. Bioinformatics. 2007;23(15):1871–4.
46. Avella MA, Baibakov B, Dean J. A single domain of the ZP2 zona pellucida protein mediates gamete recognition in mice and humans. J Cell Biol. 2014;205(6):801–9.
47. Chang MC. Fertilization capacity of spermatozoa deposited into the fallopian tubes. Nature. 1951;168(4277):697–8.
48. Austin CR. Observation on the penetration of the sperm into mammalian eggs. Aust J Biol Sci. 1951;4(4):581–96.
49. Austin C, Bishop M. Role of the rodent acrosome and perforatoruim in fertilization. Proc R Soc Lond B Biol Sci. 1958;149(935):241–8.
50. Yanagimachi R. Mammalian fertilization, the physiology of reproduction. New York: Raven Press, Ltd; 1994.
51. Green DP. Three-dimensional structure of the zona pellucida. Rev Reprod. 1997;2(3):147–56.
52. Oikawa T, Yanagimachi R, Nicolson GL. Wheat germ agglutinin blocks mammalian fertilization. Nature. 1973;241(5387):256–9.
53. Harvey M, Florman HM, Wassarman PM. O-linked oligosaccharides of mouse egg ZP3 account for its sperm receptor activity. Cell. 1985;41(1):313–24.
54. Shalgi R, Matityahu A, Nebel L. The role of carbohydrates in sperm-egg interaction in rats. Biol Reprod. 1986;34(3):446–52.
55. Florman HM, Storey BT. Mouse gamete interactions: the zona pellucida is the site of the acrosome reaction leading to fertilization in vitro. Dev Biol. 1982;91(1):121–30.
56. Wassarman PM. Zona pellucida glycoproteins. J Biol Chem. 2008;283(36):24285–9.
57. Bleil JD, Wassarman PM. Mammalian sperm-egg interaction: identification of a glycoprotein in mouse egg zonae pellucidae possessing receptor activity for sperm. Cell. 1980;20(3):873–82.
58. Bleil JD, Beall CF, Wassarman PM. Mammalian sperm-egg interaction: fertilization of mouse eggs triggers modification of the major zona pellucida glycoprotein, ZP2. Dev Biol. 1981;86 (1):189–97.
59. Bleil JD, Wassarman PM. Sperm-egg interactions in the mouse: sequence of events and induction of the acrosome reaction by a zona pellucida glycoprotein. Dev Biol. 1983;95 (2):317–24.
60. Florman HM, Wassarman PM. O-linked oligosaccharides of mouse egg ZP3 account for its sperm receptor activity. Cell. 1985;41(1):313–24.
61. Bleil JD, Wassarman PM. Galactose at the nonreducing terminus of O-linked oligosaccharides of mouse egg zona pellucida glycoprotein ZP3 is essential for the glycoprotein's sperm receptor activity. Proc Natl Acad Sci U S A. 1988;85:6778–82.

62. Bleil JD, Greve JM, Wassarman PM. Identification of a secondary sperm receptor in the mouse egg zona pellucida: role in maintenance of binding of acrosome-reacted sperm to eggs. Dev Biol. 1988;128(2):376–85.
63. Shur BD, Hall NG. A role for mouse sperm surface galactosyltransferase in sperm binding to the egg zona pellucida. J Cell Biol. 1982;95(2 Pt 1):574–9.
64. Lopez LC, Bayna EM, Litoff D, Shaper NL, Shaper JH, Shur BD. Receptor function of mouse sperm surface galactosyltransferase during fertilization. J Cell Biol. 1985;101(4):1501–10.
65. Shur BD, Hall NG. Sperm surface galactosyltransferase activities during in vitro capacitation. J Cell Biol. 1982;95(2 Pt 1):567–73.
66. Lu Q, Shur BD. Sperm from beta 1,4-galactosyltransferase-null mice are refractory to ZP3-induced acrosome reactions and penetrate the zona pellucida poorly. Development. 1997;124 (20):4121–31.
67. Rodeheffer C, Shur BD. Sperm from beta1,4-galactosyltransferase I-null mice exhibit precocious capacitation. Development. 2004;131(3):491–501.
68. Shur BD. Reassessing the role of protein-carbohydrate complementarity during sperm-egg interactions in the mouse. Int J Dev Biol. 2008;52(5–6):703–15.
69. Kinloch RA, Sakai Y, Wassarman PM. Mapping the mouse ZP3 combining site for sperm by exon swapping and site-directed mutagenesis. Proc Natl Acad Sci U S A. 1995;92(1):263–7.
70. Williams SA, Xia L, Cummings RD, McEver RP, Stanley P. Fertilization in mouse does not require terminal galactose or N-acetylglucosamine on the zona pellucida glycans. J Cell Sci. 2007;120(Pt 8):1341–9.
71. Swenson CE, Dunbar BS. Specificity of sperm-zona interaction. J Exp Zool. 1982;219(1):97–104.
72. Bedford JM. Sperm/egg interaction: the specificity of human spermatozoa. Anat Rec. 1977;188 (4):477–87.
73. Yonezawa N, Amari S, Takahashi K, Ikeda K, Imai FL, Kanai S, Kikuchi K, Nakano M. Participation of the nonreducing terminal beta-galactosyl residues of the neutral N-linked carbohydrate chains of porcine zona pellucida glycoproteins in sperm-egg binding. Mol Reprod Dev. 2005;70(2):222–7.
74. Amari S, Yonezawa N, Mitsui S, Katsumata T, Hamano S, Kuwayama M, Hashimoto Y, Suzuki A, Takeda Y, Nakano M. Essential role of the nonreducing terminal alpha-mannosyl residues of the N-linked carbohydrate chain of bovine zona pellucida glycoproteins in sperm-egg binding. Mol Reprod Dev. 2001;59(2):221–6.
75. Jin M, Fujiwara E, Kakiuchi Y, Okabe M, Satouh Y, Baba SA, Chiba K, Hirohashi N. Most fertilizing mouse spermatozoa begin their acrosome reaction before contact with the zona pellucida during in vitro fertilization. Proc Natl Acad Sci U S A. 2011;108(12):4892–6.
76. Inoue N, Satouh Y, Ikawa M, Okabe M, Yanagimachi R. Acrosome-reacted mouse spermatozoa recovered from the perivitelline space can fertilize other eggs. Proc Natl Acad Sci U S A. 2011;108(50):20008–11.
77. Kuzan FB, Fleming AD, Seidel GE Jr. Successful fertilization in vitro of fresh intact oocytes by perivitelline (acrosome-reacted) spermatozoa of the rabbit. Fertil Steril. 1984;41(5):766–70.
78. Baibakov B, Gauthier L, Talbot P, Rankin TL, Dean J. Sperm binding to the zona pellucida is not sufficient to induce acrosome exocytosis. Development. 2007;134(5):933–43.
79. Ikawa M, Inoue N, Benham AM, Okabe M. Fertilization: a sperm's journey to and interaction with the oocyte. J Clin Invest. 2010;120(4):984–94.
80. Okabe M. Sperm-egg interaction and fertilization: past, present, and future. Biol Reprod. 2018;99(1):134–46.

Chapter 17
Function of ZP

Akiko Hasegawa

Abstract The main function of mammalian ZP is to maintain growing oocytes in the ovary and to facilitate their maturation. After ovulation, ZP plays several important roles such as sperm recognition, prevention of polyspermy, and protection of preimplantation embryos. Taxonomic interactions of sperm and ZP have been studied particularly intensively. Dean's group has succeeded in establishing ZP1, ZP2, and ZP3 gene knockout mouse strains. They have demonstrated that ZP2- and ZP3-gene-targeted mice are infertile. These mice showed disrupted ovarian function and their follicles did not grow. Morphological analysis showed that ZP2 and ZP3 KO mice did not form ZP architecture. ZP1-targeted mice were fertile but fecundity was significantly lower than in wild-type. The second important role of ZP is to prevent abnormal fertilization such as heterologous sperm penetration or polyspermy. This chapter explains how ZP avoids abnormal fertilization. In the physiological process, embryos are transported in the oviduct and reach the uterus for implantation. ZP prevents direct attachment of embryonic cells to the oviductal epithelium because oviductal cells recognize embryos as a non-self-antigen and remove them. ZP protects preimplantation embryos from attack by the mother's immune system.

17.1 ZP Formation During Folliculogenesis

The mouse zona pellucida (ZP) is an extracellular matrix generally comprised of four glycoproteins that are mainly synthesized by the oocyte in oogenesis/folliculogenesis [1, 2]. A follicle is a functional unit for oocyte growth. Primordial follicles are composed of dormant oocytes surrounded by a layer of a few epithelial cells but do not express the ZP proteins. In the growth stage, the epithelial cells

A. Hasegawa (✉)
Department of Obstetrics and Gynecology, School of Medicine, Hyogo Medical University, Nishinomiya, Hyogo, Japan
e-mail: zonapel@hyo-med.ac.jp

© The Editor(s) (if applicable) and The Author(s), under exclusive license to Springer Nature Singapore Pte Ltd. 2022
H. Shibahara, A. Hasegawa (eds.), *Gamete Immunology*,
https://doi.org/10.1007/978-981-16-9625-1_17

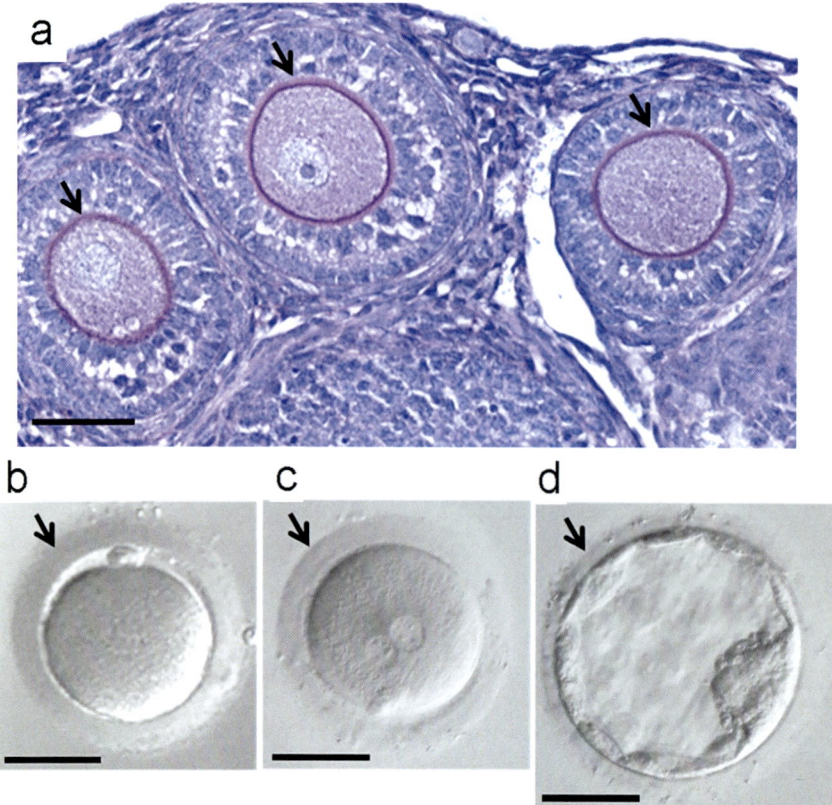

Fig. 17.1 Morphologies of zona pellucida. (**a**) Adult mouse ovarian section. Purple coloring (arrows) around the oocytes shows zona pellucida stained by PAS staining. (**b–d**) Human egg, zygote, and blastocyst surrounded by zona pellucida (arrows). Bars are 50 mm. ZP becomes thinner due to internal pressure of the blastocyst (**d**)

change morphology from flat to a cuboidal shape. This stage is called a primary follicle and the cuboidal cells are called granulosa cells. At the primary follicle stage with an oocyte surrounded by one layer of cuboidal granulosa cells, the ZP can be observed around the oocyte as a discontinuous layer. During follicle growth, the layer of ZP coalesces to form a thickening matrix [3].

The thickness is about 10–15% of the diameter of the ooplasm: 7–8 μm in mice and 15–20 μm in humans at mature stages. After ovulation, a single spermatozoon penetrates the ZP to complete fertilization. Until implantation, the ZP remains around the developing embryo to protect it during transport in the oviduct and the uterus (Fig. 17.1).

In the human ovary, ZP proteins seem to be synthesized in primordial oocytes [4]. The granulosa cells grow to form two to three layers around the oocyte, named secondary follicles. In the next stage, granulosa cells proliferate to seven to eight

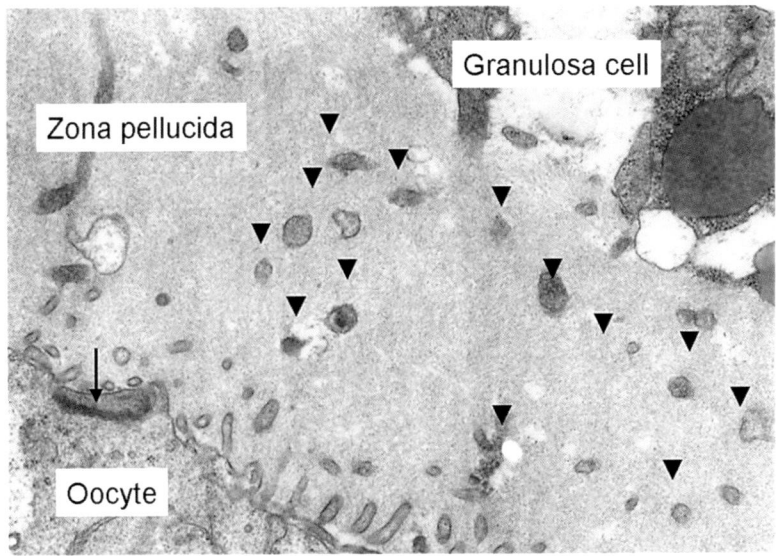

▼ : Microvilli in ZP derived from oocyte and granulosa cell

↓ : Gap junction between oocyte and granulosa cell

Fig. 17.2 Electron microscopic observations. TEM of mouse ZP in the ovary. Numerous microvilli processing from granulosa cells and oocyte can be seen in ZP. This morphology indicates that ZP supports cell-cell interactions. Occasionally gap junctions are also observed. ▼: Microvilli in ZP derived from oocyte and granulosa cell. ↓: Gap junction between oocyte and granulosa cell

layers around the oocyte, and an antrum appears in the follicle, named an antral follicle. During follicle/oocyte growth, the ZP becomes thicker until the oocyte matures (Fig. 17.1).

Figure 17.2 shows a transmission electron micrographic feature of an ultrathin section of mouse ovarian tissues. There are multiple microvilli in the area between granulosa cell and oocyte as shown by small triangles. The microvilli are penetrating through ZP from both sides, oocyte and granulosa cell. Noticeably, a gap junction is observed as shown by an arrow. Gap junctions are channels of ions, nutrient substances, and other important factors for oocyte growth [5, 6]. Nowadays, paracrine/autocrine mechanisms followed by signal transduction pathways are being investigated for research.

Since cAMP (cyclic adenosine monophosphate) is known to inhibit nuclear maturation in the oocytes, it is conceivable that keeping high concentration of cAMP is a crucial step in the maintenance of meiotic arrest of oocytes. A plausible early hypothesis was that the gap junctions allowed cAMP to pass from the granulosa cells to the oocyte. Subsequent studies showed that cAMP produced by the oocyte itself was required to maintain high intracellular cAMP. However, the hypothesis that cAMP was provided by the granulosa cells has not been entirely

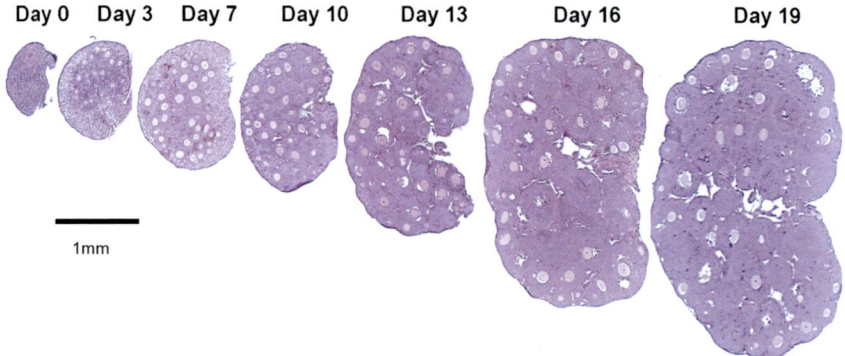

Fig. 17.3 Ovarian morphologies in the gonadotropin-independent phase of 0- to 19-day-old juvenile mice. All sections are prepared to be the maximum of the tissues. Original magnification is ×20. Each section is represented with actual proportions. Oocyte/follicles show synchronized chronological growth

ruled out [7]. cAMP transported from granulosa cells possibly maintains germinal vesicles. Following withdrawal of the microvilli, reduction of oocyte cAMP along with closing of the gap junctions induces resumption of meiosis prophase to metaphase II.

Follicle cell-oocyte gap junctions have the ability to transmit informational molecules such as cAMP between the two cell types [8]. Abundant microvilli and gap junctions promote folliculogenesis via oocyte and granulosa cell communications. Figure 17.2 shows structural evidence for the existence of interactive pathways supported by ZP. It can be said that ZP is a functional structure for oocyte growth and maturation.

More recently, gene-targeting studies have reported that under optical microscopic observations, mouse ZP2 (mZP2 $-/-$) and ZP3 (mZP3 $-/-$) null mice lack the ZP and are completely infertile due to a severe reduction in ovulated oocytes. This suggests that ovarian follicle growth was retarded in those gene-targeted mice. Hence, ZP plays essential roles in folliculogenesis [9].

In mice, the ovary at birth (day 0) is processing primordial follicle formation as shown in Fig. 17.3. At day 3 after birth, numerous primordial follicles and small primary follicles appear. At day 7, primordial follicles locate to the peripheral region of the ovary, and early secondary follicles and primary follicles are observed in the central area of the ovary. Sizes of the oocyte in the follicles as well as the whole ovary are significantly increased. At day 10, the number of secondary follicles has increased markedly. Some of them reach the late stage of "secondary follicle" with seven to eight layers of granulosa cells, but antral follicles are not observed until day 13.

At day 13, the whole ovarian tissue and each oocyte in the ovary become larger still and antral follicles also appear (as shown by arrows). On the other hand, the primordial follicle area becomes smaller. At day 16, various kinds of follicles are

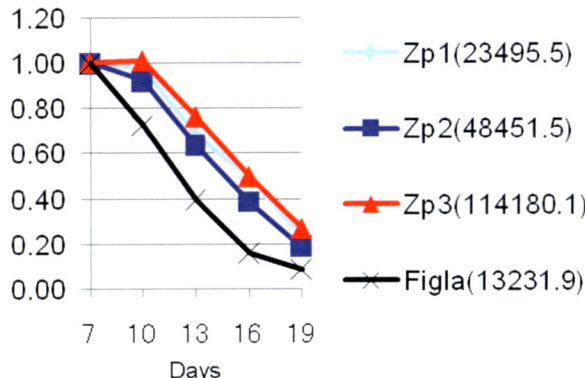

Fig. 17.4 Transcriptome analysis of mouse ZP-related genes. This figure shows relative changes during 7–19 days after birth. The figures in parentheses show intensities at day 7. The intensity of FIGLA expression decreased drastically. ZP1, 2, and 3 also decreased similarly

recognized in the ovary, also including degenerated follicles having pyknotic granulosa cells. At day 19, ovarian size and the number of granulosa cells in the follicle are further increased. Juvenile mouse ovary from day 0 to 19 changes chronologically in one cycle of folliculogenesis. This duration of mouse ovary is often used as a functional model of the ovary in large animals including humans. In mice, functional oocytes are ovulated by gonadotropin stimulation after day 21.

Figure 17.4 shows transcriptome analysis of mouse ZP-related proteins from day 7 to day 19. All of the ZP mRNAs express at a maximum at day 7 and gradually decrease until day 19. FIGLA (Factor In the GermLine, Alpha), which is a transcriptional factor for ZP proteins, also shows the same pattern of decrease. FIGLA has also been demonstrated as a central regulator of oocyte-specific genes that play roles in folliculogenesis, fertilization, and early development including ZP protein expression.

Figure 17.4 confirms ZP protein expressions are controlled by FIGLA [10]. FIGLA (Factor In the GermLine, Alpha) demonstrated its involvement in two independent developmental processes: formation of the primordial follicle and coordinated expression of zona pellucida genes. Previously, FIGLA was reported to regulate gene expression of multiple molecules related to oogenesis [11]. ZP genes are also located downstream of FIGLA [12, 13]. Also in humans, FIGLA is shown to be involved in continued oocyte survival in primordial follicles [14].

17.2   Sub-molecular Aspects of Sperm Recognition

By the mid-1990s, advances in molecular biology made it possible to produce sequenced proteins as desired. Kinloch et al. constructed a plasmid containing modified mouse ZP3 (mZP3) gene and transformed embryonic stem (ES) cells. They analyzed the resultant recombinant protein and concluded that the C-terminal region of mZP3 coded by exon 7 was essential for acrosome reaction induction as well as sperm binding [15]. The research group also succeeded in the establishment of mZP3 knockout (KO) mouse lines by mZP3 gene targeting [16]. The strain bearing heterozygous allele [mZP3 (+/−)] did not show any remarkable phenotype. Fertility was normal, although ZP thicknesses were less than in wild-type [mZP3 (+/+)] [17, 18]. On the other hand, the homozygous strain did not form ZP at all. These females were infertile, although mZP1 and mZP2 were produced normally. Furthermore, folliculogenesis was also disrupted in ZP3 KO [mZP3 (−/−)] mice [19].

Rankin et al. successfully established mZP1 KO [mZP1 (−/−)] mice, and their fecundity reduced but not to infertility [20]. This suggested that mZP1 somehow influenced reproductive systems but was not an essential factor. Detailed observation indicated ZP morphology was loosely constructed and ovarian follicles were ectopic clusters with oocytes and granulosa cells. ZP-1-targeted mice were fertile but fecundity was significantly lower than wild-type. This indicates that around a half-thickness of ZP is eventually enough to produce fertilizable oocytes in the ovary and maintain reproductive ability.

Also, Rankin et al. succeeded in establishing a mZP2 KO mouse line [mZP2 (−/−)] [21]. Optical microscopic observation showed the mice did not form ZP structures, and folliculogenesis was severely disrupted. The female mice were infertile, like the ZP3 KO [mZP3 (−/−)] mice. Although the ovarian tissues were fragile, a few oocytes reached MII stage and were ovulated by exogeneous gonadotropin stimulation similarly to the mZP3 KO line phenotype. Although the oocyte numbers collected by exogeneous gonadotropin stimulation were very small, they could fertilize and develop into the blastocyst stage. This suggests that ZP eventually affects follicle growth mediated to oocyte-granulosa cell interaction ability but does not disrupt fertilizing ability itself.

Previously, we established a number of monoclonal antibodies to determine epitopes for sperm binding sites. Some of them indicated blocking effects on sperm binding to ZP. Since a monoclonal antibody that reacted to ZP2 inhibited sperm binding to ZP, we considered the epitope in ZP2 to be a good candidate for immunocontraception (Koyama). Our own studies using ZP2 KO mice are described as follows.

Figure 17.5 shows a study of ovarian histology and compares the numbers of growing follicles between wild-type and ZP2 KO mice of the same age (8 weeks). In wild-type (a), healthy growing follicles including oocytes with ZP are observed, while in ZP2 KO mice (b), there are few growing follicles including oocytes without ZP. Instead, many degenerated follicles are recognized, shown by arrows. This suggests primordial follicles probably start to grow but most of them degenerated before fully developing. The small follicles without oocytes indicate degenerative follicles.

Figure 17.6 shows developing follicle numbers in WT and ZP2 KO mouse ovarian sections, and oocytes ovulated by gonadotropin stimulation in 8-week-old mice. Primordial follicle and ovulated oocyte numbers were significantly lower in ZP2 KO mice. However, primary, secondary, and antral follicles were not statistically significant. This means that the follicles recruited from primordial follicles frequently degenerated in the primary and secondary stages. It seemed that few follicles reach antral stages. Primary and secondary follicles are supplied from the

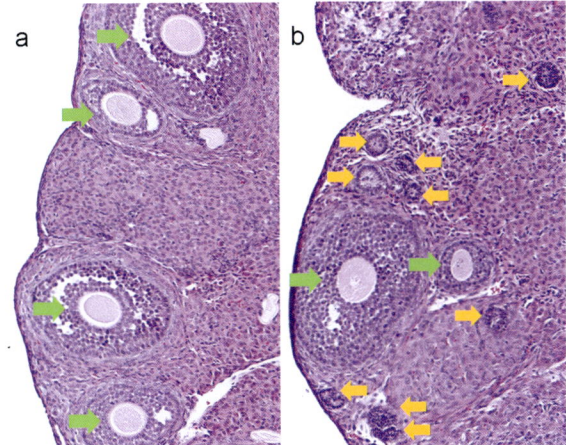

Fig. 17.5 Histological comparison of growing follicles between wild-type and ZP2 KO mice of the same ages (8 weeks). In wild type (**a**), growing follicles including oocytes with ZP are observed. Four growing follicles with small antrum are shown in green. In ZP2-KO mice however (**b**), few growing follicles are observed, as well as oocytes without ZP. Instead, many degenerated follicles are found as shown by orange arrows. This suggests that primordial follicles probably start to grow but most of them degenerate before fully developing. Actually, ZP2 KO mice can ovulate small numbers of oocytes by hyper-stimulation of gonadotropin, but not from natural cycles. The ovarian features seem to be coincident to functional ability

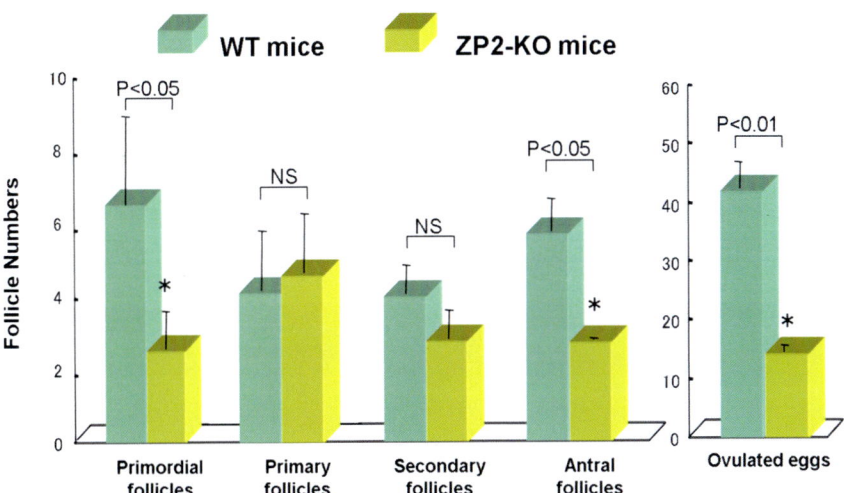

Fig. 17.6 Comparison of follicle numbers in different stages between wild-type and ZP2-KO mice in ovarian sections. Primordial follicle, antral follicle, and ovulated oocyte numbers were significantly lower in ZP2-KO mice. However, differences in primary and secondary follicle numbers were not statistically significant. This suggests that growing follicles such as primary and secondary follicles are supplied from primordial follicles after degeneration

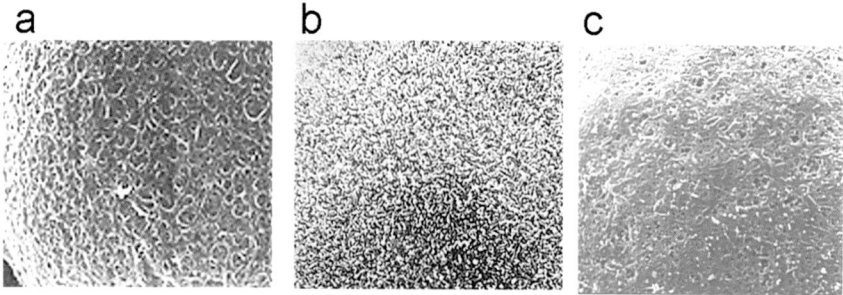

Fig. 17.7 Surface of mZP2-KO [mZP2 $(-/-)$] oocytes by scanning electron microscopy. (**a**) ZP surface of wild-type mouse oocyte. (**b**) PM surface of wild-type mouse oocyte (ZP is mechanically removed). (**c**) Oocyte surface of ZP2-KO mouse oocyte

primordial follicle pool and maintain constant numbers. However, the pool gradually becomes depleted.

Additionally, advanced-age ZP2 KO mice (20 weeks) exhibited severe ovarian disorders compared with age-matched WT mice. In normal conditions, ZP2 KO mice are infertile due to deficient follicle growth and ovulation. Actually, ZP2 KO mice do not ovulate in natural conditions even at reproductive ages (after 8 weeks old). Small numbers of oocytes can be ovulated under hyper-stimulation with gonadotropin. This reproductive ability is coincident to ovarian features in Fig. 17.5.

Figure 17.7 shows comparison of mouse oocyte surfaces by scanning electron microscopy. The ZP surface shows numerous spongelike fenestration structures. On the other hand, the surface of the vitelline membrane shows tremendous numbers of microvilli. ZP2 KO mouse oocyte, however, shows different features. Although ZP is not detected on the oocyte surface from ZP2 KO mice under optical microscopic observation, electron microscopic observation clearly revealed that very thin ZP was formed of only ZP1 and ZP3 proteins (Fig. 17.7). This very thin layer of ZP covered the vitelline membrane as shown in Fig. 17.7c. Microvilli extrude sparsely from the vitelline membrane through the thin ZP layers. This suggests that ZP1 and ZP3 may form very thin layers of ZP around the oocyte, but these are greatly insufficient for normal follicle growth.

Figure 17.8 shows transmission electron microscopic observations of oocyte and granulosa cell junctions in ZP2 KO mouse ovaries. In wild-type mice, the processes from a granulosa cell extend to the oocyte through ZP and form a gap junction. In ZP2 KO mice, granulosa cells adhere directly to the oocyte surface and form gap junctions both in preantral and antral follicles. This suggests that occasionally, granulosa cells and oocytes form gap junctions and grow to the mature follicle stage.

Due to lacking of ZP4 in mice, the roles of ZP4 could not be investigated for a long time. Instead, recently, ZP4-targeted rabbits were created to examine the roles of ZP4. The phenotype of the rabbits was similar to ZP1 KO mice. In the targeted rabbits, ovulation was not interfered with but litter size was significantly lower. This suggested that embryos surrounded by thinner ZP were impaired during

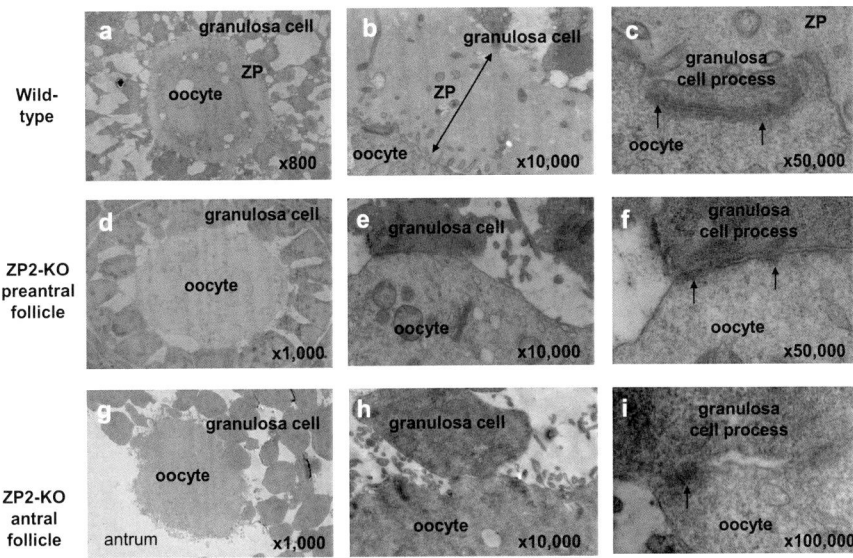

Fig. 17.8 Transmission electron microscopic observations of oocyte and granulosa cell junctions in ZP2-KO mouse ovaries. In wild-type ovarian sections, ZP structures are obviously present between oocyte and granulosa cells (**a, b**). A number of microvilli extend in the ZP (**b**) from both oocytes and granulosa cells. In ZP2-KO ovarian sections, no ZP was observed in the preantral (**d–f**) or antral sections (**g–i**). Noticeably, a small number of gap junctions formed in ZP2-KO mice, shown by → (**f, i**) as well as wild type (**c**). This suggests that ZP is not essential to build gap junctions but the numbers are extremely reduced compared with wild type. Microvilli increase surface that can make gap junctions. ZP promotes gap-junction formation followed by supporting of the microvilli. (**a–c**) Wild-type mice; (**d–f**) preantral follicles of ZP2-KO mice; (**g–i**) antral follicles of ZP2-KO mice. (**a, d, g**) Low magnifications ×800, ×1000. (**b, e, h**) Magnification ×10,000. (**c, f, i**) Magnification ×50,000, 100,000

transportation in the oviducts [22]. Together, thinner ZP from mZP1 KO [mZP1 (−/ −)] mice or ZP4 KO [ZP4 (−/−)] rabbits showed the same phenotype of reduced fecundity. However, mouse heterologous targeted ZP2 and ZP3 did not showed reduced fertility, albeit ZP was thinner than wild-type mice. This suggests that existence of three ZP proteins maintain rigid nature by cross-linking with the three proteins. It was considered that there are differences between homologous and heterologous targeted mouse ZP, if ZP is shown to be same phenotype by optical microscopic observation.

These gene-targeting studies clearly indicated that ZP plays a function in follicle growth as well as oocyte growth. They suggested that ZP physically and biochemically supports oocyte-granulosa cell interactions followed by follicle growth. Transportation of intercellular substrates and transmission of information occur via abundant microvilli extending from oocytes and granulosa cells into ZP (Fig. 17.2).

ZP-lacking oocytes from KO female mice were ovulated by ovarian stimulation with gonadotropin. The oocytes were able to be fertilized by careful intracytoplasmic sperm injection (ICSI). Further, the embryos grew to full term after implantation into pseudopregnant mice. The babies born were healthy, although the rates were low. This result indicated oocyte fertilization and development were not affected by a lack of ZP. Recently, it has been reported that ZP-lacking oocytes exist in humans, as described in Chap. 18. It was found that genetic mutation in ZP genes caused infertility.

In another aspect, Dean and his colleagues investigated species-specific binding mechanisms by gene-targeting methods. The researchers created genetically designed mice with gene-manipulated ZP that human sperm would bind to, as human sperm does not naturally bind to mouse ZP.

Rankin et al. established mZP1, mZP2, and mZP3 KO mouse lines by gene targeting [19–21]. Furthermore, they rescued mZP3 KO mice with human ZP3 to examine species-specific sperm binding to ZP. The transgenic mice showed normal fertility and conceived in vivo. Unexpectedly, human sperm did not bind to the manipulated ZP [23]. When mouse ZP2 was replaced with human ZP2, human sperm did not bind to the transgenic ZP. Even when all ZP proteins were replaced with human ZP1, ZP2, and ZP3, the genetically modified mouse ZP did not bind to human sperm [24], although mouse sperm binds to mZP2 [25]. It seemed that this was not present for human sperm binding site on human ZP2. This was a contradictory enigma to well-established observations of species-specific sperm binding to ZP.

Human ZP has four ZP proteins (ZP1, ZP2, ZP3, and ZP4) but mouse ZP contains only three proteins. The additional ZP protein, ZP4, possibly contributes to sperm binding to human ZP. Rat ZP is composed of four glycoproteins same as humans. Actually, human sperm does not bind to rat ZP as well as mouse ZP [26]. The presence of ZP4 was not related to human sperm binding to ZP. Therefore, it may be difficult to define the relationship between protein and function among animal species. In addition, a variety of carbohydrate moieties possibly contribute to species-specific sperm binding on ZP according to Wassarman's hypothesis. Dean's group claimed with the possibility that O-linked carbohydrate does not mediate sperm binding [27].

Recently, Baibakov et al. succeeded in creating genetically designed ZP in mice to which human sperm could bind [28]. They produced a transgenic mouse line in which the mouse ZP proteins (ZP1–3) were completely replaced with human ZP proteins (ZP1–3) and human ZP4 was added. The transgenic ZP was bound to human sperm mediated with human ZP2. It seems that a steric structure composed of the four ZP proteins, not single molecule interactions, makes human sperm binding to ZP possible. For extensive study, the research group divided the ZP2 gene into several sections and produced several chimera strains of mouse and human ZP2. They finally identified the human sperm binding domain as located in 51–149 aa of ZP2 [29]. They concluded that the N-terminal region of human ZP2 was the human sperm binding site.

However, this hypothesis is supported only in human sperm and ZP binding. It is not known whether or not the N-terminal region of ZP2 is always a sperm binding site in other animal species. The sperm-ZP recognition mechanisms are not always the same in different animal species; thus, it is difficult to establish a unified model.

17.3 Polyspermy Blocking

The important function of ZP is to prevent abnormal fertilization such as heterologous sperm penetration, or polyspermy. Indeed, zona-free animal ova were reported to allow penetration of different species sperm [30–32]. How ZP avoids abnormal fertilization is explained from traditional observations and cutting-edge technologies. Previously, various polyspermy-blocking mechanisms that had been reported.

Ovastacin (*ovarian astacin* family of metalloprotease) cDNA had been cloned and characterized in mice and humans in 2004. The enzyme was predominantly expressed in ovarian tissues. The function of this protease was not known [33].

However, recently, it was demonstrated that ovastacin was released from CGs and affected ZP2 cleavage into ZP2f [34]. This cleavage reaction directly caused polyspermy blocking. Indeed, recombinant ovastacin cleaved ZP2 in natural ZP. Furthermore, deletion of seven amino acids by CRISPR/Cas9 at the endogenous locus of the ovastacin gene prevents CG localization. As a result, ovastacin was released prematurely and cleaved ZP2 [35]. The gene-manipulated female mice, ovastacin KO mice, were subfertile due to deficiency of polyspermy blocking. This research demonstrated that the cleavage of ZP2 by ovastacin is essential for polyspermy blocking. ZP is the critical site for polyspermy blocking [36].

A different feature was clinically observed. A number of human sperm remained in/on ZP of the embryo after in vitro fertilization. In most cases, human sperm continued binding to ZP. After single-sperm fertilization, additional sperm could not penetrate ZP or fuse with the plasma membrane of oocytes. On the other hand, sperm bound to ZP seemed not to detach from ZP. It is not known if ZP2 cleavage by ovastacin is effective for polyspermy blocking even in humans.

As previously mentioned, we established a number of monoclonal antibodies to determine epitopes for sperm binding sites. Some of them indicated blocking effects on sperm binding to ZP [37–39]. The other monoclonal antibodies did not affect sperm binding to ZP. Surprisingly, a few monoclonal antibodies enhanced sperm binding to ZP, although the reason was not known [38]. These monoclonal antibodies possibly recognized ovastacin cleavage sites and allowed excess sperm binding to ZP in vitro.

Zinc was also reported to release sperm from the oocytes and stop their motility. There are several stages for polyspermy blocking in ZP and the oocyte plasma membrane.

Zinc incorporation into the ZP induces a number of structural and physiological changes that resemble those observed in activated egg ZPs [40]. However, recent reports have documented that glycan-independent gamete recognition triggers a zinc spark and cleaves ZP2 to ZP2f, suggesting that zinc does not affect sperm motility or ZP structure directly. Instead, zinc is a kind of cofactor of ovastacin [41, 42].

More recently, Fahrenkamp et al. described egg coat hardening for polyspermy blocking from mollusks to mammals using a novel technology, crystallographic analysis. It is interesting that crystallographic analysis predicts a common structure in abalone and mice for polyspermy blocking sites. They addressed the possibility that similar mechanisms and sites were conserved for polyspermy prevention during evolution, although thicknesses, amino acid sequences, and carbohydrate characters are different in the ZP [43].

Another candidate for a polyspermy blocking factor, plasminogen/plasmin activator (tPA), was reported [44]. This molecule was known to be present in CGs and released during ovulation [45]. Rekkas et al. provided evidence that tPA affected bovine oocyte maturation in vitro [46]. More recently, gene expression analysis showed that urokinase-type plasminogen activator (uPA) was related to oocyte maturation, fertilization, and embryo development via ZP [47].

Heparin-like glycosaminoglycans (S-GAGs) present in oviductal fluid are also suggested to contribute to the polyspermy blocking factor [48] in domestic animals. They also proposed that the observed physiological ZP hardening produced by cross-linkage of SH groups was a possible mechanism to control polyspermy [49] (Fig. 17.9).

Differently from mice, human embryos maintain sperm binding and penetration into ZP. Figure 17.10 shows human embryos such as pronuclei, 8-cells, and blastocyst stages. Extensive sperm bound to ZP are visible. This would suggest that the ovastacin system does not work in humans. Even in mice, ovastacin is not the only system for polyspermy blocking [50]. They showed that in vivo, the fertilization system tubal-utero junction regulates sperm numbers passing through the oviduct.

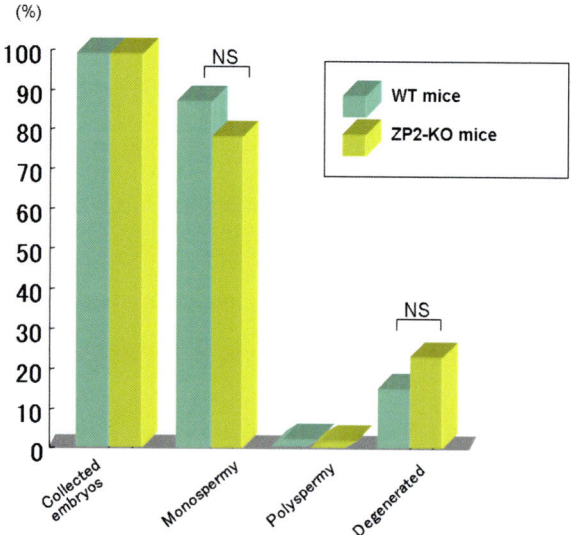

Fig. 17.9 Polyspermy frequency of in vivo-fertilized eggs from ZP2-KO mice. Although the numbers of collected embryos were different in the wild-type and mZP2-KO lines, the ratios of monospermy and polyspermy did not differ between the two groups. There were no zygotes penetrated by multiple sperm

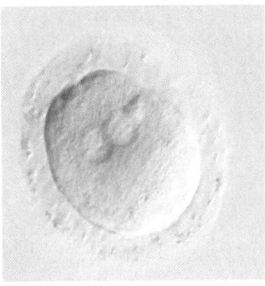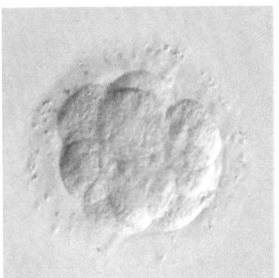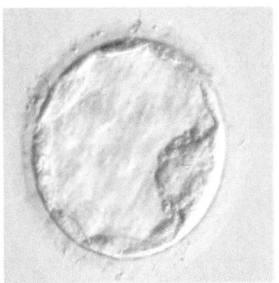

Fig. 17.10 Human sperm remaining bound to ZP. A number of human sperms remain in/on ZP of embryos after fertilization in vitro. In most cases human sperms keep binding to ZP, suggesting that ZP2 cleavage may be insufficient in humans compared with mice. After fertilization, additional sperms cannot bind to ZP but bound sperms do not detach from ZP

As described in the next section, the noticeable ZP function in vivo is to protect embryos from maternal attack rather than polyspermy blocking. The polyspermy blocking mechanism is different from species to species. Conclusively, ZP functions are different among animal species.

17.4   The Role of ZP in Embryo Development

When the ZP is removed, an embryo at the pronuclei stage can develop into the blastocyst stage under in vitro conditions (Fig. 17.11), but not in vivo. Denuded embryos transplanted into the oviduct of pseudopregnant mice did not pass through the oviduct or reach the uterus [51]. Furthermore, embryos could not be recovered from the oviduct several hours after transplantation. The embryos were probably absorbed onto the surface of the oviduct epithelium. However, embryos at the blastocyst stage could migrate in the oviducts and reach the uterus. Oviduct epithelial cells recognize early cleaved embryos as non-self-antigens and eliminate them by immunological surveillance. On the other hand, blastocyst-stage embryos are recognized as self-antigens. Therefore, the ZP protects earlier stage preimplantation embryos from attack by the mother's cells during oviduct transportation. More recently, it has been proposed that the ZP or ZP derivatives could act as intrinsic signals to the maternal immune system for implantation susceptibility [52].

Normal preimplantation embryos having ZP reach the uterus in adequate time for implantation. Cleaved embryos pass through the oviducts in 2–4 days depending on the particular stage of implantation condition. Blastocyst-stage embryos rapidly transport to the uterus after transplantation into the oviducts. It is not known how the oviduct recognizes the stages of embryos and regulates transport speed. Probably, ZP prevents direct interactions between embryo cell membranes and oviductal cells and protects impairment of embryos.

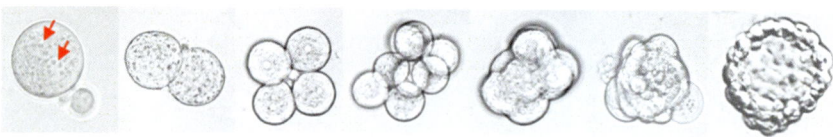

Fig. 17.11 Mouse embryo development without ZP. Embryos can develop from pronuclei to blastocyst stages in vitro

However, in some cases, ZP shows harmful effects on embryos, interfering with hatching. In reproductive medicine, AHA (assisted hatching) treatment is applied to rescue embryo hatching from ZP. In the 1990s, studies of molecular interactions between ZP and oviduct were reported. A molecule named OVGP-1 (oviduct glycoprotein 1) secreted from the oviducts facilitates fertilization of hamsters in vitro [53]. Thereafter, counterparts of this molecule were detected in various other animals including pigs, boars, and humans [54]. OVGP-1 was thought to be an oviduct-specific molecule in those days, but related molecules with a great variety of functions have since been detected in various tissues [55].

In humans, a recombinant protein of OVPG-1 (rHuOVGP1) was shown to enhance fertilization rates [56]. The mechanism of fertilization enhancement indicates that OVGP-1 binds to ZP and partially reaches the oolemma and even oocyte cytoplasm, subsequently affecting tyrosine phosphorylation [57]. Contrarily, however, there is a report that OVPG-1 inhibits sperm penetration of zona pellucida to block polyspermy [48]. Collectively, it suggests that OVPG is adequately controlled around fertilization at the zona pellucida.

Recently, as well as OVGP-1, various kinds of molecules have been reported to be secreted by oviducts bearing microexosomes [58]. Microexosomes are derived from the epithelial cells and transported to other tissues. This is a system of information transmission between tissues. Also, this system may serve to transport implantation signals from the oviducts to the uterus.

References

1. Bleil JD, Wassarman PM. Synthesis of zona pellucida proteins by denuded and follicle-enclosed mouse oocytes during culture in vitro. Proc Natl Acad Sci U S A. 1980;77(2):1029–33.
2. Shimizu S, Tsuji M, Dean J. In vitro biosynthesis of three sulfated glycoproteins of murine zonae pellucidae by oocytes grown in follicle culture. J Biol Chem. 1983;258(9):5858–63.
3. Miyano T, Hirao Y, Kato S, Kanda S. Formation of the zona pellucida in relation to the oocyte growth in the mouse ovary. Jpn J Anim Reprod. 1988;23:61–6.
4. Gook DA, Edgar DH, Borg J, Martic M. Detection of zona pellucida proteins during human folliculogenesis. Hum Reprod. 2008;23:394–40.
5. Carabatsos MJ, Sellitto C, Goodenough DA, Albertini DF. Oocyte-granulosa cell heterologous gap junctions are required for the coordination of nuclear and cytoplasmic meiotic competence. Dev Biol. 2000;226:167–79.

6. Kidder GM, Mhawi AA. Gap junctions and ovarian folliculogenesis. Reproduction. 2002;123 (5):613–20.
7. Clarke HJ. Regulation of germ cell development by intercellular signaling in the mammalian ovarian follicle. Wiley Interdiscip Rev Dev Biol. 2018;7(1):10.1002/wdev.294.
8. Anderson E, Albertini DF. Gap junctions between the oocyte and companion follicle cells in the mammalian ovary. J Cell Biol. 1976;71(2):680–6.
9. Wassarman PM, Litscher ES. Influence of the zona pellucida of the mouse egg on folliculogenesis and fertility. Int J Dev Biol. 2012;56(10–12):833–9.
10. Liang LF, Soyal SM, Dean J. FIGa, a germ cell specific transcription factor involved in the coordinate expression of the zona pellucida genes. Development. 1997;124:4939–49.
11. Joshi S, Davies H, Sims LP, Levy SE, Dean J. Ovarian gene expression in the absence of FIGLA, an oocyte-specific transcription factor. BMC Dev Biol. 2007;7:67.
12. Hu W, Gauthier L, Baibakov B, Jimenez-Movilla M, Dean J. FIGLA, a basic helix-loop-helix transcription factor, balances sexually dimorphic gene expression in postnatal oocytes. Mol Cell Biol. 2010;30(14):3661–71.
13. Li L, Zheng P, Dean J. Maternal control of early mouse development. Development. 2010;137 (6):859–8.
14. Bayne RAL, Silva SJM, Anderson RA. Increased expression of the FIGLA transcription factor is associated with primordial follicle formation in the human fetal ovary. Mol Hum Reprod. 2004;10(6):373–81.
15. Kinloch RA, Sakai Y, Wassarman PM. Mapping the mouse ZP3 combining site for sperm by exon swapping and site-directed mutagenesis. Proc Natl Acad Sci U S A. 1995;92:263–7.
16. Liu C, Litscher ES, Mortillo S, Sakai Y, Kinloch RA, Stewar CL, Wassarman PM. Targeted disruption of the mZP3 gene results in production of eggs lacking a zona pellucida and infertility in female mice. Proc Natl Acad Sci U S A. 1996;93(11):5431–6.
17. Wassarman PM, Liu C, Chen J, Qi H, Litscher ES. Ovarian development in mice bearing homozygous or heterozygous null mutations in zona pellucida glycoprotein gene mZP3. Histol Histopathol. 1998;13(1):293–300.
18. Wassarman PM, Qi H, Litscher ES. Mutant female mice carrying a single mZP3 allele produce eggs with a thin zona pellucida, but reproduce normally. Proc Biol Sci. 1997;264(1380):323–8.
19. Rankin T, Familari M, Lee E, Ginsberg A, Dwyer N, Blanchette-Mackie J, Drago J, Westphal H, Dean J. Mice homozygous for an insertional mutation in the Zp3 gene lack a zona pellucida and are infertile. Development. 1996;122(9):2903–10.
20. Rankin T, Talbot P, Lee E, Dean J. Abnormal zonae pellucidae in mice lacking ZP1 result in early embryonic loss. Development. 1999;126(17):3847–55.
21. Rankin TL, O'Brien M, Lee E, Wigglesworth K, Eppig J, Dean J. Defective zonae pellucidae in Zp2-null mice disrupt folliculogenesis, fertility and development. Development. 2001;128 (7):1119–26.
22. Lamas-Toranzo I, Fonseca Balvís N, Querejeta-Fernández A, Izquierdo-Rico MJ, González-Brusi L, Lorenzo PL, García-Rebollar P, Avilés M, Bermejo-Álvarez P. ZP4 confers structural properties to the zona pellucida essential for embryo development. Elife. 2019;8:e48904.
23. Rankin TL, Tong ZB, Castle PE, Lee E, Gore-Langton R, Nelson LM, Dean J. Human ZP3 restores fertility in Zp3 null mice without affecting order-specific sperm binding. Development. 1998;125(13):2415–24.
24. Rankin TL, Coleman JS, Epifano O, Hoodbhoy T, Turner SG, Castle PE, Lee E, Gore-Langton R, Dean J. Fertility and taxon-specific sperm binding persist after replacement of mouse sperm receptors with human homologs. Dev Cell. 2003;5(1):33–43.
25. Avella MA, Xiong B, Dean J. The molecular basis of gamete recognition in mice and humans. Mol Hum Reprod. 2013;19(5):279–89.
26. Hoodbhoy T, Joshi S, Boja ES, Williams SA, Stanleyand P, Dean J. Human sperm do not bind to rat zonae pellucidae despite the presence of four homologous glycoproteins. J Biol Chem. 2005;280(13):12721–31.

27. Hoodbhoy T, Dean J. Insights into the molecular basis of sperm–egg recognition in mammals. Reproduction. 2004;127(4):417–22.
28. Baibakov B, Boggs NA, Yauger B, Baibakov G, Dean J. Human sperm bind to the N-terminal domain of ZP2 in humanized zonae pellucidae in transgenic mice. J Cell Biol. 2012;197 (7):897–905.
29. Avella MA, Baibakov B, Dean J. A single domain of the ZP2 zona pellucida protein mediates gamete recognition in mice and humans. J Cell Biol. 2014;205(6):801–9.
30. Yanagimachi R, Yanagimachi H, Rogers BJ. The use of zona-free animal ova as a test-system for the assessment of the fertilizing capacity of human spermatozoa. Biol Reprod. 1976;15 (4):471–6.
31. Pavlok A. Penetration of hamster and pig zona-free eggs by boar ejaculated spermatozoa preincubated in vitro. Int J Fertil. 1981;26(2):101–6.
32. Binor Z, Sokoloski JE, Wolf DP. Sperm interaction with the zona-free hamster egg. J Exp Zool. 1982;222(2):187–93.
33. Quesada V, Sánchez LM, Alvarez J, López-Otín C. Identification and characterization of human and mouse ovastacin: a novel metalloproteinase similar to hatching enzymes from arthropods, birds, amphibians, and fish. J Biol Chem. 2004;279(25):26627–34.
34. Burkart AD, Xiong B, Baibakov B, Jiménez-Movilla M, Dean J. Ovastacin, a cortical granule protease, cleaves ZP2 in the zona pellucida to prevent polyspermy. J Cell Biol. 2012;197(1):37–44.
35. Vogt EJ, Tokuhiro K, Guo M, Dale R, Yang G, Shin SW, Movilla MJ, Shroff H, Dean J. Anchoring cortical granules in the cortex ensures trafficking to the plasma membrane for post-fertilization exocytosis. Nat Commun. 2019;10(1):2271.
36. Xiong B, Zhao Y, Beall S, Sadusky AB, Dean J. A unique egg cortical granule localization motif is required for ovastacin sequestration to prevent premature ZP2 cleavage and ensure female fertility in mice. PLoS Genet. 2017;13(1):e1006580.
37. Isojima S, Koyama K, Hasegawa A, Tsunoda Y, Hanada A. Monoclonal antibodies to porcine zona pellucida antigens and their inhibitory effects on fertilization. J Reprod Immunol. 1984;6 (2):77–87.
38. Koyama K, Hasegawa A, Tsuji Y, Isojima S. Production and characterization of monoclonal antibodies to cross-reactive antigens of human and porcine zonae pellucidae. J Reprod Immunol. 1985;7(3):187–98.
39. Kyurkchiev SD, Surneva-Nakova TN, Ivanova MD, Nakov LS, Dimitrova EH. Monoclonal antibodies to porcine zona pellucida that block the initial stages of fertilization. Am J Reprod Immunol Microbiol. 1988;18(1):11–6.
40. Que EL, Duncan FE, Bayer AR, Philips SJ, Roth EW, Bleher R, Gleber SC, Vogt S, Woodruff T, O'Halloran TV. Zinc sparks induce physiochemical changes in the egg zona pellucida that prevent polyspermy. Integr Biol (Camb). 2017;9(2):135–44.
41. Que EL, Duncan FE, Lee HC, Hornick JE, Vogt S, Fissore RA, O'Halloran TV, Woodruff TK. Bovine eggs release zinc in response to parthenogenetic and sperm-induced egg activation. Theriogenology. 2019;127:41–8.
42. Tokuhiro K, Dean J. Glycan-independent gamete recognition triggers egg zinc sparks and ZP2 cleavage to prevent polyspermy. Dev Cell. 2018;46:627–40.
43. Fahrenkamp E, Algarra B, Jovine L. Mammalian egg coat modifications and the block to polyspermy. Mol Reprod Dev. 2020;87(3):326–40.
44. Huarte J, Belin D, Vassalli JD. Plasminogen activator in mouse and rat oocytes: induction during meiotic maturation. Cell. 1985;43:551–8.
45. Zhang X, Rutledge J, Khamsi F, Armstrong DT. Release of tissue-type plasminogen activator by activated rat eggs and its possible role in the zona reaction. Mol Reprod Dev. 1992;32(1):28–32.
46. Rekkas CA, Besenfelder U, Havlicek V, Vainas E, Brem G. Plasminogen activator activity in cortical granules of bovine oocytes during in vitro maturation. Theriogenology. 2002;57 (7):1897–905.

47. Roldán-Olarte M, Maillo V, Sánchez-Calabuig MJ, Beltrán-Breña P, Rizos D, Gutiérrez-Adán A. Effect of urokinase type plasminogen activator on in vitro bovine oocyte maturation. Reproduction. 2017;154(3):231–40.
48. Coy P, Cánovas S, Mondéjar I, Saavedra MD, Romar R, Grullón L, Matás C, Avilés M. Oviduct-specific glycoprotein and heparin modulate sperm-zona pellucida interaction during fertilization and contribute to the control of polyspermy. Proc Natl Acad Sci U S A. 2008;105:15809–14.
49. Coy P, Grullon L, Canovas S, Romar R, Matas C, Aviles M. Hardening of the zona pellucida of unfertilized eggs can reduce polyspermic fertilization in the pig and cow. Reproduction. 2008;135(1):19–27.
50. Muro Y, Hasuwa H, Isotani A, Miyata H, Yamagata K, Ikawa M, Yanagimachi R, Okabe M. Behavior of mouse spermatozoa in the female reproductive tract from soon after mating to the beginning of fertilization. Biol Reprod. 2016;94(4):80.
51. Bronson RA, McLaren A. Transfer to the mouse oviduct of eggs with and without the zona pellucida. J Reprod Fertil. 1970;22:129–37.
52. Fujiwara H, Araki Y, Toshimori K. Is the zona pellucida an intrinsic source of signals activating maternal recognition of the developing mammalian embryo? J Reprod Immunol. 2009;81:1–8.
53. Araki Y, Kurata S, Oikawa T, Yamashita T, Hiroi M, Naiki M, Sendo F. A monoclonal antibody reacting with the zona pellucida of the oviductal egg but not with that of the ovarian egg of the golden hamster. J Reprod Immunol. 1987;11:193–208.
54. Algarra B, Maillo V, Avilés M, Gutiérrez-Adán A, Rizos D, Jiménez-Movilla M. Effects of recombinant OVGP1 protein on in vitro bovine embryo development. J Reprod Dev. 2018;64:433–43.
55. Lopera-Vásquez R, Hamdi M, Fernandez-Fuertes B, et al. Extracellular vesicles from BOEC in in vitro embryo development and quality. PLoS One. 2016;11:e0148083.
56. Zhao Y, Kan FWK. Human OVGP1 enhances tyrosine phosphorylation of proteins in the fibrous sheath involving AKAP3 and increases sperm-zona binding. J Assist Reprod Genet. 2019;36:1363–77.
57. Zhao Y, Yang X, Jia Z, Reid RL, Leclerc P, Kan FW. Recombinant human oviductin regulates protein tyrosine phosphorylation and acrosome reaction. Reproduction. 2016;152(5):561–73.
58. Almiñana C, Corbin E, Tsikis G, et al. Oviduct extracellular vesicles protein content and their role during oviduct-embryo cross-talk. Reproduction. 2017;15:153–68.

Chapter 18
Human ZP

Akiko Hasegawa

Abstract In the past, human ZP analysis was difficult due to the very small quantities available for research use. Ethical issues were also a bottleneck for the characterization and analysis of human ZP functions. Therefore, small-scale experiments and immunological methods enhancing human ZP reactivity were developed. In this chapter, human ZP structure and interaction with sperm are described, referring to animal models. Later, characterization of human ZP became possible through advanced technologies such as DNA recombination and mass spectrometry. The current aspects of human ZP function for each molecule, ZP1, ZP2, ZP3, and ZP4, are summarized. The function of carbohydrate portions, which was clarified from comparison of the recombinant proteins from *Escherichia coli* and mammalian cell culture systems, is also mentioned.

18.1 Biochemical Studies

Prior to the 1990s, it was difficult to obtain sufficient materials for human ZP studies. Research was carried out using naturally occurring pig or rabbit ZP [1–3]. Pig ZP in particular could be obtained from abattoirs in relatively abundant amounts compared to other animal species. As a human model, the genomic organization of pig ZP2 (designated as ZP1 in 1996) gene was reported, and the recombinant ZP2 protein was produced in mammalian cells [4, 5]. Immunological methods such as enzyme-linked immunosorbent assay (ELISA) and Western blotting were useful, because antibodies recognize specific antigens and enhance the reactivities to small amounts of human ZP. Many species-specific or cross-reactive monoclonal antibodies were made to examine antigenic epitopes of ZP, and the corresponding ZP antigens were

A. Hasegawa (✉)
Department of Obstetrics and Gynecology, School of Medicine, Hyogo Medical University, Nishinomiya, Hyogo, Japan
e-mail: zonapel@hyo-med.ac.jp

H. Shibahara, A. Hasegawa (eds.), *Gamete Immunology*,
https://doi.org/10.1007/978-981-16-9625-1_18

characterized [6–8]. Pig and human cross-reactive antigens were particularly focused on regarding infertility [8–10].

Also, the biological functions of human ZP were examined using immature and fertilization-failed oocytes from reproductive medicine. In 1988, Cross et al. solubilized human ZP by acidic pH and applied it to human sperm. They showed that as well as intact human ZP, the soluble ZP induced acrosome reaction in human sperm [11]. In addition, Bhandari et al. (2010) found that human ZP solubilized by heating at 70 °C for 90 min also induced acrosome reaction, suggesting that ZP is primarily a physiological inducer of acrosome reaction [12].

Since amino acid sequences had been deduced from cDNA cloning technology, antibodies against synthetic peptides of human ZP2 and ZP3 were produced [13]. The study showed a difference of molecular weights of human ZP2 at prophase I and metaphase II, but ZP3 did not show such modification. ZP2 cleavage was originated from oocyte protease, probably released during cortical reaction. They suggested that the cleavage contributes to oocyte maturation and follows sperm-egg fusion as part of polyspermy blocking. In addition, carbohydrate moieties associated with human ZP core proteins were also detected, similarly to mouse ZP.

Independently, another research group [14] produced antibodies against recombinant nonhuman primate (*Macaca radiata*) ZP. Amino acid sequence and antigenicity between human and *Macaca* were highly conserved. They produced antibodies to ZP2, ZP3, and ZP4 (designated as ZP1 in 1998), and detected counterparts of human ZP2 and ZP3 and ZP4 around 100, 53, and 60 kDa from Western blotting analysis. ZP3 and ZP4 overlapped in molecular weight in two-dimensional polyacrylamide gel electrophoresis (PAGE). They suggested the immunological cross-reactivity of ZP1 and ZP4 was due to high amino acid conformity (47% conformity between human ZP1 and human ZP4). Furthermore, they generated specific monoclonal antibodies to human ZP2, ZP3, and ZP4 and characterized their molecular weights using native human ZP [15, 16].

18.2   Characterization by Recombinant Technology

Gene cloning of human ZP proteins was performed in the following chronological order: ZP3 [17], ZP2 [18], ZP4 [19], and ZP1. ZP4 was identified as a molecule named ZP1 (ZPB) by Harris et al. in 1994. The presence of genuine ZP1 was verified by gene cloning and considering phylogenic development of other animal species as shown in Fig. 18.1. The comparison of mouse and human ZP genes is shown in Table 18.1.

Fig. 18.1 Pairwise Sequence Alignment. The alignment was made with the EMBOSS Needle pairwise alignment tool (https://www.ebi.ac.uk/Tools/psa/emboss_needle/). EMBOSS Needle creates an optimal global alignment of two sequences using the Needleman-Wunsch algorithm. (J Mol Biol 48:443 1970)

Table 18.1 Comparison of mouse and human *ZP1–4*

ZP gene		Chromosome number	Exon number	Amino acid number	Identity (similarity)
Zp1	Mouse	19	12	638	65.9% (75.5%)
	Human	11	12	623	
Zp2	Mouse	5	18	713	58.1% (72.2%)
	Human	16	19	745	
Zp3	Mouse	7	8	424	67.1% (77.5%)
	Human	7	8	424	
Zp4	Mouse	NA*	NA*	NA*	NA*
	Human	1	13	540	NA*

NA* not applicable (mouse *Zp4* is a pseudogene)

18.2.1   Human ZP1

Comprehensive study of human ZP functions was undertaken using recombinant and immunoaffinity-purified ZP proteins. Gupta's group prepared several kinds of recombinant proteins with or without carbohydrate moiety and indicated that recombinant human ZP1 derived from the baculovirus expression system bound to human capacitated sperm and induced acrosome reaction [20]. Recombinant human ZP1 of the same region from the *E coli* expression system also bound to human sperm, but did not induce acrosome reaction. This suggests the carbohydrate moiety of ZP1 does not contribute to sperm binding but plays an important role for induction of acrosome reaction [20].

18.2.2   Human ZP2

More than 20 years ago, our group demonstrated that a core peptide of recombinant human ZP2 corresponding to 1–206 amino acid residues bound to acrosome-reacted human sperm [21]. The peptide fragment bound to the mid-piece region of the sperm and antisera to the peptide inhibited human sperm binding to ZP [21–23]. Furthermore, full-length recombinant ZP2 proteins with or without carbohydrates did not bind to human sperm before acrosome reaction [24]. Native human ZP2 purified by immunoaffinity chromatography with a specific monoclonal antibody failed to bind to human sperm before acrosome reaction but bound to the sperm after acrosome reaction [25]. Collectively, it is suggesting that the N-terminal peptide portion of ZP2 has a function of acrosome-reacted human sperm binding to ZP without contribution of carbohydrates. More recently, Avella et al. supported this hypothesis using gene manipulation technology, in which the N-terminal region of mouse ZP2 was replaced with the corresponding human ZP2 region. The ZP from the transgenic mice could accept human sperm, whereas natural mouse ZP does not [26].

18.2.3   Human ZP3

As to the third ZP protein, ZP3, naturally occurring human ZP3 binds to capacitated sperm at the anterior region of the head and induces acrosome reaction [16]. Recombinant human ZP3 has been produced in several kinds of cell systems, and signal pathways in acrosome reaction induction to human sperm have been investigated. Several studies were suggested contributions of ion channel for acrosome reaction such as glycine receptor/Cl(−) channels and T-and L-type voltage-gated calcium channels [27, 28]. An analysis of kinetic parameters of ZP-treated sperm demonstrated specific motility pattern [29]. A Gi protein-coupled receptor was postulated to be a signal transducer in ZP3-mediated acrosome reaction induction [29, 30]. Details of sperm membrane and signal transduction mechanisms are not known.

Around fertilization events, human sperm bound to the recombinant ZP3 proteins only before and during acrosome reaction. In detailed analysis, the C-terminal fragment of ZP3 expressing carbohydrates increased significantly and dose dependently in acrosome reaction. The binding localization was the anterior head, similar to naturally occurring ZP3 [24, 25].

18.2.4   Human ZP4

Naturally occurring ZP4 was shown to bind to the entire head region of intact sperm before acrosome reaction [16]. Recombinant human ZP4 binds to capacitated (acrosome intact) sperm with or without carbohydrate moieties. Similar to ZP3, ZP4 including carbohydrates is significantly increased on acrosome reaction induction

[24]. ZP4 seems not to have a marked function in humans, as it seems that ZP3 with carbohydrates has sufficient function without ZP4. Instead, ZP4 may play a role in fine regulation of ZP-sperm interactions at fertilization.

To summarize, human ZP1, ZP3, and ZP4 seem to be competent to bind to sperm and induce acrosome reaction. ZP1, ZP3, and ZP4 correlate to primary binding to the capacitated sperm followed by acrosome reaction induction, and ZP2 binds to acrosome-reacted sperm as a secondary receptor. This suggests that ZP2 supports penetration of ZP by acrosome-reacted sperm. The present status of human ZP proteins has been summarized by Gupta SK [31]. As a whole, the claim supported the hypothesis established by Wassarman in mice [32]. Conclusively, since ZP has a three-dimensional architecture, total function cannot be understood by separation of each protein and segment.

References

1. Isojima S, Koyama K, Hasegawa A, Tsunoda Y, Hanada A. Monoclonal antibodies to porcine zona pellucida antigens and their inhibitory effects on fertilization. J Reprod Immunol. 1984;6 (2):77–87.
2. Drell DW, Dunbar BS. Monoclonal antibodies to rabbit and pig zonae pellucidae distinguish species-specific and shared antigenic determinants. Biol Reprod. 1984;30(2):445–57.
3. Timmons TM, Maresh GA, Bundman DS, Dunbar BS. Use of specific monoclonal and polyclonal antibodies to define distinct antigens of the porcine zonae pellucidae. Biol Reprod. 1987;36(5):1275–87.
4. Yamasaki N, Tsubamoto H, Hasegawa A, Inoue M, Koyama K. Genomic organization of the gene for pig zona pellucida glycoprotein ZP1 and its expression in mammalian cells. J Reprod Fertil Suppl. 1996;50:19–23.
5. Tsubamoto H, Yamasaki N, Hasegawa A, Koyama K. Expression of a recombinant porcine zona pellucida glycoprotein ZP1 in mammalian cells. Protein Expr Purif. 1999;17(1):8–15.
6. Maresh GA, Dunbar BS. Antigenic comparison of five species of mammalian zonae pellucidae. J Exp Zool. 1987;244(2):299–307.
7. Drell DW, Dunbar BS. Monoclonal antibodies to rabbit and pig zonae pellucidae distinguish species-specific and shared antigenic determinants. Biol Reprod. 1984;30(2):445–57. https:// doi.org/10.1095/biolreprod30.2.445.
8. Koyama K, Hasegawa A, Isojima S. Further characterization of the porcine zona pellucida antigen corresponding to monoclonal antibody 3A4-2G1 exclusively cross-reactive with porcine and human zonae pellucidae. J Reprod Immunol. 1991;19(2):131–48.
9. Mori T, Kamada M, Yamano S, Kinoshita T, Kano K, Mori T. Production of monoclonal antibodies to porcine zona pellucida and their inhibition of sperm penetration through human zona pellucida in vitro. J Reprod Immunol. 1985;8(1):1–11.
10. Mori E, et al. A monoclonal antibody cross-reactive with porcine and human zona pellucida recognized the protein's steric structure. Eur J Obstet Gynecol Reprod Biol. 1989;32(3): 213–21.
11. Cross NL, Morales P, Overstreet JW, Hanson FW. Induction of acrosome reactions by the human zona pellucida. Biol Reprod. 1988;38(1):235–44.
12. Bhandari B, Bansal P, Talwar P, Gupta SK. Delineation of downstream signaling components during acrosome reaction mediated by heat solubilized human zona pellucida. Reprod Biol Endocrinol. 2010;8:7.
13. Bauskin AR, Franken DR, Eberspaecher U, Donner P. Characterization of human zona pellucida glycoproteins. Mol Hum Reprod. 1999;5(6):534–40.

14. Gupta SK, Yurewicz EC, Sacco AG, Kaul R, Jethanandani P, Govind CK. Human zona pellucida glycoproteins: characterization using antibodies against recombinant non-human primate ZP1, ZP2 and ZP3. Mol Hum Reprod. 1998;4(11):1058–64.

15. Bukovsky A, Gupta SK, Bansal P, Chakravarty S, Chaudhary M, Svetlikova M, White RS, Copas P, Upadhyaya NB, Van Meter SE, Caudle MR. Production of monoclonal antibodies against recombinant human zona pellucida glycoproteins: utility in immunolocalization of respective zona proteins in ovarian follicles. J Reprod Immunol. 2008;78(2):102–14.

16. Chiu PC, Wong BS, Chung MK, Lam KK, Pang RT, Lee KF, Sumitro SB, Gupta SK, Yeung WS. Effects of native human zona pellucida glycoproteins 3 and 4 on acrosome reaction and zona pellucida binding of human spermatozoa. Biol Reprod. 2008;79(5):869–77.

17. Chamberlin ME, Dean J. Human homolog of the mouse sperm receptor. Proc Natl Acad Sci U S A. 1990;87(16):6014–8.

18. Liang LF, Dean J. Conservation of mammalian secondary sperm receptor genes enables the promoter of the human gene to function in mouse oocytes. Dev Biol. 1993;156(2):399–408.

19. Harris JD, Hibler DW, Fontenot GK, Hsu KT, Yurewicz EC, Sacco AG. Cloning and characterization of zona pellucida genes and cDNAs from a variety of mammalian species: the ZPA, ZPB and ZPC gene families. DNA Seq. 1994;4(6):361–93.

20. Ganguly A, Bukovsky A, Sharma RK, Bansal P, Bhandari B, Gupta SK. In humans, zona pellucida glycoprotein-1 binds to spermatozoa and induces acrosomal exocytosis. Hum Reprod. 2010;25(7):1643–56.

21. Tsubamoto H, Hasegawa A, Inoue M, Yamasaki N, Koyama K. Binding of recombinant pig zona pellucida protein 1 (ZP1) to acrosome-reacted spermatozoa. J Reprod Fertil Suppl. 1996;50:63–7.

22. Hasegawa A, Tsubamoto H, Hamada Y, Koyama K. Blocking effect of antisera to recombinant zona pellucida proteins (r-ZPA) on in vitro fertilization. Am J Reprod Immunol. 2000;44(1):59–64.

23. Tsubamoto H, Hasegawa A, Nakata Y, Naito S, Yamasaki N, Koyama K. Expression of recombinant human zona pellucida protein 2 and its binding capacity to spermatozoa. Biol Reprod. 1999;61(6):1649–54.

24. Chiu PC, Wong BS, Lee CL, Pang RT, Lee KF, Sumitro SB, Gupta SK, Yeung WS. Native human zona pellucida glycoproteins: purification and binding properties. Hum Reprod. 2008;23(6):1385–93.

25. Chakravarty S, Kadunganattil S, Bansal P, Sharma RK, Gupta SK. Relevance of glycosylation of human zona pellucida glycoproteins for their binding to capacitated human spermatozoa and subsequent induction of acrosomal exocytosis. Mol Reprod Dev. 2008;75(1):75–88.

26. Avella MA, Baibakov B, Dean J. A single domain of the ZP2 zona Pellucida protein mediates gamete recognition in mice and humans. J Cell Biol. 2014;205(6):801–9.

27. Bray C, Son JH, Kumar P, Harris JD, Meizel S. A role for the human sperm glycine receptor/Cl (−) channel in the acrosome reaction initiated by recombinant ZP3.年・ 雑誌名.

28. José O, Hernández-Hernández O, Chirinos M, González-González ME, Larrea F, Almanza A, Felix R, Darszon A, Treviño CL. Recombinant human ZP3-induced sperm acrosome reaction: evidence for the involvement of T- and L-type voltage-gated calcium channels. Biol Reprod. 2002;66(1):91–7.

29. Dong KW, Chi TF, Juan YW, Chen CW, Lin Z, Xiang XQ, Mahony M, Gibbons WE, Oehninger S. Characterization of the biologic activities of a recombinant human zona pellucida protein 3 expressed in human ovarian teratocarcinoma (PA-1) cells. Am J Obstet Gynecol. 2001;184(5):835–43.

30. Gupta SK, Bhandari B, Shrestha A, Biswal BK, Palaniappan C, Malhotra SS, Gupta N. Mammalian zona pellucida glycoproteins: structure and function during fertilization. Cell Tissue Res. 2012;349(3):665–78.

31. Gupta SK. The human egg's zona pellucida. Curr Top Dev Biol. 2018;130:379–411.

32. Bleil JD, Wassarman PM. Mammalian sperm-egg interaction: identification of a glycoprotein in mouse egg zonae pellucidae possessing receptor activity for sperm. Cell. 1980;20(3):873–82.

Part V
Anti-Zona Pellucida Antibody (AZPA): Zona Pellucida (ZP) in Clinical Inpact

Chapter 19
ZP in Assisted Reproductive Technology

Akiko Hasegawa

Abstract Assisted reproductive technology (ART) has provided many opportunities to observe normal and aberrant human ZP. Medical researchers have examined the relationship between several ZP parameters and fertilization capacity or pregnancy outcome. It seems that morphology does not reflect oocyte and embryo competence. Additionally, the status of empty follicle syndrome (EFS), in which oocytes are not retrieved during IVF programs, is described. As parameters to evaluate embryo quality, ZP thickness variation (ZPTV) and birefringence of the inner region of ZP are provided. These may be useful as noninvasive parameters. ZP features plausibly affect implantation rather than fertilization and embryo development. In ART, empty follicle syndrome (EFS) is often encountered. EFS is usually considered to be derived from inadequate stimulation of gonadotropin, or oocyte aspiration errors. Alternatively, EFS is caused by anti-ZP antibody (AZPA) and/or genetic disorders of ZP genes. The presence of ZP-free or ZP-fragile oocytes has been reported in infertile women with recurrent EFS. Reports about genetic polymorphisms of ZP proteins are currently increasing. The first report was significant in suggesting that inheritance of homologous ZP1 deletion was related to infertility. Here, case reports and preliminary meta-analysis of ZP genetic variation are summarized.

19.1 Morphological Studies of Human ZP

In the past, morphological studies were performed by electron microscopic observation. Researchers tried to detect differences in ZP structure between unfertilized and fertilized oocytes [1, 2]. Precise observations by scanning electron microscopy (SEM) showed the difference between the inner and outer surfaces of ZP [3, 4]. The

A. Hasegawa (✉)
Department of Obstetrics and Gynecology, School of Medicine, Hyogo Medical University, Nishinomiya, Hyogo, Japan
e-mail: zonapel@hyo-med.ac.jp

253

H. Shibahara, A. Hasegawa (eds.), *Gamete Immunology*,
https://doi.org/10.1007/978-981-16-9625-1_19

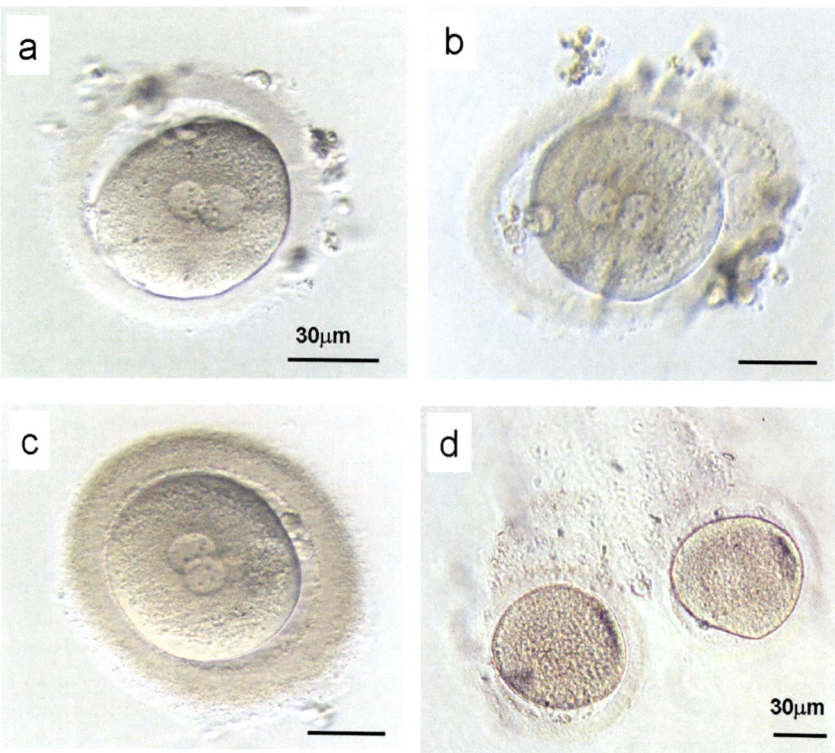

Fig. 19.1 Irregular ZP morphologies observed in ART programs. (A) Normal shape. (**b**) Oval shape. (**c**) Dense and thicker shape of ZP. (**a–c**) Two pronuclei presumably derived from egg and sperm can be seen in the oocyte cytoplasm. This suggests that oocytes were penetrated by a single sperm. (**d**) ZP partially and loosely formed. Sperm did not penetrate into the ooplasm, just binding to an amorphous area of ZP

zona pellucida showed two basically different patterns: a mesh-like, spongy structure having wide and/or close meshes and a compact, smooth surface. The smooth-surfaced zona was most commonly seen in cultured oocytes belonging to the immature and atretic groups [1]. They suggested a relationship between sperm permeability and the amorphous spongelike structure. However, another study indicated that variations in size and morphology did not correlate either with the maturity of the oocytes or with the age of the women [5]. Furthermore, a different research group reported that no correlation was detected between the morphology of ZP and success of fertilization [6].

Irregular ZP morphologies observed in our ART programs are shown in Fig. 19.1. Excepting panel D, two pronuclei presumably derived from egg and sperm can be seen in the oocyte cytoplasm. The ZP shape in D could not indicate direction to sperm going to progress. A study suggested that oocytes with a dark ZP shown in panel C in Fig. 19.1 affected the clinical outcome in IVF/ICSI cycles and appeared to contain more ultrastructural alterations [7]. Collectively, irregular ZP morphology is probably caused by the local environment of surrounding somatic cells but does not influence the sperm binding and fertilizing abilities of ZP. Recently, it was provided data that expected the correlation of ZP gene variation and ZP morphology [8].

19.2 ZP Thickness

With significant advances in ART, studies using human oocytes have increased. The relationship between ZP thickness and fertilization capacity of oocytes or pregnancy outcome has been examined. Some studies showed a positive relation between ZP thickness and pregnancy outcome [9]. Oocytes with thicker ZP tended to have better fertilization rates and embryo growth. ZP thickness was generally regarded in terms of cellular activities including respiratory status, protein synthesis, and potential of embryo growth.

Recent studies reported that ZP thickness variation (ZPTV) rather than ZP thickness is a more effective parameter to evaluate embryo quality [10–13]. However, opposite studies also documented that there is no beneficial effect in ZP measurements in in vitro fertilization treatment [14]. Recently, the relationships between ZPTV and implantation/assisted hatching were assessed using imaging software, in which the ZPTV values were obtained from 100 points in each embryo. There were no significant trends for association between ZPTV and implantation but slightly higher odds for implantation when assisted hatching was used [15].

More recently, measuring optical density showed more promise for ZP quality than thickness [16]. With further development of the polarized light microscope (PLM), precise observation of oocyte spindles and ZP became possible. Retardance observed by PLM was applied to noninvasive prediction of fertilizing ability and embryo development. Furthermore, it suggested a relationship between ZP birefringence in the inner region and implantation potential in ICSI cycles or embryo development [17].

Another study examined the correlation of oocyte maturity and ZP birefringence in the inner region of ZP. The results showed that the inner regions of ZP observed by PLM were thicker in mature oocytes than in immature oocytes. In addition, fertilizable oocytes tended to have thicker inner regions than unfertilized oocytes. These reports suggested that the thickness of the inner region of ZP could be a noninvasive predictor for fertilizing competence of the oocytes [18].

Higher and thicker birefringence in the inner area of ZP may reflect higher competences of oocyte and embryo growth, but how birefringence originates is not well understood. Probably, dense cross-linking of ZP protein filaments produces higher birefringence. Higher cross-linking of ZP filaments composed of ZP proteins is not always an advantage. In some steps for fertilization and embryo growth, higher cross-linking of ZP filaments may be unfavorable to the oocytes/embryos such as for permeation of nutrients and growth factors from outer environments as well as embryo hatching [19]. Whether PLM can be useful for the identification of good competence oocytes for ART selection is still under debate.

As another interesting predictor, mRNAs for ZP components in granulosa cells are being investigated [20]. Before the end stage of oocyte growth and maturation, mRNAs of ZP2 and ZP3 proteins were detected in granulosa cells as well as oocytes in humans [21]. Quantitative analysis of ZP3 mRNA in granulosa cells was proposed as a noninvasive predictor for fertilization capacity of oocytes, but its accuracy is as yet unknown [20]. Overall, parameters related to ZP are not promising predictors for determining fertilization and pregnancy competences at this moment.

19.3 Empty Follicle Syndrome (EFS)

The occurrence of "empty follicle syndrome" (EFS) is often experienced in reproductive medicine. EFS is usually considered to be derived from inadequate stimulation of gonadotropin or oocyte aspiration errors [22, 23]. However, anti-ZP autoantibodies can induce EFS. Our animal experiments showed that immunization with solubilized pig ZP antigens produced histological features of empty follicles composed of only granulosa cells [24]. These follicles could possibly develop into antral follicles or ZP-lacking/degenerated oocytes by high-intensity gonadotropin stimulation.

As another possibility, EFS might arise from a genetic abnormality in ZP protein production, because animal experiments demonstrated that mice knocked out for the genes encoding zona proteins ZP2 and ZP3 were completely infertile [25, 26]. ZP1 knockout mice showed significantly reduced fecundity [27]. Figure 17.5 shows the ovarian histology of a ZP2 knockout mouse. A number of small follicles lacking oocytes were observed. Therefore, in humans, aberrant expression of ZP proteins by genetic ZP mutations might cause infertility. Case reports have described that ZP-free or ZP-fragile oocytes were retrieved from women with recurrent EFS [28–31]. Such ZP abnormalities arise from mutations in the genes encoding ZP proteins.

19.4 Mutation of ZP Gene

Some studies reported the relationship between ZP morphology and ZP gene mutation. Although human ZP exhibits genetic polymorphisms, most of them do not affect sperm binding or the fertilizing ability of the oocyte. One study showed that a single nucleotide polymorphism (SNP) affected ZP structure, but fertilizing ability was normal. Furthermore, there was no evidence of any relationship between the occurrence of infertility and particular DNA sequences [7]. Collectively, irregular ZP morphology is probably caused by the local environment of surrounding somatic cells but does not influence the sperm binding and fertilizing abilities of ZP. Irregular ZP morphologies that our laboratory observed are shown in Fig. 2.3.1. Excepting panel D, two pronuclei derived from egg and sperm can be seen in the oocyte cytoplasm, which provides evidence of sperm penetration of the ZP.

However, in 2005, it was found that there were 20 ZP gene variations in patients with fertilization failure. Two variations in ZP3, one in the regulatory region (c. 1-87 T -- > G), and one in exon 6 [c. 894 G -- > A (p. K298)] existed more frequently in fertilization failure group than in normal fertile group [32].

Inheritable *ZP1* gene mutation was first reported in 2014 [33]. The researchers identified a homologous frameshift mutation in *ZP1* in six family members, of whom five were known to be infertile. The frameshift mutation produced a truncated ZP1 protein of 404 amino acids, although the normal protein consists of 638. Oocytes bearing the mutation did not form ZP at all. This phenotype is different from ZP1-targeted mice. The gene-targeting study showed that ZP1 gene knockout mice form a half-thickness ZP and show reduced fecundity but are not infertile [27]. It was noticeable that similar ZP abnormality eventually occurred in humans, although the phenotype was not the exact same.

Several reports have been published in which SNP mutation possibly resulted in ZP-deficient oocytes in humans. One of them clarified the two ZP2 pathogenic variants causing infertility whose oocytes lack ZP following truncated ZP2 protein [29]. In gene-targeting experiments, ZP2 and ZP3 null mice did not form ZP, as mentioned in the previous chapter. Recently, identifications of ZP gene mutation have increased in infertile patients with abnormal ZP [31–33]. However, zona-free oocytes possibly fertilize by ICSI and develop into blastocyst resulting in live birth [28].

Recently, several kinds of ZP mutation have often been reported to induce recurrent EFS [34–38]. In addition, it was revealed that heterologous mutations in ZP1 and ZP3 also caused EFS [37, 38]. Mouse oocytes with heterologous mutation showed to produce thinner ZP structure but did not impair the oocyte growth in the follicles. On the other hand, human oocytes with heterologous mutation are possibly damaged during follicle growth, because human reproductive period is much longer than mice. In addition, the relationship between mutations in ZP4 and endometriosis was reported for the first time [39]. As the counterpart to ZP4 is not expressed in mouse ZP, it is impossible to know the affect from this deficiency by mouse model.

Taken together, ZP gene mutation seems to impede fertility with various syndromes. Investigations of human ZP gene mutation will expand the understanding of pathogenic mechanisms for infertility.

References

1. Familiari G, Nottola SA, Micara G, Aragona C, Motta PM. Is the sperm-binding capability of the zona pellucida linked to its surface structure? A scanning electron microscopic study of human in vitro fertilization. J In Vitro Fert Embryo Transf. 1988;5(3):134–43.
2. Nikas G, Paraschos T, Psychoyos A, Handyside AH. The zona reaction in human oocytes as seen with scanning electron microscopy. Hum Reprod. 1994;9(11):2135–8.
3. Familiari G, Nottola SA, Macchiarelli G, Micara G, Aragona C, Motta PM. Human zona pellucida during in vitro fertilization: an ultrastructural study using saponin, ruthenium red, and osmium-thiocarbohydrazide. Mol Reprod Dev. 1992;32(1):51–61.

4. Familiari G, Relucenti M, Heyn R, Micara G, Correr S. Three-dimensional structure of the zona pellucida at ovulation. Microsc Res Tech. 2006;69(6):415–26.
5. Schwartz P, Hinney B, Nayudu PL, Michelmann'HW. Oocyte-sperm interaction in the course of IVF: a scanning electron microscopy analysis. Reprod Biomed Online. 2003;7(2):205–10.
6. Magerkurth C, Töpfer-Petersen E, Schwartz P, Michelmann HW. Scanning electron microscopy analysis of the human zona pellucida: influence of maturity and fertilization on morphology and sperm binding pattern. Hum Reprod. 1999;14(4):1057–66.
7. Shi W, Xu B, Wu L, Jin R, Luan H, Luo L, Zhu Q, Johansson L, Liu Y, Tong X. Oocytes with a dark zona pellucida demonstrate lower fertilization, implantation and clinical pregnancy rates in IVF/ICSI cycles. PLoS One. 2014;9(2):e89409.
8. Pökkylä RM, Lakkakorpi JT, Nuojua-Huttunen SH, Tapanainen JS. Sequence variations in human ZP genes as potential modifiers of zona pellucida architecture. Fertil Steril. 2011;95: 2669–72.
9. Bertrand E, Van den Bergh M, Englert Y. Clinical parameters influencing human zona pellucida thickness. Fertil Steril. 1996;66(3):408–11.
10. Palmstierna M, Murkes D, Csemiczky G, Andersson O, Wramsby H. Zona pellucida thickness variation and occurrence of visible mononucleated blastomers in preembryos are associated with a high pregnancy rate in IVF treatment. J Assist Reprod Genet. 1998;15(2):70–5.
11. Gabrielsen A, Bhatnager PR, Petersen K, Lindenberg S. Influence of zona pellucida thickness of human embryos on clinical pregnancy outcome following in vitro fertilization treatment. J Assist Reprod Genet. 2000;17(6):323–8.
12. Gabrielsen A, Lindenberg S, Petersen K. The impact of the zona pellucida thickness variation of human embryos on pregnancy outcome in relation to suboptimal embryo development. A prospective randomized controlled study. Hum Reprod. 2001;16(10):2166–70.
13. Sun YP, Xu Y, Cao T, Su YC, Guo YH. Zona pellucida thickness and clinical pregnancy outcome following in vitro fertilization. Int J Gynaecol Obstet. 2005;89(3):258–62.
14. Balakier H, Sojecki A, Motamedi G, Bashar S, Mandel R, Librach C. Is the zona pellucida thickness of human embryos influenced by women's age and hormonal levels? Fertil Steril. 2012;98(1):77–83.
15. Lewis EI, Farhadifar R, Farland LV, Needleman DJ, Missmer SA, Racowsky C. Use of imaging software for assessment of the associations among zona pellucida thickness variation, assisted hatching, and implantation of day 3 embryos. J Assist Reprod Genet. 2017;34(10):1261–9.
16. Pelletier C, Keefe DL, Trimarchi JR. Noninvasive polarized light microscopy quantitatively distinguishes the multilaminar structure of the zona pellucida of living human eggs and embryos. Fertil Steril. 2004;81:850–6.
17. Montag M, Schimming T, Köster M, Zhou C, Dorn C, Rösing B. Oocyte zona birefringence intensity is associated with embryonic implantation potential in ICSI cycles. Reprod Biomed Online. 2008;16(2):239–44.
18. Ebner T, Balaban B, Moser M, Shebl O, Urman B, Ata B, Tews G. Automatic user-independent zona pellucida imaging at the oocyte stage allows for the prediction of preimplantation development. Fertil Steril. 2010;94:913–20.
19. Molinari E, Evangelista F, Racca C, Cagnazzo C, Revelli A. Polarized light microscopy-detectable structures of human oocytes and embryos are related to the likelihood of conception in IVF. J Assist Reprod Genet. 2012;29:1117–22.
20. Gook DA, Edgar DH, Borg J, Martic M. Detection of zona pellucida proteins during human folliculogenesis. Hum Reprod. 2008;23:394–440.
21. Canosa S, Adriaenssens T, Coucke W, Dalmasso P, Revelli A, Benedetto C, Sm J. Zona pellucida gene mRNA expression in human oocytes is related to oocyte maturity, zona inner layer retardance and fertilization competence. Mol Hum Reprod. 2017;23(5):292–303.
22. Kim JH, Jee BC. Empty follicle syndrome. Clin Exp Reprod Med. 2012;39:132–7.
23. Vutyavanich T, Piromlertamorn W, Ellis J. Immature oocytes in "apparent empty follicle syndrome": a case report. J Case Rep Med. 2010;2010:367505.

24. Hasegawa A, Koyama K, Inoue M, Takemura T, Isojima S. Antifertility effect of active immunization with ZP4 glycoprotein family of porcine zona pellucida in hamsters. J Reprod Immunol. 1992;22:197–210.
25. Rankin TL, Tong ZB, Castle PE, Lee E, Gore-Langton R, Nelson LM, et al. Human ZP3 restores fertility in Zp3 null mice without affecting order-specific sperm binding. Development. 1998;125:2415–24.
26. Rankin TL, O'Brien M, Lee E, Wigglesworth K, Eppig J, Dean J. Defective zonae pellucidae in Zp2-null mice disrupt folliculogenesis, fertility and development. Development. 2001;128: 1119–26.
27. Rankin T, Talbot P, Lee E, Dean J. Abnormal zonae pellucidae in mice lacking ZP1 result in early embryonic loss. Development. 1999;126:3847–55.
28. Shu Y, Peng W, Zhang J. Pregnancy and live birth following the transfer of vitrified-warmed blastocysts derived from zona- and corona-cell-free oocytes. Reprod Biomed Online. 2010;21: 527–32.
29. Dai C, Hu L, Gong F, Tan Y, Cai S, Zhang S, et al. ZP2 pathogenic variants cause in vitro fertilization failure and female infertility. Genet Med. 2019;21(2):431–40.
30. Duru NK, Cincik M, Dede M, Hasimi A, Baser I. Retrieval of zona-free immature oocytes in a woman with recurrent empty follicle syndrome: a case report. J Reprod Med. 2007;52:858–63.
31. Zhou Z, Ni C, Wu L, Chen B, Xu Y, Zhang Z, Mu J, Li B, Yan Z, Fu J, Wang W, Zhao L, Dong J, Sun X, Kuang Y, Sang Q, Wang L. Novel mutations in ZP1, ZP2, and ZP3 cause female infertility due to abnormal zona pellucida formation. Hum Genet. 2019;138(4):327–37.
32. Männikkö M, Törmälä RM, Tuuri T, Haltia A, Martikainen H, Ala-Kokko L, et al. Association between sequence variations in genes encoding human zona pellucida glycoproteins and fertilization failure in IVF. Hum Reprod. 2005;20:1578–85.
33. Huang H, Lv C, Zhao Y, Li W, He X, Li P, Sha A, Tian X, Papasian CJ, Deng H, Lu G, Xiao H. Mutant ZP1 in familial infertility. N Engl J Med. 2014;370(13):1220–6.
34. Chen T, Bian Y, Liu X, Zhao S, Wu K, Yan L, et al. A recurrent missense mutation in ZP3 causes empty follicle syndrome and female infertility. Am J Hum Genet. 2017;101(3):459–65.
35. Dai C, Chen Y, Hu L, Du J, Gong F, Dai J, Zhang S, Wang M, Chen J, Guo J, Zheng W, Lu C, Wu Y, Lu G, Lin G. ZP1 mutations are associated with empty follicle syndrome: evidence for the existence of an intact oocyte and a zona pellucida in follicles up to the early antral stage. A case report. Hum Reprod. 2019;34(11):2201–7.
36. Xu Q, Zhu X, Maqsood M, Li W, Tong X, Kong S. A novel homozygous nonsense ZP1 variant causes human female infertility associated with empty follicle syndrome (EFS). Mol Genet Genomic Med. 2020;8(7):e1269.
37. Ling S, Fang X, Zhiheng Chen Z, Zhang H, Zhang Z, Zhou P, et al. Compound heterozygous ZP1 mutations cause empty follicle syndrome in infertile sisters. Hum Mutat. 2019;40(11): 2001–6.
38. Liu W, Li K, Bai D, Yin J, Tang Y, Chi F, Zhang L. Dosage effects of ZP2 and ZP3 heterozygous mutations cause human infertility. Hum Genet. 2017;136(8):975–98.
39. Zou Y, Zhou J, Guo J, Zhang Z, Luo Y, Liu F, Huang H, Wang F, He M, Wang L, Huang O. Mutation analysis of ZP1, ZP2, ZP3 and ZP4 genes in 152 Han Chinese samples with ovarian endometriosis. Mutat Res. 2019;813:46–50.

Chapter 20
Infertility and Anti-ZP Antibody (AZPA)

Akiko Hasegawa

Abstract Animal experiments showed that AZPA inhibited fertilization due to blocking of sperm binding to ZP. Whether AZPA causes infertility in humans has been debated since the 1970s. The detection frequency of AZPA in infertile women varies in individual research, depending on the detection methods used in the studies. In this chapter, methods for detecting AZPA are summarized. In the early stages of this field, AZPA was screened for using pig ZP. Immunofluorescence, ELISA, radioimmunoassay, and heme-agglutination tests were adopted for screening. Recently, dot immunoassay, in which solubilized native human ZP is applied to a nitrocellulose membrane, has been exploited as a highly specific method. Hemizona assay may also be applicable for AZPA detection. Despite application of these methods, the correlation of AZPA with infertility is still controversial. Advanced immunological studies will clarify whether or not AZPA is a pathogenic factor of infertility.

20.1 History

Early studies showed that in animal experiments, immunization with ZP or ovarian tissues produced antibodies reactive to ZP [1–4]. Serum antibodies recognized autologous ZP as well as heterologous ZP used as immunogens. This means that there are cross-reactive antigens in different animal species and that heterologous antigens produce self-reactive antibodies to ZP. Furthermore, the antibodies strongly blocked fertilization by prevention of sperm binding to ZP. Numerous studies have been carried out to analyze AZPA. Some of them focused on pathogenic factors of infertility.

A. Hasegawa (✉)
Department of Obstetrics and Gynecology, School of Medicine, Hyogo Medical University, Nishinomiya, Hyogo, Japan
e-mail: zonapel@hyo-med.ac.jp

H. Shibahara, A. Hasegawa (eds.), *Gamete Immunology*,
https://doi.org/10.1007/978-981-16-9625-1_20

Shivers et al. found AZPA in human sera [5]. In infertile women, they detected 32% with strongly positive sera and 41% with moderately positive sera, suggesting that the production of AZPA was one of the pathogenic factors in fertilization failure. After publication, many studies supporting their results followed [6–11].

However, contrary results were also reported. Sacco et al. documented that AZPA-positive reactions were found in fertile and infertile women and even in men [12]. Their results showed that the prevalence of AZPA in infertile women was a nonspecific reaction. Many studies supporting these contrary results were also reported [13–18]. Afterward, it was reported that most or all positive reactions disappeared after absorption with pig erythrocytes [13, 14]. Only a few positive cases remained after this absorption. It seemed that most of the AZPA detected in humans recognized heteroantigens (antigens from other species) because earlier studies had used pig ZP as a source of antigens for an indirect immunofluorescent assay. This method was of low sensitivity with a high background. To demonstrate the genuine existence of anti-ZP antibodies causing human infertility, improvements in assay methods were needed.

20.2 Detection Methods for AZPA

To clarify these controversial results, intensive efforts were made to develop detection methods with enhanced sensitivities and without nonspecific reactions. Several highly sensitive assays were proposed [16–22]. Quantitative radioimmunoassays revealed similar frequencies of anti-ZP antibodies in cases of unexplained infertility comparing fertile women and men [16, 17]. As another sensitive and quantitative method, the passive hemagglutination reaction showed a significantly higher incidence of anti-ZP antibodies in individuals with idiopathic infertility [21]. The authors performed careful and large-scale investigations and confirmed the existence of anti-ZP antibodies specifically reactive to human ZP proteins. They showed that the presence of anti-ZP antibodies was an etiological factor in infertility, although the incidence was very low [23]. Caudle suggested that in a given individual, anti-ZP antibodies might coexist with other fertility disorders and that it is difficult to distinguish whether these antibodies are the real cause of concurrent fertility disorders [24].

20.3 Correlation of AZPA with Infertility

Assisted reproductive technology brought an opportunity to use human ZP proteins for an anti-ZP antibody detection assay. The hemizona assay was originally exploited to predict the zona-binding potential of spermatozoa [25]. This assay was also applied to testing inhibitory effects of sperm binding to ZP by anti-ZP antibodies [26–28]. To detect ZP antibody, one hemizona was treated with test serum, and the other hemizona was treated with control serum before incubation with fertile spermatozoa. The sperm binding capacity was expected to be identical in each hemizona because they were derived from the same oocyte. Serum that reduced the numbers of spermatozoa showing tight binding to the ZP by more than 50% compared with the control was judged to be positive for anti-ZP antibodies.

For another human ZP-based assay, Takamizawa et al. developed a dot immunoassay using solubilized human ZP [29]. The prevalence of anti-ZP antibodies in the sera was not significantly higher in women with unexplained infertility compared with other parameters. However, the frequency was particularly higher in women with premature ovarian failure (POF) [29–31]. Indeed, there have been reports that POF was associated with elevated levels of anti-ZP antibodies [32, 33]. Animal studies showed that ovarian autoimmune disease could be induced by immunization with ZP proteins and that this was associated with ovarian disruption and infertility, similar to POF [34–37].

Assuming that anti-ZP antibodies are generated in some infertile women: what is the mechanism for generating such autoantibodies? One possible mechanism is immune stimulation in the oviducts and uterus during repeated ovulatory cycles or longer exposure to antigens. Aging may be one plausible factor for autoantibodies to ZP proteins [12, 16, 38]. Another possibility is that these women might have been sensitized to foreign antigens that cross-reacted with their own ZP antigens. It was reported that anti-ZP antibodies are frequently detected in mycoplasma-infected patients [39]. Also several studies showed that anti-ZP antibodies cross-react with thyroid microsomal antigen [33, 40], although further investigations are necessary to clarify whether autoimmunity to thyroid antigens is an etiological factor in infertility. In another report, it was shown that the presence of mycoplasma was significantly correlated with the presence of AZPA [41].

There have been reports showing an association between elevated anti-ZP antibody titers and poor fertilization rates in conventional in vitro fertilization (IVF). Some studies focused on follicular fluid rather than blood sera and documented a high detection frequency of anti-ZP antibodies in cases of unexplained infertility [42, 43]. The presence of anti-ZP antibodies in follicular fluid is a plausible cause of poor fertilization because follicle fluid encloses the oocyte before ovulation. Some clinicians have recommended intracytoplasmic sperm injection (ICSI) rather than conventional IVF, because repeated follicular puncture tends to increase the levels of circulating anti-ZP antibodies [44, 45].

In spite of numerous studies, no significant association of anti-ZP antibodies with idiopathic human infertility has been established conclusively, and the topic is still controversial [24, 46, 47]. The ZP has many kinds of highly immunogenic antigens, but usually antibody production has been suppressed by the development of immune tolerance. Nevertheless, ZP antigens or cross-reactive foreign antigens could be recognized as altered self-antigens resulting in autoantibodies reactive to native ZP.

References

1. Gwatkin RB, Williams DT, Carlo DJ. Immunization of mice with heat-solubilized hamster zonae: production of anti-zona antibody and inhibition of fertility. Fertil Steril. 1977;28:871–7.
2. Yanagimachi R, Winkelhake J, Nicolson GL. Immunological block to mammalian fertilization: survival and organ distribution of immunoglobulin which inhibits fertilization in vivo. Proc Natl Acad Sci U S A. 1976;73:2405–8.
3. Tsunoda Y, Chang MC. Effects of antisera against eggs and zonae pellucidae on fertilization and development of mouse eggs in vivo and in culture. J Reprod Fertil. 1978;54:233–7.
4. Sacco AG. Inhibition of fertility in mice by passive immunization with antibodies to isolated zonae pellucidae. J Reprod Fertil. 1979;56:533–7.
5. Shivers CA, Dunbar BS. Autoantibodies to zona pellucida: a possible cause for infertility in women. Science. 1977;197:1082–4.
6. Mori T, Nishimoto T, Kohda H, Takai I, Nishimura T, Oikawa T. A method for specific detection of autoantibodies to the zona pellucida in infertile women. Fertil Steril. 1979;32: 67–72.
7. Singh J, Mhaskar AM. Enzyme-linked immunosorbent determination of autoantibodies to zona pellucida as a possible cause of infertility in women. J Immunol Methods. 1985;79:133–41.
8. Kamada M, Daitoh T, Mori K, Maeda N, Hirano K, Irahara M, et al. Etiological implication of autoantibodies to zona pellucida in human female infertility. Am J Reprod Immunol. 1992;28: 104–9.
9. Hovav Y, Almagor M, Benbenishti D, Margalioth EJ, Kafka I, Yaffe H. Immunity to zona pellucida in women with low response to ovarian stimulation, in unexplained infertility and after multiple IVF attempts. Hum Reprod. 1994;9:643–5.
10. Mikulíková L, Veselský L, Cerný V, Martínek J, Malbohan I, Fialová L. Immunofluorescence detection of porcine anti-zona pellucida antibodies in sera of infertile women. Acta Univ Carol Med (Praha). 1989;35:63–8.
11. Buckshee K, Mhaskar A. Status of autoantibodies to zona pellucida in human reproduction. Int J Fertil. 1985;30:13–7.
12. Sacco AG, Moghissi KS. Anti-zona pellucida activity in human sera. Fertil Steril. 1979;31:503–6.
13. Dakhno FV, Hjort T, Grischenko VI. Evaluation of immunofluorescence on pig zona pellucida for detection of anti-zona antibodies in human sera. J Reprod Immunol. 1980;2:281–91.
14. Nayudu PL, Freemann LE, Trounson AO. Zona pellucida antibodies in human sera. J Reprod Fertil. 1982;65:77–84.
15. Caudle MR, Shivers CA, Wild RA. Clinical significance of naturally occurring anti-zona pellucida antibodies in infertile women. Am J Reprod Immunol Microbiol. 1987;15:119–21.
16. Dietl J, Knop G, Mettler L. The frequency of serological anti-zona pellucida activity in males, females and children. J Reprod Immunol. 1982;4:123–31.
17. Kurachi H, Wakimoto H, Sakumoto T, Aono T, Kurachi K. Specific antibodies to porcine zona pellucida detected by quantitative radioimmunoassay in both fertile and infertile women. Fertil Steril. 1984;41:265–9.
18. Gerrity M, Niu E, Dunbar BS. A specific radioimmunoassay for evaluation of serum antibodies to zona pellucida antigens. J Reprod Immunol. 1981;3:59–70.
19. Sacco AG, Subramanian MG, Yurewicz EC. Application of a radioimmunoassay (RIA) for monitoring immune response to porcine zonae pellucidae. Proc Soc Exp Biol Med. 1981;167: 318–26.
20. Noda Y, Yamada I, Hayashi S, Takai I, Watanabe E. A new method for detection of anti-zona activity in human sera using latex agglutination reaction. Experientia. 1983;39:926–8.
21. Kamada M, Hasebe H, Irahara M, Kinoshita T, Naka O, Mori T. Detection of anti-zona pellucida activities in human sera by the passive hemagglutination reaction. Fertil Steril. 1984;41:901–6.

22. Czuppon AB, Dietl J, Tripatzis I. Solid-phase antigen immunoassay for the detection of anti-zona pellucida-antibodies in clinically defined sera. J Clin Chem Clin Biochem. 1987;25:87–90.

23. Kamada T, Maegawa M, Daitoh T, Mori K, Yamamoto S, et al. Sperm-zona pellucida interaction and immunological infertility. Reprod Med Biol. 2006;5:95–104.

24. Caudle MR, Shivers CA. Current status of anti-zona pellucida antibodies. Am J Reprod Immunol. 1989;21:57–60.

25. Burkman LJ, Coddington CC, Franken DR, Kruger TF, Rosenwaks Z, Hodgen GD. The hemizona assay (HZA): development of a diagnostic test for the binding of human spermatozoa to the human hemizona pellucida to predict fertilization potential. Fertil Steril. 1988;49:688–97.

26. Govind CK, Hasegawa A, Koyama K, Gupta SK. Delineation of a conserved B cell epitope on bonnet monkey (Macaca radiata) and human zona pellucida glycoprotein-B by monoclonal antibodies demonstrating inhibition of sperm-egg binding. Biol Reprod. 2000;62:67–75.

27. Afzalpurkar A, Shibahara H, Hasegawa A, Koyama K, Gupta SK. Immunoreactivity and in-vitro effect on human sperm-egg binding of antibodies against peptides corresponding to bonnet monkey zona pellucida-3 glycoprotein. Hum Reprod. 1997;12:2664–70.

28. Sivapurapu N, Upadhyay A, Hasegawa A, Koyama K, Gupta SK. Native zona pellucida reactivity and in-vitro effect on human sperm-egg binding with antisera against bonnet monkey ZP1 and ZP3 synthetic peptides. J Reprod Immunol. 2002;56:77–91.

29. Takamizawa S, Shibahara H, Shibayama T, Suzuki M. Detection of antizona pellucida antibodies in the sera from premature ovarian failure patients by a highly specific test. Fertil Steril. 2007;88:925–32.

30. Koyama K, Hasegawa A, Mochida N, Calongos G. Follicular dysfunction induced by autoimmunity to zona pellucida. Reprod Biol. 2005;5:269–78.

31. Calongos G, Hasegawa A, Komori S, Koyama K. Harmful effects of anti-zona pellucida antibodies in folliculogenesis, oogenesis, and fertilization. J Reprod Immunol. 2009;79:148–55.

32. Tuohy VK, Altuntas CZ. Autoimmunity and premature ovarian failure. Curr Opin Obstet Gynecol. 2007;19:366–9.

33. Kelkar RL, Meherji PK, Kadam SS, Gupta SK, Nandedkar TD. Circulating auto-antibodies against the zona pellucida and thyroid microsomal antigen in women with premature ovarian failure. J Reprod Immunol. 2005;66:53–67.

34. Rhim SH, Millar SE, Robey F, Luo AM, Lou YH, Yule T, Tung KS. Autoimmune disease of the ovary induced by a ZP3 peptide from the mouse zona pellucida. J Clin Invest. 1992;89:28–35.

35. Tung K, Agersborg S, Bagavant H, Garza K, Wei K. Autoimmune ovarian disease induced by immunization with zona pellucida (ZP3) peptide. Curr Protoc Immunol. 2002;Chapter 15:Unit 15-17.

36. Lloyd ML, Papadimitriou JM, O'Leary S, Robertson SA, Shellam GR. Immunoglobulin to zona pellucida 3 mediates ovarian damage and infertility after contraceptive vaccination in mice. J Autoimmun. 2010;35:77–85.

37. Bagavant H, Sharp C, Kurth B, Tung KS. Induction and immunohistology of autoimmune ovarian disease in cynomolgus macaques (Macaca fascicularis). Am J Pathol. 2002;160:141–9.

38. Nishimoto T, Mori T, Yamada I, Nishimura T. Autoantibodies to zona pellucida in infertile and aged women. Fertil Steril. 1980;34:552–6.

39. Al-Daghistani HI, Fram KM. Incidence of anti-zona pellucida and anti-sperm antibodies among infertile Jordanian women and its relation to mycoplasma. East Mediterr Health J. 2009;15: 1263–71.

40. Twig G, Shina A, Amital H, Shoenfeld Y. Pathogenesis of infertility and recurrent pregnancy loss in thyroid autoimmunity. J Autoimmun. 2012;38:J275–81.

41. Al-Daghistani HI, Fram KM. Incidence of anti-zona pellucida and anti-sperm antibodies among infertile Jordanian women and its relation to mycoplasmas. East Mediterr Health J. 2009;15(5): 1263–71.

42. Mantzavinos T, Dalamanga N, Hassiakos D, Dimitriadou F, Konidaris S, Zourlas PA. Assessment of autoantibodies to the zona pellucida in serum and follicular fluid in in-vitro fertilization patients. Clin Exp Obstet Gynecol. 1993;20:111–5.
43. Papale ML, Grillo A, Leonardi E, Giuffrida G, Palumbo M, Palumbo G. Assessment of the relevance of zona pellucida antibodies in follicular fluid of in-vitro fertilization (IVF) patients. Hum Reprod. 1994;9:1827–31.
44. Ulcová-Gallová Z, Mardesic T. Does in vitro fertilization (IVF) influence the levels of sperm and zona pellucida (ZP) antibodies in infertile women? Am Reprod Immunol. 1996;36:216–9.
45. Mardesic T, Ulcova-Gallova Z, Huttelova R, Muller P, Voboril J, Mikova M, et al. The influence of different types of antibodies on in vitro fertilization results. Am J Reprod Immunol. 2000;43:1–5.
46. Van Voorhis BJ, Stovall DW. Autoantibodies and infertility: a review of the literature. J Reprod Immunol. 1997;33:239–56.
47. Forges T, Monnier-Barbarino P, Faure GC, Béné MC. Autoimmunity and antigenic targets in ovarian pathology. Hum Reprod Update. 2004;10:163–75.

Part VI
Anti-Zona Pellucida Antibody: ZP-Based Immuno-contraceptive Vaccine Studies

Chapter 21
Development of Immunocontraceptives Using ZP-Related Antigens

Akiko Hasegawa

Abstract Several decades ago, AZPA was reported to inhibit fertilization in vitro and in vivo in animal experiments. ZP has thus been considered a promising candidate for an immunocontraceptive vaccine. However, vaccination by various kinds of antigen interfered with pregnancy but concomitantly provoked undesirable side effects such as ovarian impairment. This was thought to be caused by contamination of somatic antigens. Recombinant ZP proteins were expected to be superior candidates as immunocontraceptive vaccine targets because the immunogens can be produced in large amounts and contamination from somatic tissues is excluded. Despite intensive efforts, most human model experiments provoked ovarian impairment. In the next stage, researchers designed various antigenic molecules using DNA technology. Artificial sequences such as a tandemly repeated peptide ZP epitope, or ZP peptide linked to a foreign epitope, were engineered. In addition, DNA vaccines automatically proliferate and continuously stimulate an immune reaction in inoculated animals. Recently, fragments termed scFv that directly bind to ZP as antibodies have also been focused on. In the future, an artificial contraceptive vaccine might be designed using multiple antigens to interfere with fertilization without undesirable side effects. However, the current general consensus is that an immunocontraceptive vaccine using ZP antigens could not be applied to humans, but only for animal population control.

21.1 Natural ZP as a Vaccine Candidate

The zona pellucida (ZP) is an extracellular matrix surrounding ovarian oocytes, ovulated eggs and preimplantation embryos (Fig. 21.1). It plays several important roles at different stages of reproduction. Its constituent glycoproteins are expressed

A. Hasegawa (✉)
Department of Obstetrics and Gynecology, School of Medicine, Hyogo Medical University, Nishinomiya, Hyogo, Japan
e-mail: zonapel@hyo-med.ac.jp

H. Shibahara, A. Hasegawa (eds.), *Gamete Immunology*,
https://doi.org/10.1007/978-981-16-9625-1_21

269

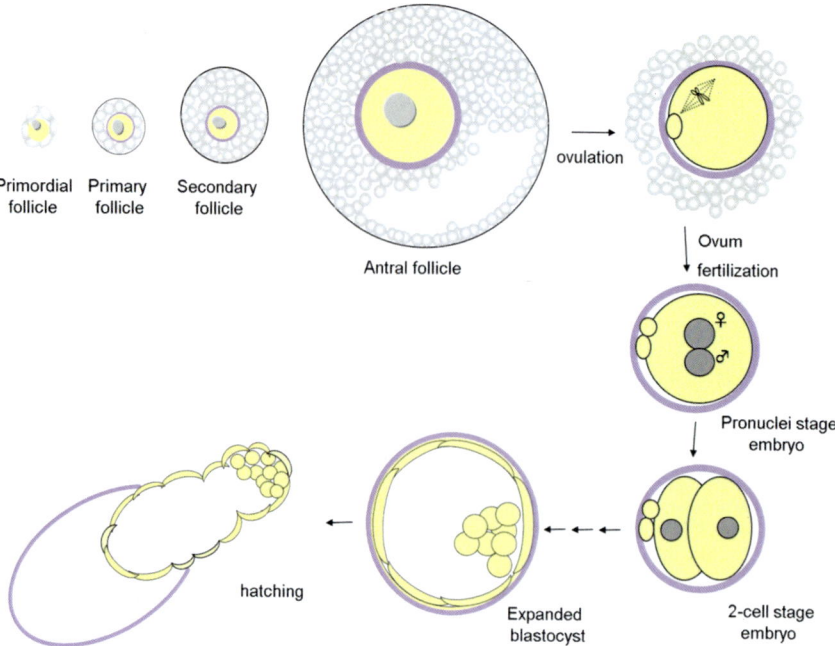

Fig. 21.1 Schematic drawing of ZP presence during follicle/oocyte growth, egg ovulation, and embryo preimplantation. ZP forms around the oocyte at the primary follicle stage and the thickness increases during follicle/oocyte growth. After ovulation, ZP functions as a sperm barrier as well as in sperm acrosome reaction induction and embryo protection. The zona derivative after hatching may have some function for pregnancy maintenance. ZP is indicated by a purple line

in a tissue-specific manner. It is thus possible to produce self-reactive antibodies to ZP antigens that interfere with reproductive functions including folliculogenesis, fertilization, and implantation. This chapter describes the historical development of ZP-based contraceptive vaccine studies.

Since the 1970s, ZP had been assumed to be a promising candidate for immunocontraceptive vaccines [1–3], because anti-ZP antibody (AZPA) blocked fertilization by prevention of sperm binding to ZP in vitro. Indeed, ZP plays several important functional roles such as sperm recognition, polyspermy blocking, and protection of preimplantation embryos. As mentioned previously, biochemical and immunological studies revealed that ZP is composed of glycoproteins and possesses strong antigenicity. Furthermore, ZP is limitedly localized in female reproductive tissues. This means that ZP antigen is tissue specifically expressed in the ovary and oviduct and that sperm-ZP interaction is a highly specific event in the oviducts. Thus, it was considered that ZP could surely satisfy the conditions for effective contraceptive vaccines.

Contraceptive vaccines for human use should be highly effective and reversible as well as safe. Unfortunately, early animal models showed that infertility by ZP immunization resulted from ovarian impairment rather than by inhibition of sperm

Table 21.1 Technologies in contraceptive vaccine studies

Naturally occurring ZP glycoprotein and deglycosylated ZP proteins
Recombinant and synthetic ZP proteins/peptides
Engineered multiple epitopes including promiscuous T-cell epitope
Self-proliferating DNA vaccine (delivery system)

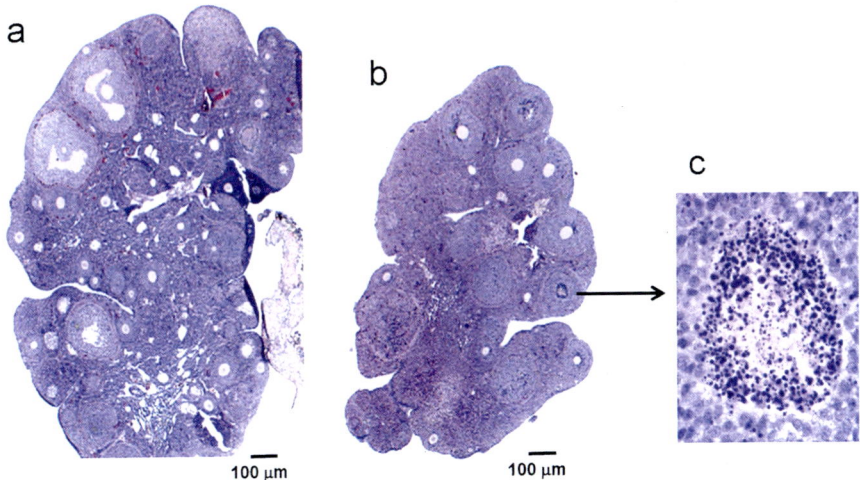

Fig. 21.2 Ovarian histology of hamster immunized with pig ZP. (**a**) Nonimmunized and age-matched control ovary. Various stages of follicles are observed, including from primordial to antral follicles. (**b**) Pig ZP-immunized ovarian section. The ovary is shrunken and there are few growing follicles. (**c**) Higher magnification of a degenerated follicle in **b**. There are many granulosa cells with fragmented nuclei

binding to ZP. There was conflict between sufficient contraceptive efficacy and adverse ovarian failure. In order to avoid ovarian impairment, much effort was made to isolate ideal antigens. Progress of ZP-based immunocontraceptive vaccine studies is summarized in Table 21.1.

As a first strategy, naturally occurring pig ZP was used, as pig ZP was relatively easily obtained from slaughterhouses and contained antigens cross-reactive among mammalian animals. Active immunization with pig ZP antigens produced not only anti-pig ZP antibodies but also self-reactive antibodies against ZP of immunized animals. As expected, fecundity was reduced, but undesired effects such as ovarian disruption also occurred in mice [4, 5], rabbits [6], dogs [7], and monkeys [8, 9].

In our experiments, the hamster ovary was severely damaged following immunization with naturally occurring pig ZP glycoproteins (Fig. 21.2). In the control sections (Fig. 21.2a), various stages of follicular growth can be observed ranging from primordial to antral follicles. In contrast, few growing follicles can be seen in the immunized ovary in which follicular reserves including primordial follicles were very limited. Granulosa cell nuclei were densely stained and fragmented (Fig. 21.2b, c). This ovarian pathology was considered to result from antibodies against somatic cell components which had contaminated in pig ZP during the vaccine preparation.

Isolation of pure pig ZP antigens was attempted with the aim of achieving the blocking of sperm binding to ZP without leading to ovarian pathology. However, these studies failed to identify any ideal antigenic protein. Even highly purified pig ZP antigens also still elicited ovarian pathology after immunization. It was considered that antigen materials still concomitantly included somatic antigens due to carbohydrate moieties, because the carbohydrate moiety is often cross-reactive with somatic tissues [2, 5]. Then, deglycosylated ZP proteins were prepared for immunization. However, these trials did not succeed in preventing ovarian pathology. Probably, ZP protein itself contains antigens which cause ovarian pathology [6–9].

At present, because the mechanism is mainly ovarian dysfunction, application of immunocontraceptive vaccines with ZP has been limited to wildlife population control [10–12], such as in feral horses [13, 14], white-tailed deer [15], African elephants [16], gray seals [17], and marsupials [18, 19]. Most of the studies used native pig ZP as an antigen. In those cases, the fecundity reduction mechanism was ovarian impairment.

21.2   Application of Recombinant ZP Proteins

As a next strategy of ZP-based immunocontraceptive vaccine development, recombinant technology was adopted (Table 21.1). Recombinant ZP3 was produced in an *Escherichia coli* expression system and used for vaccination. Application of recombinant ZP proteins could not avoid ovarian pathology in homologous and heterologous immunizations, despite numerous efforts [20–27]. Paterson et al. showed that recombinant human ZP3 and human peptides produced antibodies cross-reactive to marmoset ZP and induced ovarian failure in the animals [28]. The homologous primate model could not avoid ovarian failure either [29, 30]. These studies documented that contraception by immunization with ZP3 peptide antigens is associated with undesired side effects in nonhuman primates.

From another point of view, we designed an experiment using hamster ZP proteins. In the experiments, two recombinant hamster ZP proteins, ZP2 (rec-hamZP2) and ZP3 (rec-hamZP3), were produced by an *E. coli* expression system and used for immunization. The results are shown in Fig. 21.3b and c. Immunization with both proteins, i.e., rec-hamZP2 and rec-hamZP3, caused ovarian impairment, although primordial follicles remained.

Although the AZPA raised by immunization with recombinant ZP proteins inhibited sperm binding to the ZP in vitro, actual fecundity-reduction effects in vivo were caused by ovarian disruption. This indicated that cellular immunity targeting ZP antigens induced cytotoxic effects followed by ovarian disruption. Our question was whether or not AZPA could block sperm penetration into ZP in vivo

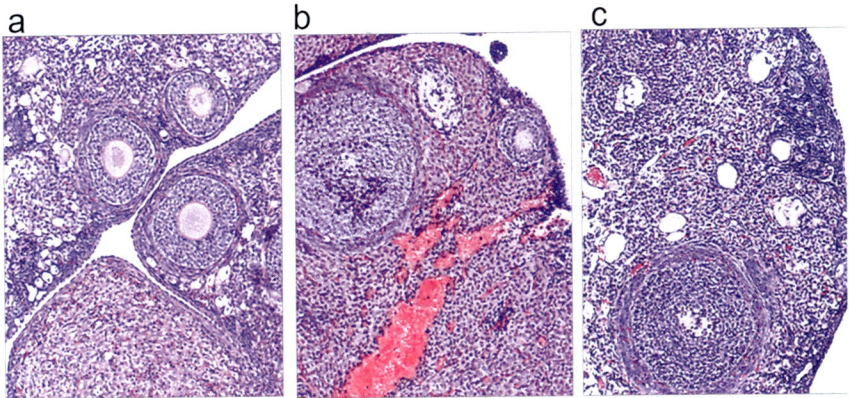

Fig. 21.3 Ovarian histology of hamsters immunized with homologous recombinant ZP proteins. (**a**) Nonimmunized control section. (**b** and **c**) are ovarian sections from hamsters immunized with rec-hamZP2 and rec-hamZP3, respectively. **b** and **c** were markedly impaired compared with **a**

Table 21.2 Comparison of active and passive immunizations with recombinant hamster ZP2 in homologous immunization

	Ovarian disorders	Ovulation inhibition	Fertility inhibition
Active immunization[a]	Yes	Yes	Yes
Passive immunization[b]	No	No	Yes

[a]Recombinant hamster ZP2 was prepared in an *E. coli* expression system and immunized in hamsters with Freund's complete adjuvant
[b]Antiserum produced by active immunization was inoculated in non-treated hamsters

without ovarian impairment. To make this clear, an experiment applying passive immunization was conducted, in which effects from cellular immunity were completely excluded.

In this study, AZPA was prepared by immunization with rec-hamZP2 and injected into non-treated female hamsters as a passive immunization. The hamsters receiving passive immunization had significantly reduced litter sizes, but ovarian disorders were not observed (Table 21.2) (unpublished data).

It was confirmed that AZPA against rec-hamZP2 react to oocyte ZP in the ovary as shown in Fig. 21.4. Then, how was the animals' fecundity reduced? To examine the precise mechanism of litter size reduction, the oocytes were collected from oviducts 1 day after mating. Surprisingly, the collected oocytes developed into two-cell stage embryos and were partially or completely extruded from ZP (Fig. 21.5). This was interpreted as showing that AZPA did not block fertilization but disturbed embryo growth by binding to ZP. Reduced fecundity resulted from embryo disruption rather than sperm binding and penetration into ZP.

The reason that embryos escaped from AZPA-bound ZP could be explained in that necessary factors such as nutrient sources and essential growth factors did not sufficiently permeate inside ZP. Embryos were probably extruded from ZP through

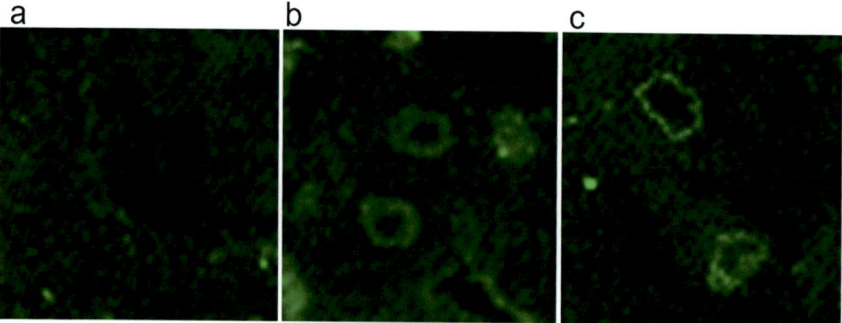

Fig. 21.4 Detection of self ZP-reactive antibodies by immunofluorescent staining after passive immunization. (**a–c**) are ovarian sections inoculated with control sera (**a**), rec-hamZP2 antisera (**b**), and rec-hamZP3 antisera (**c**). **b** and **c** show positive reactions with ZP, indicating that inoculated antisera bound to the ovarian ZP

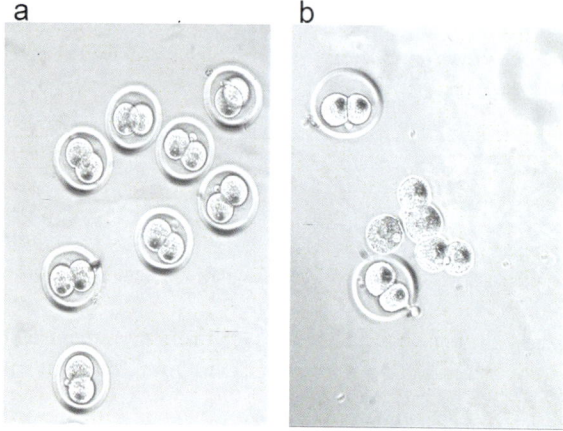

Fig. 21.5 Effect of passive immunization with rec-hamZP2 on two-cell-stage embryos. Hamsters were inoculated intraperitoneally with 1 mL antisera as a passive immunization. They were treated with gonadotropin for superovulation and mated with a fertile male. Embryos were collected from the oviducts 2 days later. (**a** and **b**) show the embryos from inoculation with control serum and rec-hamZP2 antisera, respectively. The embryos in B did not survive in the oviduct

the sperm penetration tunnel as shown. As previously mentioned, denuded embryos do not survive in the oviduct without ZP [31]. ZP protects embryos before preimplantation until they reach the uterus. Actually, no evidence of implantation was detected in the uterus. The results indicated that the reduced litter size by AZPA was due to embryo denudation from ZP in the oviduct, where maternal immunological systems may attack denuded embryos. It is concluded that humoral antibody, AZPA, cannot block sperm binding to ZP in vivo. Initial observation in vitro that sperm binding to ZP was inhibited by AZPA might not be the true mechanism of reduced fecundity. In the future, if AZPA production were engineered by advanced technology without immunological mediation, it could be possible to control fertility without ovarian impairment.

21.3 Application of Engineered Synthetic Peptides

Development of a safe contraceptive vaccine for human use continued with a search for epitopes. The focus of studies moved onto a sequential peptide epitope consisting of 10–20 amino acids that inhibited fertilization without producing any adverse side effects (Table 21.1).

Millar et al. demonstrated that a synthetic peptide based on ZP3 could produce a fertilization-blocking antibody without adverse side effects in the ovary [32]. Subsequently, the T-cell epitope causing ovarian pathology was identified, and immunization with the epitope concomitantly induced infertility [33]. In addition, a ZP3 peptide vaccine which induced antibodies and reversible infertility without ovarian dysfunction was reported [34]. These reports encouraged researchers to segregate fertilization-blocking epitopes from ovarian pathology-producing epitopes. Based on these reports, numerous studies have attempted to identify the fertilization-blocking B-cell epitope that does not induce cytotoxic T-cell activity. B-cell and T-cell epitopes were presumed to produce desired and undesired effects, respectively.

For B-cell epitope mapping, numerous monoclonal antibodies against ZP proteins were produced to identify a sequential peptide for a vaccine candidate [35–39]. We generated monoclonal antibodies to block human spermatozoa from binding to human ZP [40, 41]. Among them, a monoclonal antibody recognized the amino-terminal region of ZP2 and produced a strong inhibition of sperm binding. For a model study, synthetic peptides corresponding to the monoclonal antibody were prepared and conjugated with T-cell epitopes. No contraceptive effect was observed by immunization with this antigen, although sperm binding to ZP was inhibited in vitro [42–44]. Taken together, a single B-cell epitope included in ZP is insufficient for contraception if self-reactive ZP antibody is produced.

Next, contraceptive vaccine development was conducted to engineer multiple B-cell epitope constructs. The T-cell epitope does not necessarily involve the ZP peptide (Table 21.1). Tetanus toxoid (amino acid residues 830–844) [45] and circumsporozoite proteins (amino acid residues 378–398) [46] were available as promiscuous T-cell epitopes. Artificial DNA constructs coding promising peptides were designed with promiscuous T-cell epitopes. Among them, a chimeric peptide, i.e., several tandemly linked epitope peptides based on bonnet monkey ZP and T-cell epitopes, was reported as a feasible candidate. The generated antibodies blocked human sperm binding to ZP [45, 47, 48].

Furthermore, a fusion protein with foreign peptide for vaccine antigens was proposed. Two studies designed DNA constructs for producing fusion proteins comprising ZP3, GnRH, and promiscuous T-cell epitopes [49, 50]. The first study examined the contraceptive potential of an *E. coli*-expressed fusion protein, i.e.,

T-cell epitope (tetanus toxoid), dog ZP3, T-cell epitope (bovine RNase), GnRH, and T-cell epitope (circumsporozoite protein) and GnRH. It was revealed that immunization with the fusion protein exhibited high contraceptive efficacy in a mating study as well as an in vitro fertilization study [49]. The second study examined the effects of immunization with a fusion protein, i.e., dog ZP3 and GnRH followed by ovalbumin as a T-cell epitope. The antibodies against ZP3 and GnRH have different fertilization-blocking mechanisms [50]. These antigen combination strategies might produce an ideal vaccine for ZP-based immunocontraception [51].

Another interesting approach was reported by Gupta's research group. They evaluated a recombinant fusion protein, i.e., promiscuous T-cell epitope (tetanus toxoid), a fragment of ZP3, and a fragment of Izumo with space in each component comprising several amino acids, in mice. Izumo is known to be an essential transmembrane factor on acrosome-reacted sperm at the oocyte-sperm fusion event. Although the raised antibody recognized acrosome-reacted sperm, and antibody titers to ZP 3 increased, contraceptive potential could not be further enhanced by the inclusion of Izumo [52].

21.4   DNA Strategy for Contraceptive Vaccines

As a DNA vaccine (Table 21.1), a self-proliferating plasmid was adopted. Antigen was expected to be spontaneously produced in inoculated animals. Several DNA vaccines encoding ZP peptides have been developed [53–55]. Successful constructs that rendered infertility were reported. In mice, for example, triple epitopes were tandemly linked for a vaccine design [55]. This strategy is advantageous because immune stimulation happens continuously without booster injections. Immunogens are automatically produced from the inoculated DNA vaccine in the animal's body. The immunity may be enhanced by a combination of promiscuous T-cell epitopes such as ovalbumin, bovine RNase, and tetanus toxoid. Efficacy should be confirmed before application. For fertility management of wildlife, pets, and stray animals, permanent sterility may be permitted.

Finally, as a future strategy, a single chain fragment for variable regions (scFv) was recently proposed. A research group constructed a plasmid DNA comprising dog ZP3, scFv reactive to dog ZP3, and dendric cell receptor (DEC-201), which was effective as an antigen causing cell stimulation. Mice immunized with the construct showed reduced litter size [56]. If this strategy can achieve effective contraception, there is no need to immunize with an antigen. AZPA designed as an effective scFv may directly be expressed in the inoculated animal's body without the contribution of B-cell and T-cell epitopes.

In conclusion, intensive studies and technological advances have not found ideal targets for ZP-based immunocontraceptive vaccines at this moment. AZPA blocked sperm penetration into ZP in vitro but not in vivo. We are anticipating future technological developments.

References

1. Gwatkin RB, Williams DT, Carlo DJ. Immunization of mice with heat-solubilized hamster zonae: production of anti-zona antibody and inhibition of fertility. Fertil Steril. 1977;28:871–7.
2. Yanagimachi R, Winkelhake J, Nicolson GL. Immunological block to mammalian fertilization: survival and organ distribution of immunoglobulin which inhibits fertilization in vivo. Proc Natl Acad Sci U S A. 1976;73:2405–8.
3. Tsunoda Y, Chang MC. Effects of antisera against eggs and zonae pellucidae on fertilization and development of mouse eggs in vivo and in culture. J Reprod Fertil. 1978;54:233–7.
4. Sacco AG. Inhibition of fertility in mice by passive immunization with antibodies to isolated zonae pellucidae. J Reprod Fertil. 1989;56:533–7.
5. Sacco AG, Subramanian MG, Yurewicz EC. Active immunization of mice with porcine zonae pellucidae: immune response and effect on fertility. J Exp Zool. 1981;218:405–18.
6. Keenan JA, Sacco AG, Subramanian MG, Kruger M, Yurewicz EC, Moghissi KS. Endocrine response in rabbits immunized with native versus deglycosylated porcine zona pellucida antigens. Biol Reprod. 1991;44:150–6.
7. Mahi-Brown CA, Yanagimachi R, Nelson ML, Yanagimachi H, Palumbo N. Ovarian histopathology of bitches immunized with porcine zonae pellucidae. Am J Reprod Immunol Microbiol. 1988;18:94–103.
8. Sacco AG, Subramanian MG, Yurewicz EC, DeMayo FJ, Dukelow WR. Heteroimmunization of squirrel monkeys (Saimiri sciureus) with a purified porcine zona antigen (PPZA): immune response and biologic activity of antiserum. Fertil Steril. 1983;39:350–8.
9. Paterson M, Koothan PT, Morris KD, O'Byrne KT, Braude P, Williams A, et al. Analysis of the contraceptive potential of antibodies against native and deglycosylated porcine ZP3 in vivo and in vitro. Biol Reprod. 1992;46:523–34.
10. Kirkpatrick JF, Lyda RO, Frank KM. Contraceptive vaccines for wildlife: a review. Am J Reprod Immunol. 2011;66:40–50.
11. Barber MR, Fayrer-Hosken RA. Possible mechanisms of mammalian immunocontraception. J Reprod Immunol. 2000;46(2):103–24.
12. Joonè CJ, Schulman ML, Bertschinger HJ. Ovarian dysfunction associated with zona pellucida-based immunocontraceptive vaccines. Theriogenology. 2017;89:329–37.
13. Kirkpatrick JF, Liu IM, Turner JW Jr, Naugle R, Keiper R. Long-term effects of porcine zonae pellucidae immunocontraception on ovarian function in feral horses (Equus caballus). J Reprod Fertil. 1992;94:437–44.
14. Nolan MB, Schulman ML, Botha AE, Human AM, Roth R, Crampton MC, Bertschinger HJ. Serum antibody immunoreactivity and safety of native porcine and recombinant zona pellucida vaccines formulated with a non-Freund's adjuvant in horses. Vaccine. 2019;37(10):1299–306.
15. Curtis PD, Richmond ME, Miller LA, Quimby FW. Pathophysiology of white-tailed deer vaccinated with porcine zona pellucida immunocontraceptive. Vaccine. 2007;25:4623–30.
16. Fayrer-Hosken RA, Grobler D, Van Altena JJ, Bertschinger HJ, Kirkpatrick JF. Immunocontraception of African elephants. Nature. 2000;407:149.
17. Brown RG, Bowen WD, Eddington JD, Kimmins WC, Mezei M, Parsons JL, et al. Evidence for a long-lasting single administration contraceptive vaccine in wild grey seals. J Reprod Immunol. 1997;35:43–51.
18. Kitchener AL, Kay DJ, Walters B, Menkhorst P, McCartney CA, Buist JA, et al. The immune response and fertility of koalas (Phascolarctos cinereus) immunised with porcine zonae pellucidae or recombinant brushtail possum ZP3 protein. J Reprod Immunol. 2009;82:40–7.
19. Kitchener AL, Harman A, Kay DJ, McCartney CA, Mate KE, Rodger JC. Immunocontraception of Eastern Grey kangaroos (Macropus giganteus) with recombinant brushtail possum (Trichosurus vulpecula) ZP3 protein. J Reprod Immunol. 2009;79:156–62.

20. VandeVoort CA, Schwoebel ED, Dunbar BS. Immunization of monkeys with recombinant complimentary deoxyribonucleic acid expressed zona pellucida proteins. Fertil Steril. 1995;64: 838–47.
21. Kaul R, Sivapurapu N, Afzalpurkar A, Srikanth V, Govind CK, Gupta SK. Immunocontraceptive potential of recombinant bonnet monkey (Macaca radiata) zona pellucida glycoprotein-C expressed in Escherichia coli and its corresponding synthetic peptide. Reprod Biomed Online. 2001;2:33–9.
22. Govind CK, Srivastava N, Gupta SK. Evaluation of the immunocontraceptive potential of Escherichia coli expressed recombinant non-human primate zona pellucida glycoproteins in homologous animal model. Vaccine. 2002;21:78–88.
23. Srivastava N, Santhanam R, Sheela P, Mukund S, Thakral SS, Malik BS, et al. Evaluation of the immunocontraceptive potential of Escherichia coli-expressed recombinant dog ZP2 and ZP3 in a homologous animal model. Reproduction. 2002;123:847–57.
24. Skinner SM, Mills T, Kirchick HJ, Dunbar BS. Immunization with zona pellucida proteins results in abnormal ovarian follicular differentiation and inhibition of gonadotropin-induced steroid secretion. Endocrinology. 1984;115(6):2418–32.
25. Hasegawa A, Koyama K, Inoue M, Takemura T, Isojima S. Antifertility effect of active immunization with ZP4 glycoprotein family of porcine zona pellucida in hamsters. J Reprod Immunol. 1992;22(2):197–210.
26. Luo AM, Garza KM, Hunt D, Tung KS. Antigen mimicry in autoimmune disease sharing of amino acid residues critical for pathogenic T cell activation. J Clin Invest. 1993;92:2117–23.
27. Sacco AG, Yurewicz EC, Subramanian MG, Lian Y, Dukelow WR. Immunological response and ovarian histology of squirrel monkeys (Saimiri sciureus) immunized with porcine zona pellucida ZP3 (M(r) = 55,000) macromolecules. Am J Primatol. 1991;24(1):15–28.
28. Paterson M, Wilson MR, Morris KD, van Duin M, Aitken RJ. Evaluation of the contraceptive potential of recombinant human ZP3 and human ZP3 peptides in a primate model: their safety and efficacy. Am J Reprod Immunol. 1998;40:198–209.
29. Paterson M, Wilson MR, Jennings ZA, van Duin M, Aitken RJ. Design and evaluation of a ZP3 peptide vaccine in a homologous primate model. Mol Hum Reprod. 1999;5:342–52.
30. Paterson M, Jennings ZA, Wilson MR, Aitken RJ. The contraceptive potential of ZP3 and ZP3 peptides in a primate model. J Reprod Immunol. 2002;53:99–107.
31. Bronson RA, McLaren A. Transfer to the mouse oviduct of eggs with and without the zona pellucida. J Reprod Fertil. 1970;22:129–37.
32. Millar SE, Chamow SM, Baur AW, Oliver C, Robey F, Dean J. Vaccination with a synthetic zona pellucida peptide produces long-term contraception in female mice. Science. 1989;246: 935–8.
33. Lou Y, Ang J, Thai H, McElveen F, Tung KS. A zona pellucida 3 peptide vaccine induces antibodies and reversible infertility without ovarian pathology. J Immunol. 1995;155:2715–20.
34. Sun W, Lou YH, Dean J, Tung KS. A contraceptive peptide vaccine targeting sulfated glycoprotein ZP2 of the mouse zona pellucid. Biol Reprod. 1999;60:900–7.
35. Isojima S, Koyama K, Hasegawa A, Tsunoda Y, Hanada A. Monoclonal antibodies to porcine zona pellucida antigens and their inhibitory effects on fertilization. Am J Reprod Immunol. 1984;6:77–87.
36. Drell DW, Dunbar BS. Monoclonal antibodies to rabbit and pig zonae pellucidae distinguish species-specific and shared antigenic determinants. Biol Reprod. 1984;30:445–57.
37. Koyama K, Hasegawa A, Tsuji Y, Isojima S. Production and characterization of monoclonal antibodies to cross-reactive antigens of human and porcine zonae pellucidae. J Reprod Immunol. 1985;7:187–98.
38. Dean J, East IJ. Effects of anti-zona pellucida monoclonal antibodies on fertilization and early development. Adv Exp Med Biol. 1986;207:37–53.
39. Gupta SK, Yurewicz EC, Afzalpurkar A, Rao KV, Gage DA, Wu H, et al. Localization of epitopes for monoclonal antibodies at the N-terminus of the porcine zona pellucida glycoprotein pZPC. Mol Reprod Dev. 1995;42:220–5.

40. Koyama K, Hasegawa A, Inoue M, Isojima S. Blocking of human sperm-zona interaction by monoclonal antibodies to a glycoprotein family (ZP4) of porcine zona pellucid. Biol Reprod. 1991;45:727–35.
41. Koyama K, Hasegawa A, Isojima S. Further characterization of the porcine zona pellucida antigen corresponding to monoclonal antibody 3A4-2G1 exclusively cross-reactive with porcine and human zonae pellucidae. J Reprod Immunol. 1991;19:131–48.
42. Hasegawa A, Tsubamoto H, Hamada Y, Koyama K. Blocking effect of antisera to recombinant zona pellucida proteins (r-ZPA) on in vitro fertilization. Am J Reprod Immunol. 2000;44:59–64.
43. Hasegawa A, Hamada Y, Shigeta M, Koyama K. Contraceptive potential of a synthetic peptide of zona pellucida protein (ZPA). J Reprod Immunol. 2002;53:91–8.
44. Koyama K, Hasegawa A, Gupta SK. Prospect for immunocontraception using the NH2-terminal recombinant peptide of human zona pellucida protein (hZPA). Am J Reprod Immunol. 2002;47:303–10.
45. Gupta N, Shrestha A, Panda AK, Gupta SK. Production of tag-free recombinant fusion protein encompassing promiscuous T cell epitope of tetanus toxoid and dog zona pellucida glycoprotein-3 for contraceptive vaccine development. Mol Biotechnol. 2013;54:853–62.
46. Sivapurapu N, Upadhyay A, Hasegawa A, Koyama K, Gupta SK. Native zona pellucida reactivity and in-vitro effect on human sperm-egg binding with antisera against bonnet monkey ZP1 and ZP3 synthetic peptides. J Reprod Immunol. 2002;56(1–2):77–91.
47. Sivapurapu N, Upadhyay A, Hasegawa A, Koyama K, Gupta SK. Efficacy of antibodies against E. coli expressed chimeric recombinant protein encompassing multiple epitopes of zona pellucida glycoproteins to inhibit in vitro human sperm-egg binding. Mol Reprod Dev. 2003;65:309–17.
48. Sivapurapu N, Hasegawa A, Gahlay GK, Koyama K, Gupta SK. Efficacy of antibodies against a chimeric synthetic peptide encompassing epitopes of bonnet monkey (Macaca radiata) zona pellucida-1 and zona pellucida-3 glycoproteins to inhibit in vitro human sperm-egg binding. Mol Reprod Dev. 2005;70:247–54.
49. Arukha AP, Minhas V, Shrestha A, Gupta SK. Contraceptive efficacy of recombinant fusion protein comprising zona pellucida glycoprotein-3 fragment and gonadotropin releasing hormone. J Reprod Immunol. 2016;114:18–26.
50. Wang Y, Li Y, Zhang B, Zhang F. The preclinical evaluation of immunocontraceptive vaccines based on canine zona pellucida 3 (cZP3) in a mouse model. Reprod Biol Endocrinol. 2018;16 (1):47.
51. Gupta SK, Srinivasan VA, Suman P, Rajan S, Nagendrakumar SB, Gupta N, Shrestha A, Joshi P, Panda AK. Contraceptive vaccines based on the zona pellucida glycoproteins for dogs and other wildlife population management. Am J Reprod Immunol. 2011;66(1):51–62.
52. Shrestha A, Wadhwa N, Gupta SK. Evaluation of recombinant fusion protein comprising dog zona pellucida glycoprotein-3 and Izumo and individual fragments as immunogens for contraception. Vaccine. 2014;32(5):564–71.
53. Rath A, Choudhury S, Hasegawa A, Koyama K, Gupta SK. Antibodies generated in response to plasmid DNA encoding zona pellucida glycoprotein-B inhibit in vitro human sperm-egg binding. Mol Reprod Dev. 2002;62:525–33.
54. Choudhury S, Ganguly A, Chakrabarti K, Sharma RK, Gupta SK. DNA vaccine encoding chimeric protein encompassing epitopes of human ZP3 and ZP4: immunogenicity and characterization of antibodies. J Reprod Immunol. 2009;79:137–47.
55. Yu MF, Fang WN, Xiong GF, Yang Y, Peng JP. Evidence for the inhibition of fertilization in vitro by anti-ZP3 antisera derived from DNA vaccine. Vaccine. 2011;29:4933–9.
56. Wang Y, Zhang B, Li J, Aipire A, Li Y, Zhang F. Enhanced contraception of canine zona pellucida 3 DNA vaccine via targeting DEC-205 in mice. Theriogenlogy. 2018;113:56–62.

Printed by Books on Demand, Germany